ANOVA and ANCOVA

ANOVA and ANCOVA
A GLM Approach

Second Edition

ANDREW RUTHERFORD
Keele University
School of Psychology
Staffordshire, United Kingdom

Published by John Wiley & Sons, Inc., Hoboken, New Jersey
Published simultaneously in Canada

For general information on our other products and services or for technical support, please contact our Customer Care Department within the United States at (800) 762-2974, outside the United States at (317) 572-3993 or fax (317) 572-4002.

Wiley also publishes its books in a variety of electronic formats. Some content that appears in print may not be available in electronic formats. For more information about Wiley products, visit our web site at www.wiley.com.

Library of Congress Cataloging-in-Publication Data:

Rutherford, Andrew, 1958-
ANOVA and ANCOVA : a GLM approach / Andrew Rutherford. – 2nd ed.
p. cm.
Includes bibliographical references and index.
ISBN 978-0-470-38555-5 (cloth)
1. Analysis of variance. 2. Analysis of covariance. 3. Linear models (Statistics) I. Title.
QA279.R879 2011
519.5′38–dc22

2010018486

10 9 8 7 6 5 4 3 2 1

Contents

Acknowledgments

I'd like to thank Dror Rom and Juliet Shaffer for their generous comments on the topic of multiple hypothesis testing. Special thanks go to Dror Rom for providing and naming Shaffer's R test—any errors in the presentation of this test are mine alone. I also want to thank Sol Nte for some valuable mathematical aid, especially on the enumeration of possibly true null hypotheses!

I'd also like to extend my thanks to Basir Syed at SYSTAT Software, UK and to Supriya Kulkarni at SYSTAT Technical Support, Bangalore, India. My last, but certainly not my least thanks go to Jacqueline Palmieri at Wiley, USA and to Sanchari Sil at Thomson Digital, Noida for all their patience and assistance.

A. R.

CHAPTER 1

An Introduction to General Linear Models: Regression, Analysis of Variance, and Analysis of Covariance

1.1 REGRESSION, ANALYSIS OF VARIANCE, AND ANALYSIS OF COVARIANCE

Regression and analysis of variance (ANOVA) are probably the most frequently applied of all statistical analyses. Regression and analysis of variance are used extensively in many areas of research, such as psychology, biology, medicine, education, sociology, anthropology, economics, political science, as well as in industry and commerce.

There are several reasons why regression and analysis of variance are applied so frequently. One of the main reasons is they provide answers to the questions researchers ask of their data. Regression allows researchers to determine if and how variables are related. ANOVA allows researchers to determine if the mean scores of different groups or conditions differ. Analysis of covariance (ANCOVA), a combination of regression and ANOVA, allows researchers to determine if the group or condition mean scores differ after the influence of another variable (or variables) on these scores has been equated across groups. This text focuses on the analysis of data generated by psychology experiments, but a second reason for the frequent use of regression and ANOVA is they are applicable to experimental, quasi-experimental, and non-experimental data, and can be applied to most of the designs employed in these studies. A third reason, which should not be underestimated, is that appropriate regression and ANOVA statistical software is available to analyze most study designs.

ANOVA and ANCOVA: A GLM Approach, Second Edition. By Andrew Rutherford.

1.2 A POCKET HISTORY OF REGRESSION, ANOVA, AND ANCOVA

Historically, regression and ANOVA developed in different research areas to address different research questions. Regression emerged in biology and psychology toward the end of the nineteenth century, as scientists studied the relations between people's attributes and characteristics. Galton (1886, 1888) studied the height of parents and their adult children, and noticed that while short parents' children usually were shorter than average, nevertheless, they tended to be taller than their parents. Galton described this phenomenon as "regression to the mean." As well as identifying a basis for predicting the values on one variable from values recorded on another, Galton appreciated that the degree of relationship between some variables would be greater than others. However, it was three other scientists, Edgeworth (1886), Pearson (1896), and Yule (1907), applying work carried out about a century earlier by Gauss (or Legendre, see Plackett, 1972), who provided the account of regression in precise mathematical terms. (See Stigler, 1986, for a detailed account.)

The t-test was devised by W.S. Gosset, a mathematician and chemist working in the Dublin brewery of Arthur Guinness Son & Company, as a way to compare the means of two small samples for quality control in the brewing of stout. (Gosset published the test in *Biometrika* in 1908 under the pseudonym "Student," as his employer regarded their use of statistics to be a trade secret.) However, as soon as more than two groups or conditions have to be compared more than one t-test is needed. Unfortunately, as soon as more than one statistical test is applied, the Type 1 error rate inflates (i.e., the likelihood of rejecting a true null hypothesis increases—this topic is returned to in Sections 2.1 and 3.6.1). In contrast, ANOVA, conceived and described by Ronald A. Fisher (1924, 1932, 1935b) to assist in the analysis of data obtained from agricultural experiments, was designed to compare the means of any number of experimental groups or conditions without increasing the Type 1 error rate. Fisher (1932) also described ANCOVA with an approximate adjusted treatment sum of squares, before describing the exact adjusted treatment sum of squares a few years later (Fisher, 1935b, and see Cox and McCullagh, 1982, for a brief history). In early recognition of his work, the F-distribution was named after him by G.W. Snedecor (1934).

ANOVA procedures culminate in an assessment of the ratio of two variances based on a pertinent F-distribution and this quickly became known as an F-test. As all the procedures leading to the F-test also may be considered as part of the F-test, the terms "ANOVA" and "F-test" have come to be used interchangeably. However, while ANOVA uses variances to compare means, F-tests *per se* simply allow two (independent) variances to be compared without concern for the variance estimate sources.

In subsequent years, regression and ANOVA techniques were developed and applied in parallel by different groups of researchers investigating different research topics, using different research methodologies. Regression was applied most often to data obtained from correlational or non-experimental research and came to be regarded only as a technique for describing, predicting, and assessing the relations between predictor(s) and dependent variable scores. In contrast, ANOVA was applied to experimental data beyond that obtained from agricultural experiments

(Lovie, 1991a), but still it was considered only as a technique for determining whether the mean scores of groups differed significantly. For many areas of psychology, particularly experimental psychology, where the interest was to assess the average effect of different experimental manipulations on groups of subjects in terms of a particular dependent variable, ANOVA was the ideal statistical technique. Consequently, separate analysis traditions evolved and have encouraged the mistaken belief that regression and ANOVA are fundamentally different types of statistical analysis. ANCOVA illustrates the compatibility of regression and ANOVA by combining these two apparently discrete techniques. However, given their histories it is unsurprising that ANCOVA is not only a much less popular analysis technique, but also one that frequently is misunderstood (Huitema, 1980).

1.3 AN OUTLINE OF GENERAL LINEAR MODELS (GLMs)

The availability of computers for statistical analysis increased hugely from the 1970s. Initially statistical software ran on mainframe computers in batch processing mode. Later, the statistical software was developed to run in a more interactive fashion on PCs and servers. Currently, most statistical software is run in this manner, but, increasingly, statistical software can be accessed and run over the Web.

Using statistical software to analyze data has had considerable consequence not only for analysis implementations, but also for the way in which these analyses are conceived. Around the 1980s, these changes began to filter through to affect data analysis in the behavioral sciences, as reflected in the increasing number of psychology statistics texts that added the general linear model (GLM) approach to the traditional accounts (e.g., Cardinal and Aitken, 2006; Hays, 1994; Kirk, 1982, 1995; Myers, Well, and Lorch, 2010; Tabachnick and Fidell, 2007; Winer, Brown, and Michels, 1991) and an increasing number of psychology statistics texts that presented regression, ANOVA, and ANCOVA exclusively as instances of the GLM (e.g., Cohen and Cohen, 1975, 1983; Cohen et al., 2003; Hays, 1994; Judd and McClelland, 1989; Judd, McClelland, and Ryan, 2008; Keppel and Zedeck, 1989; Maxwell and Delaney, 1990, 2004; Pedhazur, 1997).

A major advantage afforded by computer-based analyses is the easy use of matrix algebra. Matrix algebra offers an elegant and succinct statistical notation. Unfortunately, however, human matrix algebra calculations, particularly those involving larger matrices, are not only very hard work but also tend to be error prone. In contrast, computer implementations of matrix algebra are not only very efficient in computational terms, but also error free. Therefore, most computer-based statistical analyses employ matrix algebra calculations, but the program output usually is designed to concord with the expectations set by traditional (scalar algebra) calculations.

When regression, ANOVA, and ANCOVA are expressed in matrix algebra terms, a commonality is evident. Indeed, the same matrix algebra equation is able to summarize all three of these analyses. As regression, ANOVA, and ANCOVA can be described in an identical manner, clearly they share a common pattern. This common pattern is the GLM. Unfortunately, the ability of the same matrix algebra equation to describe regression, ANOVA, and ANCOVA has resulted in the inaccurate identification of the matrix algebra equation as the GLM. However, just as a particular language provides a means of expressing an idea, so matrix algebra provides only one notation for expressing the GLM.

Tukey (1977) employed the GLM conception when he described data as

$$\text{Data} = \text{Fit} + \text{Residual} \tag{1.1}$$

The same GLM conception is employed here, but the fit and residual component labels are replaced with the more frequently applied labels, model (i.e., the fit) and error (i.e., the residual). Therefore, the usual expression of the GLM conception is that data may be accommodated in terms of a model plus error

$$\text{Data} = \text{Model} + \text{Error} \tag{1.2}$$

In equation (1.2), the model is a representation of our understanding or hypotheses about the data, while the error explicitly acknowledges that there are other influences on the data. When a full model is specified, the error is assumed to reflect all influences on the dependent variable scores not controlled in the experiment. These influences are presumed to be unique for each subject in each experimental condition. However, when less than a full model is represented, the score component attributable to the omitted part(s) of the full model also is accommodated by the error term. Although the omitted model component increments the error, as it is neither uncontrolled nor unique for each subject, the residual label would appear to be a more appropriate descriptor. Nevertheless, many GLMs use the error label to refer to the error parameters, while the residual label is used most frequently in regression analysis to refer to the error parameter estimates. The relative sizes of the full or reduced model components and the error components also can be used to judge how well the particular model accommodates the data. Nevertheless, the tradition in data analysis is to use regression, ANOVA, and ANCOVA GLMs to express different types of ideas about how data arises.

1.3.1 Regression

Simple linear regression examines the degree of the linear relationship (see Section 1.5) between a single predictor or independent variable and a response or dependent variable, and enables values on the dependent variable to be predicted from the values recorded on the independent variable. Multiple linear regression does the same, but accommodates an unlimited number of predictor variables.

In GLM terms, regression attempts to explain data (the dependent variable scores) in terms of a set of independent variables or predictors (the model) and a residual component (error). Typically, the researcher applying regression is interested in predicting a quantitative dependent variable from one or more quantitative independent variables and in determining the relative contribution of each independent variable to the prediction. There is also interest in what proportion of the variation in the dependent variable can be attributed to variation in the independent variable(s).

Regression also may employ categorical (also known as nominal or qualitative) predictors-the use of independent variables such as gender, marital status, and type of teaching method is common. As regression is an elementary form of GLM, it is possible to construct regression GLMs equivalent to any ANOVA and ANCOVA GLMs by selecting and organizing quantitative variables to act as categorical variables (see Section 2.7.4). Nevertheless, throughout this chapter, the convention of referring to these particular quantitative variables as categorical variables will be maintained.

1.3.2 Analysis of Variance

Single factor or one-way ANOVA compares the means of the dependent variable scores obtained from any number of groups (see Chapter 2). Factorial ANOVA compares the mean dependent variable scores across groups with more complex structures (see Chapter 5).

In GLM terms, ANOVA attempts to explain data (the dependent variable scores) in terms of the experimental conditions (the model) and an error component. Typically, the researcher applying ANOVA is interested in determining which experimental condition dependent variable score means differ. There is also interest in what proportion of variation in the dependent variable can be attributed to differences between specific experimental groups or conditions, as defined by the independent variable(s).

The dependent variable in ANOVA is most likely to be measured on a quantitative scale. However, the ANOVA comparison is drawn between the groups of subjects receiving different experimental conditions and is categorical in nature, even when the experimental conditions differ along a quantitative scale. As regression also can employ categorical predictors, ANOVA can be regarded as a particular type of regression analysis that employs only categorical predictors.

1.3.3 Analysis of Covariance

The ANCOVA label has been applied to a number of different statistical operations (Cox and McCullagh, 1982), but it is used most frequently to refer to the statistical technique that combines regression and ANOVA. As ANCOVA is the combination of these two techniques, its calculations are more involved and time consuming than either technique alone. Therefore, it is unsurprising that an increase in ANCOVA applications is linked to the availability of computers and statistical software.

Fisher (1932, 1935b) originally developed ANCOVA to increase the precision of experimental analysis, but it is applied most frequently in quasi-experimental research. Unlike experimental research, the topics investigated with quasi-experimental methods are most likely to involve variables that, for practical or ethical reasons, cannot be controlled directly. In these situations, the statistical control provided by ANCOVA has particular value. Nevertheless, in line with Fisher's original conception, many experiments may benefit from the application of ANCOVA.

As ANCOVA combines regression and ANOVA, it too can be described in terms of a model plus error. As in regression and ANOVA, the dependent variable scores constitute the data. However, as well as experimental conditions, the model includes one or more quantitative predictor variables. These quantitative predictors, known as covariates (also concomitant or control variables), represent sources of variance that are thought to influence the dependent variable, but have not been controlled by the experimental procedures. ANCOVA determines the covariation (correlation) between the covariate(s) and the dependent variable and then removes that variance associated with the covariate(s) from the dependent variable scores, prior to determining whether the differences between the experimental condition (dependent variable score) means are significant. As mentioned, this technique, in which the influence of the experimental conditions remains the major concern, but one or more quantitative variables that predict the dependent variable are also included in the GLM, is labeled ANCOVA most frequently, and in psychology is labeled ANCOVA exclusively (e.g., Cohen et al., 2003; Pedhazur, 1997, cf. Cox and McCullagh, 1982). An important, but seldom emphasized, aspect of the ANCOVA method is that the relationship between the covariate(s) and the dependent variable, upon which the adjustments depend, is determined empirically from the data.

1.4 THE "GENERAL" IN GLM

The term "general" in GLM simply refers to the ability to accommodate distinctions on quantitative variables representing continuous measures (as in regression analysis) and categorical distinctions representing groups or experimental conditions (as in ANOVA). This feature is emphasized in ANCOVA, where variables representing both quantitative and categorical distinctions are employed in the same GLM.

Traditionally, the label *linear modeling* was applied exclusively to regression analyses. However, as regression, ANOVA, and ANCOVA are but particular instances of the GLM, it should not be surprising that consideration of the processes involved in applying these techniques reveals any differences to be more apparent than real. Following Box and Jenkins (1976), McCullagh and Nelder (1989) distinguish four processes in linear modeling: (1) model selection, (2) parameter estimation, (3) model checking, and (4) the prediction of future values. (Box and Jenkins refer to model identification rather than model selection, but McCullagh and Nelder resist this terminology, believing it to imply that a correct model can be known with certainty.) While such a framework is useful heuristically, McCullagh and Nelder acknowledge that in reality these four linear modeling processes are not so distinct and that the whole, or parts, of the sequence may be iterated before a model finally is selected and summarized.

Usually, prediction is understood as the forecast of new, or independent values with respect to a new data sample using the GLM already selected. However, McCullagh and Nelder include Lane and Nelder's (1982) account of prediction, which unifies conceptions of ANCOVA and different types of standardization. Lane and Nelder consider prediction in more general terms and regard the values fitted by the GLM (graphically, the values intersected by the GLM line or hyper plane) to be instances of prediction and part of the GLM summary. As these fitted values are often called predicted values, the distinction between the types of predicted value is not always obvious, although a greater standard error is associated with the values forecast on the basis of a new data sample (e.g., Cohen et al., 2003; Kutner et al., 2005; Pedhazur, 1997).

With the linear modeling process of prediction so defined, the four linear modeling processes become even more recursive. For example, when selecting a GLM, usually the aim is to provide a best fit to the data with the least number of predictor variables (e.g., Draper and Smith, 1998; McCullagh and Nelder, 1989). However, the model checking process that assesses best fit employs estimates of parameters (and estimates of error), so the processes of parameter estimation and prediction must be executed within the process of model checking.

The misconception that this description of general linear modeling refers only to regression analysis is fostered by the effort invested in the model selection process with correlational data obtained from non-experimental studies. Usually in non-experimental studies, many variables are recorded and the aim is to identify the GLM that best predicts the dependent variable. In principle, the only way to select the best GLM is to examine every possible combination of predictors. As it takes relatively few potential predictors to create an extremely large number of possible GLM selections, a number of predictor variable selection procedures, such as all-possible regressions, forward stepping, backward stepping, and ridge regression (e.g., Draper and Smith, 1998; Kutner et al., 2005) have been developed to reduce the number of GLMs that need to be considered.

Correlations between predictors, termed *multicollinearity* (but see Pedhazur, 1997; Kutner et al., 2005; and Section 11.7.1) create three problems that affect the processes of GLM selection and parameter estimation. These are (i) the substantive interpretation of partial coefficients (if calculated simultaneously, correlated predictors' partial coefficients are reduced), (ii) the sampling stability of partial coefficients (different data samples do not provide similar estimates), and (iii) the accuracy of the calculation of partial coefficients and their errors (Cohen et al., 2003). The reduction of partial coefficient estimates is due to correlated predictor variables accommodating similar parts of the dependent variable variance. Because correlated predictors share association with the same part of the dependent variable, as soon as a correlated predictor is included in the GLM, all of the dependent variable variance common to the correlated predictors is accommodated by this first correlated predictor, so making it appear that the remaining correlated predictors are of little importance.

When multicollinearity exists and there is interest in the contribution to the GLM of sets of predictors or individual predictors, an incremental regression analysis can be adopted (see Section 5.4). Essentially, this means that predictors (or sets of predictors) are entered into the GLM cumulatively in a principled order (Cohen et al., 2003). After each predictor has entered the GLM, the new GLM may be compared with the previous GLM, with any changes attributable to the predictor just included. Although there is similarity between incremental regression and forward stepping procedures, they are distinguished by the, often theoretical, principles employed by incremental regression to determine the entry order of predictors into the GLM. Incremental regression analyses also concord with Nelder's (McCullagh and Nelder, 1989; Nelder, 1977) approach to ANOVA and ANCOVA, which attributes variance to factors in an ordered manner, accommodating the marginality of factors and their interactions (also see Bingham and Fienberg, 1982).

After selection, parameters must be estimated for each GLM and then model checking engaged. Again, due to the nature of non-experimental data, model checking may detect problems requiring remedial measures. Finally, the nature of the issues addressed by non-experimental research make it much more likely that the GLMs selected will be used to forecast new values.

A little consideration reveals identical GLM processes underlying a typical analysis of experimental data. For experimental data, the GLM selected is an expression of the experimental design. Moreover, most experiments are designed so that the independent variables translate into independent (i.e., uncorrelated) predictors, so avoiding multicollinearity problems. The model checking process continues by assessing the predictive utility of the GLM components representing the experimental effects. Each significance test of an experimental effect requires an estimate of that experimental effect and an estimate of a pertinent error term. Therefore, the GLM process of parameter estimation is engaged to determine experimental effects, and as errors represent the mismatch between the predicted and the actual data values, the calculation of error terms also engages the linear modeling process of prediction. Consequently, all four GLM processes are involved in the typical analysis of experimental data. The impression of concise experimental analyses is a consequence of the experimental design acting to simplify the processes of GLM selection, parameter estimation, model checking, and prediction.

1.5 THE "LINEAR" IN GLM

To explain the distinctions required to appreciate model linearity, it is necessary to describe a GLM in more detail. This will be done by outlining the application of a simple regression GLM to data from an experimental study. This example of a regression GLM also will be useful when least square estimates and regression in the context of ANCOVA are discussed.

Consider a situation where the relationship between study time and memory was examined. Twenty-four subjects were divided equally between three study time groups and were asked to memorize a list of 45 words. Immediately after studying the words for 30 seconds (s), 60 s, or 180 s, subjects were given 4 minutes to free recall and write down as many of the words they could remember. The results of this study are presented in Figure 1.1, which follows the convention of plotting

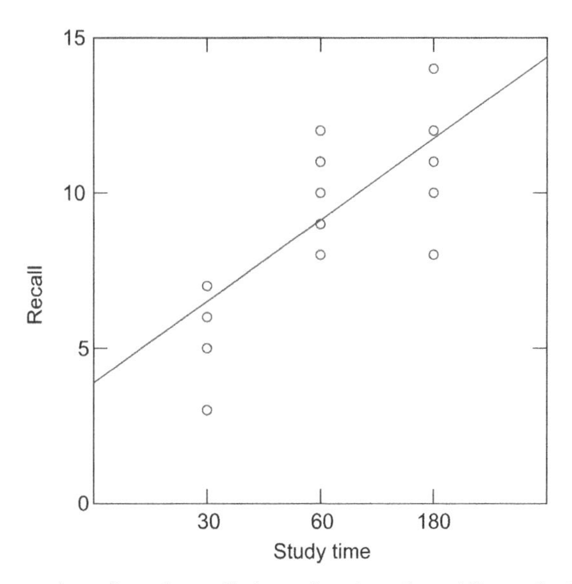

Figure 1.1 The number of words recalled as a function of word list study time. (NB. Some plotted data points depict more than one score.)

independent or predictor variables on the *X*-axis and dependent variables on the *Y*-axis.

Usually, regression is applied to non-experimental situations where the predictor variable can take any value and not just the three time periods defined by the experimental conditions. Indeed, regression usually does not accommodate categorical information about the experimental conditions. Instead, it assesses the linearity of the relationship between the predictor variable (study time) and the dependent variable (free recall score) across all of the data. The relationship between study time and free recall score can be described by the straight line in Figure 1.1 and in turn, this line can be described by equation (1.3)

$$\widehat{Y}_i = \beta_0 + \beta_1 X_i \tag{1.3}$$

where the subscript *i* denotes values for the *i*th subject (ranging from $i = 1, 2, \ldots, N$), $\widehat{Y}_i$ is the predicted dependent variable (free recall) score for the *i*th subject, the parameter β_0 is a constant (the intercept on the *Y*-axis), the parameter β_1 is a regression coefficient (equal to the slope of the regression line), and X_i is the value of the predictor variable (study time) recorded for the same *i*th subject.

As the line describes the relationship between study time and free recall, and equation (1.3) is an algebraic version of the line, it follows that equation (1.3) also describes the relationship between study time and free recall. Indeed, the terms $(\beta_0 + \beta_1 X_1)$ constitute the *model* component of the regression GLM applicable to this data. However, the full GLM equation also includes an error component. The error represents the discrepancy between the scores predicted by the model, through which the regression line passes, and the actual data values. Therefore, the full regression GLM equation that describes the data is

$$Y_i = \beta_0 + \beta_1 X_i + \varepsilon_i \tag{1.4}$$

where Y_i is the observed score for the ith subject and ε_i is the random variable parameter denoting the error term for the same subject. Note that it is a trivial matter of moving the error term to right-hand side of equation (1.4) to obtain the formula that describes the predicted scores

$$\widehat{Y}_i = (Y_i - \varepsilon_i) = \beta_0 + \beta_1 X_i \tag{1.5}$$

Now that some GLM parameters and variables have been specified, it makes sense to say that GLMs can be described as being linear with respect to both their parameters and predictor variables. Linear in the parameters means no parameter is multiplied or divided by another, nor is any parameter above the first power. Linear in the predictor variables also means no variable is multiplied or divided by another, nor is any above the first power. However, as shown below, there are ways around the variable requirement.

For example, equation (1.4) above is linear with respect to both parameters and variables. However, the equation

$$Y_i = \beta_0 + \beta_1^2 X_i + \varepsilon_i \tag{1.6}$$

is linear with respect to the variables, but not to the parameters, as β_1 has been raised to the second power. Linearity with respect to the parameters also would be violated if any parameters were multiplied or divided by other parameters or appeared as exponents. In contrast, the equation

$$Y_i = \beta_0 + \beta_1 X_i^2 + \varepsilon_i \tag{1.7}$$

is linear with respect to the parameters, but not with respect to the variables, as X_i^2 is X_i raised to the second power. However, it is very simple to define $Z_i = X_i^2$ and to substitute Z_i in place of X_i^2. Therefore, models such as described by equation (1.7) continue to be termed linear, whereas such as those described by equation (1.6) do not. In short, linearity is presumed to apply only to the parameters. Models that are not linear with respect to their parameters are described specifically as nonlinear. As a result, models can be assumed to be linear with respect to their parameters, unless specified otherwise, and frequently the term linear is omitted.

Nevertheless, the term "linear" in GLM often is misunderstood to mean that the relation between any data and any predictor variable must be described by a straight line. Although GLMs can describe straight-line relationships, they are capable of much more. Through the use of transformations and polynomials, GLMs can describe many complex curvilinear relations between the data and the predictor variables (e.g., Draper and Smith, 1998; Kutner et al., 2005).

1.6 LEAST SQUARES ESTIMATES

Parameters describe or apply to populations. However, it is rare for data from whole populations to be available. Much more available are samples of these populations. Consequently, parameters usually are estimated from sample data. A standard form of distinction is to use Greek letters, such as α and β, to denote parameters and to place a hat on them (e.g., $\widehat{\alpha}$, $\widehat{\beta}$), when they denote parameter estimates. Alternatively, the ordinary letter equivalents, such as a and b, may be used to represent the parameter estimates.

The parameter estimation method underlying all of the analyses presented in Chapters 2–11 is that of *least squares*. Some alternative parameter estimation methods are discussed briefly in Chapter 12. Although these alternatives are much more computationally demanding than least squares, their use has increased with greater availability and access to computers and relevant software. Nevertheless, least squares remains by far the most frequently applied parameter estimation method.

The least squares method identifies parameter estimates that minimize the sum of the squared discrepancies between the predicted and the observed values. From the GLM equation

$$Y_i = \beta_0 + \beta_1 X_i + \varepsilon_i \qquad (1.4, \text{rptd})$$

the sum of the squared deviations may be described as

$$\sum_{i=1}^{N} \varepsilon_i^2 = \sum_{i=1}^{N} (Y_i - \beta_0 - \beta_1 X_1)^2 \qquad (1.8)$$

The estimates of β_0 and β_1 are chosen to provide the smallest value of $\sum_{i=1}^{N} \varepsilon_i^2$. By differentiating equation (1.8) with respect to each of these parameters, two (simultaneous) normal equations are obtained. (More GLM parameters require more differentiations and produce more normal equations.) Solving the normal equations for each parameter provides the formulas for calculating their least squares estimates and in turn, all other GLM (least squares) estimates.

Least squares estimates have a number of useful properties. Employing an estimate of the parameter β_0 ensures that the residuals sum to zero. Given that the error terms also are uncorrelated with constant variance, the least squares estimators will be unbiased and will have the minimum variance of all unbiased linear estimators. As a result they are termed the *best linear unbiased estimators* (BLUE). However, for conventional significance testing, it is also necessary to assume that the errors are distributed normally. (Checks of these and other assumptions are considered in Chapter 10. For further details of least squares estimates, see Kutner et al., 2005; Searle, 1987.) However, when random variables are employed in GLMs, least squares estimation requires the application of restrictive constraints (or assumptions) to allow the normal equations to be solved. One way to escape from these constraints is to employ a different method of parameter estimation. Chapter 12 describes the use of some different parameter estimation methods, especially restricted maximum likelihood (REML), to estimate parameters in repeated measures designs where subjects are accommodated as levels of a random factor. Current reliance on computer-based maximum likelihood parameter estimation suggests this is a recent idea but, in fact, it is yet another concept advanced by Fisher (1925, 1934), although it had been used before by others, such as Gauss, Laplace, Thiele, and Edgeworth (see Stigler, 2002).

1.7 FIXED, RANDOM, AND MIXED EFFECTS ANALYSES

Fixed, random, and mixed effects analyses refer to different sampling situations. Fixed effects analyses employ only fixed variables in the GLM *model* component, random effects analyses employ only random variables in the GLM *model* component, while mixed effects analyses employ both fixed and random variables in the GLM *model* component.

When a fixed effects analysis is applied to experimental data, it is assumed that all the experimental conditions of interest are included in the experiment. This assumption is made because the inferences made on the basis of a fixed effects analysis apply fully only to the conditions included in the experiment. Therefore, the experimental conditions used in the original study are fixed in the sense that exactly the same conditions must be employed in any replication of the study. For most genuine experiments, this presents little problem. As experimental conditions usually are chosen deliberately and with some care, so fixed effects analyses are appropriate for most experimental data (see Keppel and Wickens, 2004, for a brief discussion). However, when ANOVA is applied to data obtained from non-experimental studies, care should be exercised in applying the appropriate form of analysis. Nevertheless, excluding estimates of the magnitude of experimental effects, it is not until factorial designs are analyzed that differences between the estimates of fixed and random effects are apparent.

Random effects analyses consider those experimental conditions employed in the study to be only a random sample of a population of experimental conditions and so, inferences drawn from the study may be applied to the wider population of conditions. Consequently, study replications need not be restricted to exactly the same experimental conditions. As inferences from random effects analyses can be generalized more widely than fixed effects inferences, all else being equal, more conservative assessments are provided by random effects analyses.

In psychology, mixed effects analyses are encountered most frequently with respect to related measures designs. The measures are related by virtue of arising from the same subject (repeated measures designs) or from related subjects (matched samples designs, etc.) and accommodating the relationship between these related scores makes it possible to identify effects uniquely attributable to the repeatedly measured subjects or the related subjects. This subject effect is represented by a random variable in the GLM model component, while the experimental conditions continue as fixed effects. It is also possible to define a set of experimental conditions as levels of a random factor and mix these with other sets of experimental conditions defined as fixed factors in factorial designs, with or without a random variable representing subjects. However, such designs are rare in psychology.

Statisticians have distinguished between regression analyses, which assume fixed effects, and correlation analyses, which do not. Correlation analyses do not distinguish between predictor and dependent variables. Instead, they study the degree of relation between random variables and are based on bivariate-normal models. However, it is rare for this distinction to be maintained in practice. Regression is applied frequently to situations where the sampling of predictor variables is random and where replications employ predictors with values different to those used in the original study. Indeed, the term regression now tends to be interpreted simply as an analysis that predicts one variable on the basis of one or more other variables, irrespective of their fixed or random natures (Howell, 2010). Supporting this approach is the demonstration that provided the other analysis assumptions are tenable, the least square parameter estimates and F-tests of significance continue to apply even with random predictor and dependent variables (Kmenta, 1971; Snedecor and Cochran, 1980; Wonnacott and Wonnacott, 1970).

All of the analyses described in this book consider experimental conditions to be fixed. However, random effects are considered with respect to related measures designs and some consideration is given to the issue of fixed and random predictor variables in the context of ANCOVA assumptions. Chapter 12 also presents recent mixed model approaches to repeated measures designs where maximum likelihood is used to estimate a fixed experimental effect parameter and a random subject parameter.

1.8 THE BENEFITS OF A GLM APPROACH TO ANOVA AND ANCOVA

The pocket history of regression and ANOVA described their separate development and the subsequent appreciation and utilization of their communality, partly as a consequence of computer-based data analysis that promoted the use of their common matrix algebra notation. However, the single fact that the GLM subsumes regression, ANOVA, and ANCOVA seems an insufficient reason to abandon the traditional manner of carrying out these analyses and adopt a GLM approach. So what is the motivation for advocating the GLM approach?

The main reason for adopting a GLM approach to ANOVA and ANCOVA is that it provides conceptual and practical advantages over the traditional approach. Conceptually, a major advantage is the continuity the GLM reveals between regression, ANOVA, and ANCOVA. Rather than having to learn about three apparently discrete techniques, it is possible to develop an understanding of a consistent modeling approach that can be applied to different circumstances. A number of practical advantages also stem from the utility of the simply conceived and easily calculated error terms. The GLM conception divides data into model and error, and it follows that the better the model explains the data, the less the error. Therefore, the set of predictors constituting a GLM can be selected by their ability to reduce the error term.

Comparing a GLM of the data that contains the predictor(s) under consideration with a GLM that does not, in terms of error reduction, provides a way of estimating effects that is both intuitively appreciable and consistent across regression, ANOVA, and ANCOVA applications. Moreover, as most GLM assumptions concern the error terms, residuals-the error term estimates, provide a common means by which the assumptions underlying regression, ANOVA, and ANCOVA can be assessed. This also opens the door to sophisticated statistical techniques, developed primarily to assist linear modeling/regression error analysis, to be applied to both ANOVA and ANCOVA. Recognizing ANOVA and ANCOVA as instances of the GLM also provides connection to an extensive and useful literature on methods, analysis strategy, and related techniques, such as structural equation modeling, multilevel analysis (see Chapter 12) and generalized linear modeling, which are pertinent to experimental and non-experimental analyses alike (e.g., Cohen et al., 2003; Darlington, 1968; Draper and Smith, 1998; Gordon, 1968; Keppel and Zedeck, 1989; McCullagh and Nelder, 1989; Mosteller and Tukey, 1977; Nelder, 1977; Kutner et al., 2005; Pedhazur, 1997; Rao, 1965; Searle,1979, 1987, 1997; Seber, 1977).

1.9 THE GLM PRESENTATION

Several statistical texts have addressed the GLM and presented its application to ANOVA and ANCOVA. However, these texts differ in the kinds of GLM they employ to describe ANOVA and ANCOVA and how they present GLM calculations. ANOVA and ANCOVA have been expressed as cell mean GLMs (Searle, 1987) and regression GLMs (e.g., Cohen et al., 2003; Judd, McClelland, and Ryan, 2008; Keppel and Zedeck, 1989; Pedhazur, 1997). Each of these expressions has some merit. (See Chapter 2 for further description and consideration of experimental design, regression and cell mean GLMs.) However, the main focus in this text is experimental design GLMs, which also may be known as structural models or effect models.

Irrespective of the form of expression, GLMs may be described and calculated using scalar or matrix algebra. However, scalar algebra equations become increasingly unwieldy and opaque as the number of variables in an analysis increases. In contrast, matrix algebra equations remain relatively succinct and clear. Consequently, matrix algebra has been described as concise, powerful, even elegant, and as providing better appreciation of the detail of GLM operations than scalar algebra. These may seem peculiar assertions given the difficulties people experience doing matrix algebra calculations, but they make sense when a distinction between theory and practice is considered. You may be able to provide a clear theoretical description of how to add numbers together, but this will not eliminate errors if you have very many numbers to add. Similarly, matrix algebra can summarize succinctly and clearly matrix relations and manipulations, but the actual laborious matrix calculations are best left to a computer. Nevertheless, while there is much to recommend matrix algebra for expressing GLMs, unless you have some serious mathematical expertise, it is likely to be an unfamiliar notation. As it is expected that many readers of this text will not be well versed in matrix algebra, primarily scalar algebra and verbal descriptions will be employed to facilitate comprehension.

1.10 STATISTICAL PACKAGES FOR COMPUTERS

Most commercially available statistical packages have the capability to implement regression, ANOVA, and ANCOVA. The interfaces to regression and ANOVA programs reflect their separate historical developments. Regression programs require the specification of predictor variables, and so on, while ANOVA requires the specification of experimental independent variables or factors, and so on. ANCOVA interfaces tend to replicate the ANOVA approach, but with the additional requirement that one or more covariates are specified. Statistical software packages offering GLM programs are common (e.g., GENSTAT, MINITAB, STATISTICA, SYSTAT) and indeed, to carry out factorial ANOVAs with SPSS requires the use of its GLM program.

All of the analyses and graphs presented in this text were obtained using the statistical package, SYSTAT. (For further information on SYSTAT, see Appendix A.) Nevertheless, the text does not describe how to conduct analyses using SYSTAT or any other statistical package. One reason for taking this approach is that frequent upgrades to statistical packages soon makes any reference to statistical software obsolete. Another reason for avoiding implementation instructions is that in addition to the extensive manuals and help systems accompanying statistical software, there are already many excellent books written specifically to assist users in carrying out analyses with the major statistical packages and it is unlikely any instructions provided here would be as good as those already available. Nevertheless, despite the absence of implementation instructions, it is hoped that the type of account presented in this text will provide not only an appreciation of ANOVA and ANCOVA in GLM terms but also an understanding of ANOVA and ANCOVA implementation by specific GLM or conventional regression programs.

CHAPTER 2

Traditional and GLM Approaches to Independent Measures Single Factor ANOVA Designs

2.1 INDEPENDENT MEASURES DESIGNS

The type of experimental design determines the particular form of ANOVA that should be applied. A wide variety of experimental designs and pertinent ANOVA procedures are available (e.g., Kirk, 1995). The simplest of these are independent measures designs. The defining feature of independent measures designs is that the dependent variable scores are assumed to be statistically independent (i.e., uncorrelated). In practice, this means that subjects are selected randomly from the population of interest and then allocated to only one of the experimental conditions on a random basis, with each subject providing only one dependent variable score.

Consider the independent measures design with three conditions presented in Table 2.1. Here, the subjects' numbers indicate their chronological allocation to conditions. Subjects are allocated randomly with the *proviso* that one subject has been allocated to all of the experimental conditions before a second subject is allocated to any experimental condition. When this is done, a second subject is allocated randomly to an experimental condition and only after two subjects have been allocated randomly to the other two experimental conditions is a third subject allocated randomly to one of the experimental conditions, and so on. This is a simple allocation procedure that distributes any subject (or subject-related) differences that might vary over the time course of the experiment randomly across conditions. It is useful generally, but particularly if it is anticipated that the experiment will take a considerable time to complete. In such circumstances, it is possible that subjects recruited at the start of the experiment may differ in relevant and so important ways from subjects recruited toward the end of the experiment. For example, consider an experiment being

ANOVA and ANCOVA: A GLM Approach, Second Edition. By Andrew Rutherford.

Table 2.1 Subject Allocation for an Independent Measures Design with Three Conditions

Condition A	Condition B	Condition C
Subject 3	Subject 2	Subject 1
Subject 5	Subject 6	Subject 4
Subject 8	Subject 9	Subject 7
Subject 12	Subject 10	Subject 11

run over a whole term at a university, where student subjects participate in the experiment to fulfill a course requirement. Those students who sign up to participate in the experiment at the beginning of the term are likely to be well-motivated and organized students. However, students signing up toward the end of the term may be those who do so because time to complete their research participation requirement is running out. These students are likely to be motivated differently and may be less organized. Moreover, as the end-of-term examinations approach, these students may feel time pressured and be less than positive about committing the time to participate in the experiment. The different motivations, organization, and emotional states of those subjects recruited at the start and toward the end of the experiment may have some consequence for the behavior(s) measured in the experiment. Nevertheless, the allocation procedure just described ensures that subjects recruited at the start and at the end of the experiment are distributed across all conditions. Although any influence due to subject differences cannot be removed, they are prevented from being related systematically to conditions and confounding the experimental manipulation(s).

To analyze the data from this experiment using *t*-tests would require the application of, at least, two *t*-tests. The first might compare Conditions A and B, while the second would compare Conditions B and C. A third *t*-test would be needed to compare Conditions A and C. The problem with such a *t*-test analysis is that the probability of a Type 1 error (i.e., rejecting the null hypothesis when it is true) increases with the number of hypotheses tested. When one hypothesis test is carried out, the likelihood of a Type 1 error is equal to the significance level chosen (e.g., 0.05), but when two independent hypothesis tests are applied, it rises to nearly double the tabled significance level, and when three independent hypothesis tests are applied, it rises to nearly three times the tabled significance level. (In fact, as three *t*-tests applied to this data would be related, although the Type 1 error inflation would be less than is described for three independent tests, it still would be greater than 0.05—see Section 3.6.)

In contrast, ANOVA simultaneously examines for differences between any number of conditions while holding the Type 1 error at the chosen significance level. In fact, ANOVA may be considered as the *t*-test extension to more than two conditions that holds Type 1 error constant. This may be seen if ANOVA is applied to compare two conditions. In such situations, the relationship between *t*- and *F*-values is

$$t^2_{(df)} = F_{(1,df)} \tag{2.1}$$

where *df* is the denominator degrees of freedom. Yet despite this apparently simple relationship, there is still room for confusion. For example, imagine data obtained from an experiment assessing a directional hypothesis, where a one-tailed *t*-test is applied. This might provide

$$t_{(20)} = 1.725, \ p = 0.05$$

However, if an ANOVA was applied to exactly the same data, in accordance with equation (2.1) the *F*-value obtained would be

$$F_{(1,20)} = 2.976, \ p = 0.100$$

Given the conventional significance level of 0.05, the one-tailed *t*-value is significant, but the *F*-value is not. The reason for such differences is that the *F*-value probabilities reported by tables and computer output are always two-tailed.

Directional hypotheses can be preferable for theoretical and statistical reasons. However, MacRae (1995) emphasizes that one consequence of employing directional hypotheses is any effect in the direction opposite to that predicted must be interpreted as a chance result—irrespective of the size of the effect. Few researchers would be able, or willing, to ignore a large and significant effect, even when it is in the direction opposite to their predictions. Nevertheless, this is exactly what all researchers should do if a directional hypothesis is tested. Therefore, to allow further analysis of such occurrences, logic dictates that nondirectional hypotheses always should be tested.

2.2 BALANCED DATA DESIGNS

The example presented in Table 2.1 assumes a balanced data design. A balanced data design has the same number of subjects in each experimental condition. There are three reasons why this is a good design practice.

First, generalizing from the experiment is easier if the complication of uneven numbers of subjects in experimental conditions (i.e., unbalanced data) is avoided. In ANOVA, the effect of each experimental condition is weighted by the number of subjects contributing data to that condition. Giving greater weight to estimates derived from larger samples is a consistent feature of statistical analysis and is entirely appropriate when the number of subjects present in each experimental condition is unrelated to the nature of the experimental conditions. However, if the number of subjects in one or more experimental conditions is related to the nature of these conditions, it may be appropriate to replace the conventional weighted means analysis with an unweighted means analysis (e.g., Winer, Brown, and Michels, 1991). Such an analysis gives the same weight to all condition effects, irrespective of the number of subjects contributing data in each condition. In the majority of experimental studies, the number of subjects present in each experimental condition is unrelated to the nature of the experimental conditions. However, this issue needs to be given greater consideration when more applied or naturalistic studies are conducted or intact groups are employed. The second reason why it is a good design practice to employ balanced data is due to terms accommodating the different numbers per group canceling out, the mathematical formulas for ANOVA with equal numbers of subjects in each experimental condition simplify with a reduction in the computational requirement. This makes the ANOVA formulas much easier to understand, apply, and interpret. The third reason why it is good design practice to employ balanced data is ANOVA is robust with respect to certain assumption violations (i.e., distribution normality and variance homogeneity) when there are equal numbers of subjects in each experimental condition (see Sections 10.4.1.2 and 10.4.1.4).

The benefits of balanced data outlined above are such that it is worth investing some effort to achieve. In contrast, McClelland (1997) argues that experimental design power should be optimized by increasing the number of subjects allocated to key experimental conditions. As most of these optimized experimental designs are also unbalanced data designs, McClelland takes the view that it is worth abandoning the ease of calculation and interpretation of parameter estimates, and the robust nature of ANOVA with balanced data to violations of normality and homogeneity of variance assumptions, to obtain an optimal experimental design (see Section 4.7.4). Nevertheless, all of the analyses presented in this chapter employ balanced data and it would be wrong to presume that unbalanced data analyzed in exactly the same way would provide the same results and allow the same interpretation. Detailed consideration of unbalanced designs may be found in Searle (1987).

2.3 FACTORS AND INDEPENDENT VARIABLES

In the simple hypothetical experiment above, the same number of subjects was allocated to each of the three experimental conditions, with each condition receiving a different amount of time to study the same list of 45 words. Shortly after, all of the subjects were given 4 minutes to free recall and write down as many of these words as they could remember (see Section 1.5).

The experimental conditions just outlined are distinguished by quantitative differences in the amount of study time available and so one way to analyze the experimental data would be to conduct a regression analysis similar to that reported in Section 1.5. This certainly would be the preferred form of analysis if the theory under test depended upon the continuous nature of the study time variable (e.g., Cohen, 1983; Vargha et al., 1996). However, where the theory tested does not depend on the continuous nature of the study time, it makes sense to treat the three different study times as experimental conditions (i.e., categories) and compare across the conditions without regard for the size of the time differences between the conditions.

Although experimental condition study times are categorical, it still is reasonable to label the independent variable as *Study time*. Nevertheless, when categorical comparisons are applied generally, the experimenter needs to keep the actual differences between the experimental conditions in mind. For example, Condition A could be changed to one in which some auditory distraction is presented. Obviously, this would invalidate the independent variable label *Study time*, but it would not invalidate exactly the same categorical comparisons of memory performance under these three different conditions. The point here is to draw attention to the fact that the levels of a qualitative factor may involve multidimensional distinctions between conditions. While there should be some logical relation between the levels of any factor, they may not be linked in such a continuous fashion as is suggested by the term *independent variable*. So, from now on, the label, *Factor*, will be used in preference.

2.4 AN OUTLINE OF TRADITIONAL ANOVA FOR SINGLE FACTOR DESIGNS

ANOVA is employed in psychology most frequently to address the question—are there significant differences between the mean scores obtained in the different experimental conditions? As the name suggests, ANOVA operates by comparing the *sample* score variation observed between groups with the *sample* score variation observed within groups. If the experimental manipulations exert a real influence, then subjects' scores should vary more between the experimental conditions than within the experimental conditions. ANOVA procedures specify the calculation of an F-value, which is the ratio of between groups to within groups variation. Between groups variation depends on the difference between the group (experimental condition) means, whereas the within groups variation depends on the variation of the individual scores around their group (experimental condition) means. When there are no differences between the group (experimental condition) means, the estimates of between group and within group variation will be equal and so their ratio, the calculated F-value, will equal 1. When differences between experimental condition means increase, the between groups variation increases, and provided the within groups variation remains fairly constant, the size of the calculated F-value will increase.

The purpose of calculating an F-value is to determine whether the differences between the experimental condition means are significant. This is accomplished by comparing the calculated F-value with the sampling distribution of the F-statistic. The F-statistic sampling distribution reflects the probability of different F-values occurring when the null hypothesis is true. The null hypothesis states that no differences exist between the means of the experimental condition *populations*. If the null hypothesis is true and the *sample* of subjects and their scores accurately reflect the *population* under the null hypothesis, then between group and within group variation estimates will be equal and the calculated F-value will equal 1. However, due to chance *sampling* variation (sometimes called sampling error), it is possible to observe differences between the experimental condition means of the data *samples*.

The sampling distribution of the F-statistic can be established theoretically and empirically (see Box 2.1). Comparing the calculated F-value with the pertinent F-distribution (i.e., the distribution with equivalent *dfs*) provides the probability of observing an F-value equal to or greater than that calculated from randomly sampled data collected under the null hypothesis. If the probability of observing this F-value under the null hypothesis is sufficiently low, then the null hypothesis is rejected and

BOX 2.1

The F-distribution for the three-condition experiment outlined in Table 2.1 can be established empirically under the null hypothesis in the following way.

Assume a normally distributed population of 1000 study scores and identify the population mean and standard deviation. (The mean and standard deviation fully describe a normal distribution, so on this basis it is possible to identify the 1000 scores.) Take 1000 ping-pong balls and write a single score on each of the 1000 ping-pong balls and put all of the ping-pong balls in a container. Next, randomly select a ball and then randomly, place it into one of the three baskets, labeled Condition A, B, and C. Do this repeatedly until you have selected and placed 12 balls, with the constraint that you must finish with 4 balls in each condition basket. When complete, use the scores on the ping-pong balls in each of the A, B, and C condition baskets to calculate an F-value and plot the calculated F-value on a frequency distribution. Replace all the balls in the container. Next, randomly sample and allocate the ping-pong balls just as before, calculate an F-value based on the ball scores just as before and plot the second F-value on the frequency distribution. Repeat tens of thousands of times. The final outcome will be the sampling distribution of the F-statistic under the null hypothesis when the numerator has two *dfs* (numerator *dfs* = number of groups − 1) and the denominator has three *dfs* (denominator *dfs* = number of scores per group − 1). This empirical distribution has the same shape as the distribution predicted by mathematical theory. It is important to appreciate that the score values do not influence the shape of the sampling distribution of the F-statistic, i.e., whether scores are distributed around a mean of 5 or 500 does not affect the sampling distribution of the F-statistic. The only influences on the sampling distribution of the F-statistic are the numerator and denominator *dfs*. As might be expected, the empirical investigation of statistical issues has moved on a pace with developments in computing and these empirical investigations often are termed Monte Carlo studies.

the experimental hypothesis is accepted. The convention is that sufficiently low probabilities begin at $p = 0.05$. The largest 5% of F-values—the most extreme 5% of F-values in the right-hand tail of the F-distribution under the null hypothesis—have probabilities of ≤ 0.05 (see Figure 2.1). In a properly controlled experiment, the only reason for differences between the experimental condition means should be the experimental manipulation. Therefore, if the probability of the difference(s) observed occurring due to sampling variation is less than the criterion for significance, then it is reasonable to conclude that the differences observed were caused by the experimental manipulation. (For an introduction to the logic of experimental design and the relationship between scientific theory and experimental data, see Hinkelman and Kempthorne, 2008; Maxwell and Delaney, 2004.)

Kirk (1995, p. 96) briefly describes the F-test as providing "a one-tailed test of a nondirectional null hypothesis because MSBG, which is expected to be greater than or

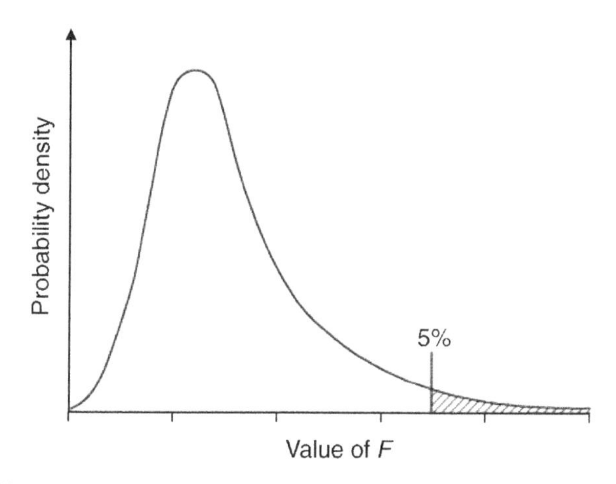

Figure 2.1 A typical distribution of F under the null hypothesis.

approximately equal to MSWG, is always in the numerator of the F statistic." (MSBG and MSWG denote the mean squares of between and within groups variance, respectively, and the F-ratio is the ratio of these two mean square estimates. Mean square estimation is described in Section 2.5.) Although perfectly correct, Kirk's description can cause confusion and obscure the reason for the apparently different t- and F-test results mentioned in Section 2.1. As Kirk says, the F-statistic in ANOVA is one-tailed because MSBG, which reflects experimental effects, is always the numerator. MSBG is always the numerator because *when the null hypothesis is false* MSBG should be greater than MSWG and the calculated F-statistic should be >1. (MSBG and MSWG are expected to be equal and $F = 1$ only *when the null hypothesis is true*.) As $F = 1$ when the influence of the experimental manipulation is zero and any influence of the experimental manipulation should provide $F > 1$, only the right-hand tail of the F-distribution needs to be examined. Consequently, the F-test is one-tailed, but *not* because it tests a directional hypothesis. In fact, the nature of the F-test numerator (MSBG) ensures the F-test always assesses a nondirectional hypothesis. The MSBG is obtained from the sum of the squared differences between the condition means, but squaring the differences between the means gives the same positive valence to all of the mean differences. Consequently, the directionality of the differences between mean is lost and so the F-test is nondirectional.

2.5 VARIANCE

Variance or variation is a vital concept in ANOVA and many other statistical techniques. Nevertheless, it can be a puzzling notion, particularly the concept of total variance. Variation measures how much the observed or calculated scores deviate from something. However, while between group variance reflects the deviation amongst condition means and within group variance reflects the deviation of scores from their condition means, it is less obvious what total variance reflects. In fact, the total variance reflects the deviation of all the observed scores from the mean of all these scores.

Before this can be illustrated, some definitions are necessary. The most frequently employed measure of central tendency is the arithmetic average or mean ($\overline{Y}$). This is defined as

$$\overline{Y} = \frac{\sum_{i=1}^{N} Y_i}{N} \tag{2.2}$$

where Y_i is the ith subject's score, $\sum_{i=1}^{N} Y_i$ is the sum of all of the subjects' scores, and N is the total number of subjects. The subscript i indexes the individual subjects and in this instance it takes the values from 1 to N. The $\sum_{i=1}^{N}$ indicates that summation occurs over all the i subject scores, from 1 to N. In turn, the population variance (σ^2) is defined as

$$\sigma^2 = \frac{\sum_{i=1}^{N} (Y_i - \overline{Y})^2}{N} \tag{2.3}$$

Therefore, variance reflects the average of the squared deviations from the mean. In other words, the variance reflects the square of the average extent to which scores differ from the mean. Equation (2.3) defines the population variance. However, it provides a biased estimate—an underestimate—of the variance of a sample drawn from a population. (This is due to the loss of a *df* from the denominator because the mean, which is based on the same set of scores, is used in this calculation—see Section 2.6). An unbiased estimate of the sample variance (s^2) is given by

$$s^2 = \frac{\sum_{i=1}^{N} (Y_i - \overline{Y})^2}{N - 1} \tag{2.4}$$

Nevertheless, while formulas (2.3) and (2.4) reveal the nature of variance quite well, they do not lend themselves to easy calculation. A useful formula for calculating sample variance (s^2) is

$$s^2 = \frac{\sum_{i=1}^{N} Y_i^2 - \left[\left(\sum_{i=1}^{N} Y_i\right)^2 / N\right]}{N - 1} \tag{2.5}$$

The standard deviation also is a very useful statistic and is simply the square root of the variance. Consequently, the population standard deviation (σ) is given by

$$\sigma = \sqrt{\frac{\sum_{i=1}^{N} (Y_i - \overline{Y})^2}{N}} \tag{2.6}$$

and the sample standard deviation (s) is given by

$$s = \sqrt{\frac{\sum_{i=1}^{N} Y_i^2 - \left[\left(\sum_{i=1}^{N} Y_i\right)^2 / N\right]}{N - 1}} \tag{2.7}$$

2.6 TRADITIONAL ANOVA CALCULATIONS FOR SINGLE FACTOR DESIGNS

The relation between total variance, between group variance, and within group variance can be illustrated using the hypothetical experimental data presented in Chapter 1. Figure 2.2 and Table 2.2 present the data provided by the 24 subjects over the three experimental conditions. (*Note*: In Table 2.2, the subject number does not reflect the chronological assignment to groups as in Table 2.1. Instead, the convention is to label the subjects in each experimental condition from 1 to N_j, where N_j denotes the number of subjects participating in each condition.)

Table 2.2 Hypothetical Experimental Data and Summary Statistics

Subjects	30 s	Subjects	60 s	Subjects	180 s
s1	7	s1	7	s1	8
s2	3	s2	11	s2	14
s3	6	s3	9	s3	10
s4	6	s4	11	s4	11
s5	5	s5	10	s5	12
s6	8	s6	10	s6	10
s7	6	s7	11	s7	11
s8	7	s8	11	s8	12
Total $\sum_{i=1}^{N_j} Y_{ij}$	48	$\sum Y_i$	80	$\sum_{i=1}^{N} Y_i$	88
$\sum Y_{ij}^2$	304	$\sum Y_i^2$	814	$\sum_{i=1}^{N} Y_i^2$	990
Mean ($\overline{Y}_j$)	6.00	$\overline{Y}_j$	10.00	$\overline{Y}_j$	11.00
Sample SD (s)	1.51	s	1.41	s	1.77
Sample variance (s^2)	2.29	s^2	1.99	s^2	3.13

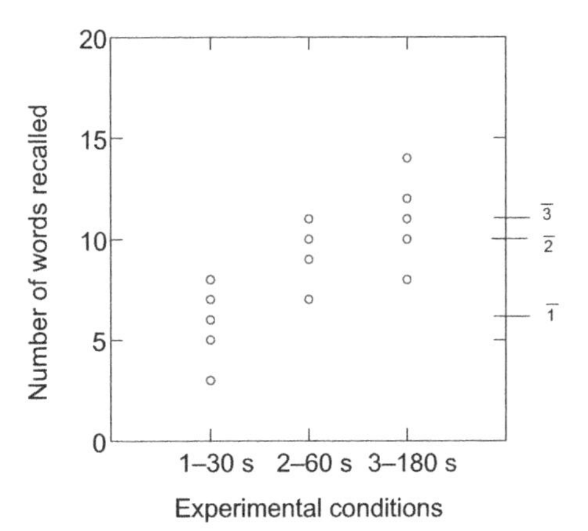

Figure 2.2 Dependent variable scores (number of words free recalled) by experimental condition.

Figure 2.2 presents subjects' dependent variable (free recall) scores plotted against the experimental condition under which the scores were recorded. The means of conditions 1 (30 s study time), 2 (60 s study time), and 3 (180 s study time) are marked on the right Y-axis by $\overline{1}$, $\overline{2}$, and $\overline{3}$. Table 2.2 provides summary statistics for the same data.

The vertical spread of the scores in Figure 2.2 provides some sense of the between groups and within groups variance. Between groups variance reflects the differences between the means of the experimental conditions (see distances between $\overline{1}$, $\overline{2}$, and $\overline{3}$ to the right of the graph). Within groups variance reflects the (average) spread of individual scores within each of the experimental conditions. Examination of Figure 2.2 suggests that the 180 s condition has the greatest within group variation and the 30 s condition has greater within group variation than the 60 s condition. These assessments are confirmed by the statistics presented in Table 2.2.

A numerical subscript is used to identify the experimental conditions in formulas, etc. The subscript j indexes the experimental conditions, and ranges from 1 to p, where p is the number of experimental conditions. In this example, the j indicator values are 1, 2, and 3, indexing, respectively, the 30 s, 60 s, and 180 s conditions.

The use of the N (or $N-1$) as the major denominator in formulae (2.2)–(2.7) reveals that these variance estimates reflect the amount of deviation from the mean, averaged across all of the subjects' scores. However, the initial variance estimates calculated in ANOVA, known as sums of squares (SS), are not averaged in this fashion.

Total sum of squares is conceived as the deviation of all the observed scores from the general mean (the mean of the experimental condition means). For this calculation, the only modification to formula (2.5) is to exclude the denominator

$$\begin{aligned} SS_{\text{total}} &= \sum_{i=1}^{N} Y_i^2 - \frac{\left(\sum_{i=1}^{N} Y_i\right)^2}{N} \\ &= 2108 - \frac{(216)^2}{24} \\ SS_{\text{total}} &= 164.000 \end{aligned} \tag{2.8}$$

Within group variance is conceived in terms of the average of the separate spreads of scores from the mean in each of the experimental conditions. The separate experimental condition variance estimates averaged over all of the scores in each experimental condition are provided in Table 2.2. These estimates can be obtained by applying formulae (2.5) and (2.7) to the scores in each of the experimental conditions. Again the components of the *within group sum of squares* can be calculated using the numerator of equation (2.5), as shown in equation (2.9) below.

$$\sum_{i=1}^{N_j} Y_{ij}^2 - \left[\frac{\left(\sum_{i=1}^{N_j} Y_{ij}\right)^2}{N_j}\right] \tag{2.9}$$

Notice that as the group or experimental condition under which the dependent variable scores were obtained now is relevant, the j subscript is included in the equation. (N denotes the total number of subjects participating in the experiment and the subscript j is used here to refer to the number of subjects in each condition. An alternative is to use n to denote the number of subjects per condition.)

For the 30 s experimental condition, this gives

$$\sum_{i=1}^{N_j} Y_{ij}^2 - \left[\frac{\left(\sum_{i=1}^{N_j} Y_{ij}\right)^2}{N_j}\right] \tag{2.10}$$

$$304 - \left[\frac{48^2}{8}\right]$$

In fact, this value is simply

$$(N_j - 1)\,(s^2) = 7\,(2.29) = 16.03 \text{ (discrepancy is due to rounding error)} \tag{2.11}$$

Similar calculations for the 60 s and 180 s experimental conditions provide the other components. The *within groups sum of squares* is calculated by summing the separate experimental condition sum of squares estimates

$$\begin{aligned} \text{SS}_{\text{within groups}} &= SS_1 + SS_2 + SS_3 \\ &= 16.000 + 14.000 + 22.000 \\ &= 52.000 \end{aligned}$$

This estimate of error variance is known as a pooled error variance estimate. As well as being an average of the experimental condition error variances, equation (2.11) reveals that the pooled error variance estimate is weighted by the individual experimental condition *dfs*. As each individual experimental condition *dfs* is equal to the sample size minus 1, it often is said the pooled error variance estimate is weighted by sample size. It is also possible to estimate an unpooled error variance, which simply averages the individual experimental condition sample variance error estimates. However, the calculation of unpooled error variance estimate *dfs* employ Satterthwaite's (1946) solution, which is a little involved. Of course, with balanced designs, both pooled and unpooled error variance estimates are identical. However, the use of pooled error variance estimates becomes less tenable as the experimental condition sample variance error estimates diverge.

Between groups variance is conceived in terms of the average of the differences among the means of the experimental conditions. However, these differences are not as simple as the mean of the 30 s condition minus the mean of the 60 s condition, and so on. Instead, the variance attributable to the differences between the condition means and the general mean are estimated. Although experimental condition means are obtained by averaging over all the subjects in the particular condition, each experimental condition mean is regarded as the score each subject in that experimental condition would record if error variation was eliminated. Consequently, in each experimental condition, there would be N_j experimental condition mean scores.

$$\begin{aligned} SS_{\text{between groups}} &= \sum_{j=1}^{p} N_j(\overline{Y}_j - \overline{Y}_G)^2 \\ &= 8(6-9)^2 + 8(10-9)^2 + 8(11-9)^2 \\ SS_{\text{between groups}} &= 112.000 \end{aligned} \tag{2.12}$$

The fact that

$$SS_{\text{total}} = SS_{\text{between groups}} + SS_{\text{within groups}} \tag{2.13}$$

can be verified easily by substituting any two of the estimates calculated for two of the terms above

$$164.000 = 112.000 + 52.000$$

The sum of squares calculations estimate the variation attributable to between groups and within groups sources. The next step in traditional ANOVA calculation is to estimate the average variance arising from between groups and within groups sources. This step requires SS denominators to provide the averages. The denominators are termed degrees of freedom and the averages they provide are termed mean squares (MS).

Degrees of freedom represent how many of the data points employed in constructing the estimate are able to take different values. For example, when sample variance is calculated using N as a denominator, underestimates of variance are obtained because in fact, there are not N *dfs*, but only $(N-1)$ *dfs*. When the correct *dfs* are used as the denominator, an accurate estimate of sample variance is obtained.

The reason there are $(N-1)$ and not N *dfs* is one *df* is lost from the sample variance because a mean is used in the sample variance calculation. Once a mean is determined for a group of scores, it is always possible to state the value of the "last" score in that group. Internal consistency demands that this "last" score takes the value that provides the appropriate sum of scores, which, when divided by the number of scores, gives the previously calculated mean. For example, for the set of scores 4, 6, 4, 6, and 5, the mean is $25/5 = 5$. If we know there are five scores, the mean is 25 and that four of the scores are 4, 6, 4, and 6 (which add to 20), it stands to reason that the other score from the set must be 5. As variance estimate calculations also use the previously calculated mean and the individual scores, the "last" score is not free to vary—it must have the value that provides the previously calculated mean. Therefore, only $(N-1)$ scores are really free to vary and so, there are $(N-1)$ *dfs*.

For the between groups SS, although three experimental condition means are involved, it is their variation around a general mean that is determined. With balanced data, the general mean is the average of the three experimental condition means. In this situation, the means are the data points contributing to the general mean and for the reasons described above, one *df* is lost. Consequently

$$\begin{aligned}\text{Between groups } df &= p - 1\\ &= 3 - 1\\ &= 2\end{aligned}$$

The within groups SS is comprised of the variation of scores from the experimental condition mean, over the three different conditions. As a separate mean is employed in each condition, a *df* will be lost in each condition.

$$\begin{aligned} df_{\text{experimental condition 1}} &= (N_1 - 1) = (8 - 1) = 7\\ df_{\text{experimental condition 2}} &= (N_2 - 1) = (8 - 1) = 7\\ df_{\text{experimental condition 3}} &= (N_3 - 1) = (8 - 1) = 7\end{aligned}$$

$$\begin{aligned}\text{Within groups } df &= 3(N_j - 1)\\ &= 3(7)\\ &= 21\end{aligned}$$

Armed with the sums of squares and *dfs*, the mean squares can be calculated by dividing the former by the latter. The ratio of the between groups MS to the within groups MS provides the F-statistic. The last item provided in the ANOVA summary table (Table 2.3) is the probability of the calculated F-value being obtained by chance under the null hypothesis given the data analyzed. This p-value associated with the calculated F-value is provided by most statistical packages. However, if the ANOVA is calculated by hand, or the statistical software used does not output the associated p-values, the table of critical F-values for $\alpha = 0.25$, 0.10, 0.05, and 0.01, provided in Appendix B may be employed. As the probability associated with the between groups F-value is less than 0.05, the null hypothesis (H_0), which states that there are no differences between experimental condition means can be rejected, and the experimental hypothesis (H_E), which states some experimental condition means differ, can be accepted (also see Section 2.3.2). Further tests are required to identify exactly which experimental condition means differ (see Chapter 3).

Table 2.3 ANOVA Summary Table

Source	SS	*df*	MS	*F*	*p*
Between groups	112.000	2	56.000	22.615	<0.001
Within groups	52.000	21	2.476		
Total	164.000	23			

2.7 CONFIDENCE INTERVALS

Confidence intervals also are useful statistics and with ANOVA now described, the description of confidence intervals is much easier. A confidence interval specifies the range within which it is expected that a particular estimate will occur. For instance, it can be stated with different probabilities (or degrees of confidence) that any population experimental condition mean will be within a specified range. Usually, a 95% confidence interval is set. A 95% confidence interval denotes that 95% of the time, the range will include the population experimental condition mean and that the population experimental condition mean will not fall within the specified range 5% of the time.

As well as the estimate of the population experimental condition mean (i.e., the mean of the experimental condition based on the sample data—the data in the experiment), two other statistics also are required to determine a confidence interval. The first is the critical F-value for $\alpha = 0.05$ with numerator $df = 1$ and denominator $dfs = (N_j - 1)$. The second statistic required is an estimate of the standard error. The sample standard error (se) can be defined as

$$\text{se} = \frac{s}{\sqrt{N_j}} \tag{2.14}$$

where s is the sample standard deviation (see Section 2.5). The standard error is a variance estimate and so it may be calculated in two ways: by assuming homogeneity of variance across experimental conditions or by employing separate estimates of the experimental condition variances. The data presented in Table 2.2 indicate variance homogeneity and so the standard error calculation also will assume homogeneity of variance. In such circumstances, the best estimate of standard error is given by

$$\text{se} = \frac{\text{MSe}}{\sqrt{N_j}}$$

Therefore, the confidence interval of any of the experimental condition means is given by

$$\text{CI} \pm \sqrt{F_{(\alpha;1,N_j-1)}}\sqrt{\frac{\text{MSe}}{\sqrt{N_j}}} \tag{2.15}$$

$$\text{CI} \pm \sqrt{F_{(0.05;1,7)}}\sqrt{\frac{2.476}{\sqrt{8}}}$$

$$\text{CI} \pm \sqrt{5.59}\sqrt{0.875}$$

$$\text{CI} \pm 2.21$$

The 95% CI for the 30 s experimental condition means is 6 ± 2.21, so there is a 95% likelihood that the population experimental condition mean will fall somewhere between 3.79 and 8.21. The 95% CI for the 60 s population experimental condition mean is 10 ± 2.21, so there is a 95% likelihood that the population experimental condition mean will fall somewhere between 7.79 and 12.21. As the 95% CI for the 180 s experimental condition means is 11 ± 2.21, there is a 95% likelihood that the population experimental condition mean will fall somewhere between 8.79 and 13.21.

In addition to CIs for individual means, it is also possible to apply the same rationale to differences between experimental condition means. In such circumstances when variance homogeneity is assumed, the CI is defined exactly as it was above for experimental condition means and so, the differences between means will have exactly the same CI

$$\text{CI} \pm \sqrt{F_{(\alpha;1,N_j-1)}}\sqrt{\frac{\text{MSe}}{\sqrt{N_j}}} \qquad (2.15, \text{rptd})$$
$$\text{CI} \pm 2.21.$$

Therefore, there is a 95% likelihood that the population difference between the mean of the 30 s experimental condition and the mean of the 60 s experimental condition is 4 ± 2.21, or between 1.79 and 6.21. There is a 95% likelihood that the population difference between the mean of the 60 s experimental condition and the mean of the 180 s experimental condition is 1 ± 2.21, or between -1.21 and 3.21. As this range includes 0, it implies that zero difference is a likely outcome and so this difference cannot be significant. With regard to the last comparison, there is a 95% likelihood that the population difference between the mean of the 30 s and the mean of the 60 s experimental condition means is 5 ± 2.21, or between 2.79 and 7.21. (A potential problem caused by this sort of use of CIs is considered in Section 3.6.4.1.)

2.8 GLM APPROACHES TO SINGLE FACTOR ANOVA

2.8.1 Experimental Design GLMs

GLM equations for ANOVA have become common sights in statistical texts, even when a traditional approach to ANOVA is applied. However, when a GLM equation is provided in the context of traditional ANOVA, the labels structural model, experimental effects model, or experimental design model can be employed. The equation

$$Y_{ij} = \mu + \alpha_j + \varepsilon_{ij} \qquad (2.16)$$

describes the GLM underlying the independent measures design ANOVA carried out on the data presented in Table 2.2. Y_{ij} is the ith subject's dependent variable score in the jth experimental condition, the parameter μ is the general mean of the experimental condition population means that underlies all subjects' dependent variable scores, the parameter α_j is the effect of the jth experimental condition and the random variable, ε_{ij}, is the error term, which reflects variation due to any uncontrolled source. Therefore, equation (2.16) is actually a summary of a set or system of equations, where each equation describes a single dependent variable score.

Predicted scores are based on the *model* component of GLM equations. Therefore, inspection of equation (2.17) reveals predicted scores $(\hat{Y}_{ij})$ to be given by

$$\hat{Y}_{ij} = \mu + \alpha_j \qquad (2.17)$$

As μ is a constant, the only variation in prediction can come from the effect of the j experimental conditions. Consequently, experimental design GLMs predict only as many different scores as there are experimental conditions, so every subject's score within an experimental condition ($\widehat{Y}_{ij}$) is predicted to be the mean score for that experimental condition ($\overline{Y}_j$)

$$\widehat{Y}_{ij} = \overline{Y}_j = \mu_j \tag{2.18}$$

For the data listed in Table 2.2, the μ_j estimates are

$$\overline{Y}_1 = 6,\ \overline{Y}_2 = 10,\ \text{and}\ \overline{Y}_3 = 11$$

Equation (2.16) defines the general mean as

$$\mu = \frac{\sum_{j=1}^{p} \mu_j}{p} \tag{2.19}$$

This reveals μ as the mean of the separate experimental condition means. Of course, with balanced data, this is also the mean of all dependent variable scores. Applying equation (2.19) to the data in Table 2.2 provides the estimate of the general mean, $\overline{Y}_G$

$$\overline{Y}_G = \frac{6 + 10 + 11}{3} = 9$$

which is identified by the dotted horizontal line in Figure 2.3. The estimate of the effect of a particular experimental condition, $\widehat{\alpha}_j$, is defined as

$$\widehat{\alpha}_j = \overline{Y}_j - \overline{Y}_G \tag{2.20}$$

Equation (2.20) reveals the effect attributable to each experimental condition to be the difference between the mean of the particular experimental condition and the general mean.

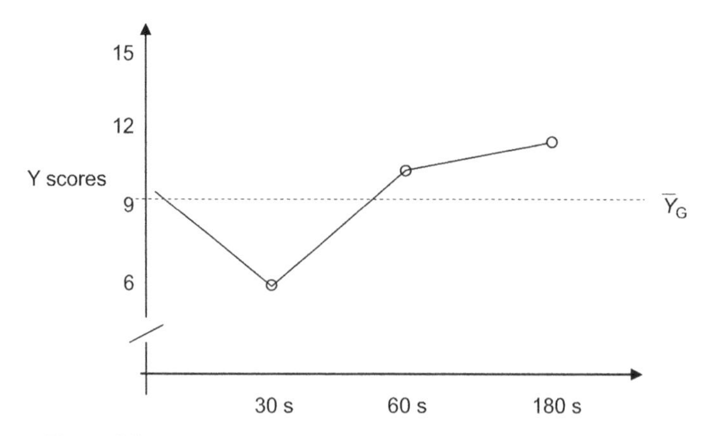

Figure 2.3 Dependent variable scores by experimental condition.

In Chapter 1, the regression GLM fit to the data was illustrated by the regression line that followed the path through all of the predicted scores. (NB. the *regression* predicted scores are not necessarily the means of the three experimental conditions.) The fit of experimental design GLMs also can be illustrated by plotting a line through all of the predicted scores – the experimental condition means. Figure 2.2 plots the experimental design GLM line that follows a path through these estimates – the experimental condition means. (As convention dictates, the dependent variable is plotted on the Y-axis and the factor levels, i.e., the experimental conditions, are plotted on the X-axis.)

However, the experimental design GLM line differs in important ways from the regression GLM line presented in Figure 1.1. To begin with, the factor levels (i.e., the experimental conditions) have a qualitative (i.e., categorical) rather than a quantitative representation in the experimental design GLM. This reflects the interest in the differences between the factor levels rather than the way in which the quantitative value of the predictor and dependent variables vary together in a linear fashion. The regression GLM estimates only two parameters – the Y-axis intercept, β_0, and the regression coefficient, β_1, and both of these estimates apply to all of the data. The experimental design GLM described by equation (2.16) may appear to estimate only two parameters, μ and α_j, but, in fact, α_j describes each experimental condition mean in terms of its deviation from, μ, so α_j actually summarizes as many effect estimates as there are experimental conditions. Moreover, as each effect is estimated as a deviation between each experimental condition mean and the general mean, the order in which the factor levels (i.e., the experimental conditions) are plotted along the X-axis is inconsequential. Usually, experimental conditions are presented in an order consistent with the theoretical or practical account, but, in contrast to the regression GLM, there is nothing in the experimental design GLM that prescribes the order of factor levels. As the shape of the experimental design GLM line is determined by an arbitrary order of experimental conditions, it emphasizes the point that the term linear in GLM does not necessarily refer to a *straight line*.

Given equations (2.19) and (2.20) and balanced data, it follows that

$$\sum_{j=1}^{p} \widehat{\alpha_j} = 0. \tag{2.21}$$

As can be seen in Table 2.4, applying equation (2.18) to the data listed in Table 2.2, provides the estimates of the three experimental condition effects and confirms equation (2.19). However, when unbalanced data is analyzed, the different numbers of subjects in the experimental conditions must be accommodated. Consequently, equation (2.21) becomes

$$\sum_{j=1}^{p} (N_i \widehat{\alpha_j}) = 0 \tag{2.22}$$

Experimental effects summing to 0 is more than just a logical outcome of the calculations. In fact, it is a mathematical side condition required to allow unique

Table 2.4 Estimates of the Three Experimental Condition Effects and Their Sum

$\widehat{\alpha}_1$	=	$6-9$	=	-3
$\widehat{\alpha}_2$	=	$10-9$	=	1
$\widehat{\alpha}_3$	=	$11-9$	=	2
$\sum_{j=1}^{p}\widehat{\alpha}_j$			=	0

estimation of the experimental design GLM parameters. This constraint is required because the experimental design GLM is overparameterized—it contains more parameters (μ, α_1, α_2, and α_3) than there are experimental condition means (30, 60, and 180 s) from which to estimate these parameters. One way of dealing with the problems caused by overparameterization is to set a constraint or side condition that all experimental effects sum to zero. Indeed, ANOVA may be defined as the special case of multiple regression that includes the side condition that experimental effects sum to 0.

Equation (2.16) summarizes a set of GLM equations that predict each subjects' score. Bearing in mind that each subject in each experimental condition is predicted to obtain the same score (2.17), the predicted experimental condition means can be described by the GLM equations

$$\widehat{Y}_{.,1} = 9 + (-3) = 6$$
$$\widehat{Y}_{.,2} = 9 + (1) = 10$$
$$\widehat{Y}_{.,3} = 9 + (2) = 11$$

In contrast to predicted scores, the estimated ε_{ij} terms ($\widehat{\varepsilon}_{ij}$) representing the discrepancy between actual and predicted scores may be different for each subject (see Table 2.5)

$$\widehat{\varepsilon}_{ij} = Y_{ij} - \overline{Y}_j \tag{2.23}$$

The average of all of these experimental errors squared provides the Mean Square error (MSe)

$$\text{MSe} = \frac{\sum_{i=1}^{N}(\widehat{\varepsilon}_{ij}^{\,2})}{p(N_j - 1)} \tag{2.24}$$

where N_j is the number of subjects in each experimental condition and p is the number of experimental conditions. The denominator of equation (2.24), $p(N_j - 1)$, gives the degrees of freedom for the ANOVA MSe

$$52/3(8-1) = 52/21 = 2.476$$

Table 2.5 Calculation of Error Terms, Their Squares, and Sums

$\hat{\varepsilon}_{ij}$		$Y_{ij} - \overline{Y}_{ij}$		$\hat{\varepsilon}_{ij}$	$(\hat{\varepsilon}_{ij})^2$
$\hat{\varepsilon}_{1,1}$	=	7 − 6	=	1	1
$\hat{\varepsilon}_{2,1}$	=	3 − 6	=	−3	9
$\hat{\varepsilon}_{3,1}$	=	6 − 6	=	0	0
$\hat{\varepsilon}_{4,1}$	=	6 − 6	=	0	0
$\hat{\varepsilon}_{5,1}$	=	5 − 6	=	−1	1
$\hat{\varepsilon}_{6,1}$	=	8 − 6	=	2	4
$\hat{\varepsilon}_{7,1}$	=	6 − 6	=	0	0
$\hat{\varepsilon}_{8,1}$	=	7 − 6	=	1	1
$\hat{\varepsilon}_{9,2}$	=	7 − 10	=	−3	9
$\hat{\varepsilon}_{10,2}$	=	11 − 10	=	1	1
$\hat{\varepsilon}_{11,2}$	=	9 − 10	=	−1	1
$\hat{\varepsilon}_{12,2}$	=	11 − 10	=	1	1
$\hat{\varepsilon}_{13,2}$	=	10 − 10	=	0	0
$\hat{\varepsilon}_{14,2}$	=	10 − 10	=	0	0
$\hat{\varepsilon}_{15,2}$	=	11 − 10	=	1	1
$\hat{\varepsilon}_{16,2}$	=	11 − 10	=	1	1
$\hat{\varepsilon}_{17,3}$	=	8 − 11	=	−3	9
$\hat{\varepsilon}_{18,3}$	=	14 − 11	=	3	9
$\hat{\varepsilon}_{19,3}$	=	10 − 11	=	−1	1
$\hat{\varepsilon}_{20,3}$	=	11 − 11	=	0	0
$\hat{\varepsilon}_{21,3}$	=	12 − 11	=	1	1
$\hat{\varepsilon}_{22,3}$	=	10 − 11	=	−1	1
$\hat{\varepsilon}_{23,3}$	=	11 − 11	=	0	0
$\hat{\varepsilon}_{24,3}$	=	12 − 11	=	1	1
$\sum$				0	52

The experimental design GLM can be used as a basis for partitioning variance in traditional ANOVA. Employing equations (2.20) and (2.23) allows equation (2.17) to be rewritten as

$$Y_{ij} = \overline{Y}_{G} + \left(\overline{Y}_{j} - \overline{Y}_{G}\right) + \left(Y_{ij} - \overline{Y}_{j}\right) \tag{2.25}$$

Moving the general mean $\left(\overline{Y}_{G}\right)$ to the left-hand side of equation (2.25) gives

$$\left(Y_{ij} - \overline{Y}_{G}\right) = \left(\overline{Y}_{j} - \overline{Y}_{G}\right) + \left(Y_{ij} - \overline{Y}_{j}\right) \tag{2.26}$$

Equation (2.26) defines the variation between the dependent variable scores (Y_{ij}) and the general mean ($\overline{Y}_{G}$) as comprising variation due to experimental conditions

Table 2.6 Calculation of Variation Due to Experimental Conditions

Conditions		$(\overline{Y}_j - \overline{Y}_G)^2$		$(\overline{Y}_j - \overline{Y}_G)^2$		$(\overline{Y}_j - \overline{Y}_G)^2$
30 s	=	$(6-9)^2$	=	$(-3)^2$	=	9
60 s	=	$(10-9)^2$	=	1^2	=	1
180 s	=	$(11-9)^2$	=	2^2	=	4
				$\sum_{j=1}^{p}(\overline{Y}_j - \overline{Y}_G)^2$	=	14

$(\overline{Y}_j - \overline{Y}_G)$ and variation due to errors $(Y_{ij} - \overline{Y}_j)$. However, to obtain accurate variance estimates the number of scores contributing to each estimate must be included. Although experimental condition means are obtained by averaging over all the subjects in the particular condition, each condition mean is regarded as the score each subject in that experimental condition would record if error variation was eliminated. Consequently, in each experimental condition there would be N_j scores equal to the mean. Therefore, the SS partition is

$$\sum_{j=1}^{p}\sum_{i=1}^{N}(Y_{ij} - \overline{Y}_G)^2 = \sum_{j=1}^{p} N_j(\overline{Y}_j - \overline{Y}_G)^2 + \sum_{j=1}^{p}\sum_{i=1}^{N}(Y_{ij} - \overline{Y}_j)^2 \tag{2.27}$$

Equation (2.27) and the account of variation due to experimental conditions should seem familiar, as exactly the same argument was applied to the estimation of the traditional ANOVA between groups SS. The calculation of the variation due to experimental conditions for the data in Table 2.2 is presented in Table 2.6

$$\text{Experimental conditions SS} = \sum_{j=1}^{p} N_j(\overline{Y}_j - \overline{Y}_G)^2 \tag{2.28}$$

So,

$$\text{Experimental conditions SS} = 8(-3)^2 + 8(1^2) + 8(2^2) = 8(14) = 112$$

Therefore, the sum of squares due to the experimental conditions (i.e., the experimental effects) is also the *average* of the square of the differences between unique pairs of experimental condition means.

The sum of squares calculations for the experimental effect above provides a value identical to that calculated for traditional ANOVA presented in Table 2.3. As degrees of freedom can be defined for the experimental design GLM as they were for traditional ANOVA, the mean square values also will be identical to those calculated for traditional ANOVA and presented in Table 2.3

$$\begin{aligned}\text{Experimental conditions MS} &= \text{Experimental effect SS}/df \\ &= 112/2 \\ &= 56\end{aligned} \quad (2.29)$$

With the error sum of squares and MSe calculated previously, it is clear that the components of equation (2.26), based on the experimental design GLM, and the components of the traditional ANOVA equation (2.15) are equivalent

$$SS_{\text{total}} = SS_{\text{between groups}} + SS_{\text{within groups}} \quad (2.15, \text{rptd})$$

It is left to the reader to confirm that

$$\sum_{j=1}^{p}\sum_{i=1}^{N}(Y_{ij} - \overline{Y}_{\text{G}})^2 = 112 + 52 = 164$$

2.8.2 Estimating Effects by Comparing Full and Reduced Experimental Design GLMs

In Chapter 1, it is explained that the same statistical procedures underlie regression and ANOVA, but that concise experimental analyses are a consequence of the experimental design acting to simplify the processes of GLM selection, parameter estimation, model checking, and prediction. In conventional regression or linear modeling, an aim is to try and find a minimal set of predictors that accommodates the maximal amount of dependent variable variance. More predictors always can accommodate some more variance—even when the predictors have no significant or meaningful relationship with the dependent variable measure. Therefore, a tension exists between finding a minimal set of predictors and accommodating the maximal amount of dependent variable variance and a whole set of linear modeling procedures and criteria have been established to aid the comparison of different GLMs of the same data and identify the best GLM (Draper and Smith, 1998).

As much of conventional linear modeling involves comparing different GLMs of the same data, it follows that comparing of full and reduced GLMs to estimate experimental effects is more in the spirit of conventional linear modeling than any of the other methods of experimental effect estimation so far described. However, just as simplified linear modeling processes are used to apply a GLM to experimental data, so the comparison of full and reduced experimental design GLMs to estimate experimental effects applies a distilled form of linear modeling processes to analyze experimental data.

In Chapter 1, the GLM conception was described as

$$\text{Data} = \text{Model} + \text{Error} \quad (1.2, \text{rptd})$$

Linear modeling processes attempt to identify the "best" GLM of the data by comparing different linear models. The GLMs are assessed in terms of the relative proportions of data variance attributed to the model and the error components. With a fixed data set, the sum of model and error components is a constant (i.e., the data variance), so any increase in variance accommodated by the model component will result in an equivalent decrease in the error component.

Consider the experimental design GLM for the independent single factor experiment

$$Y_{ij} = \mu + \alpha_j + \varepsilon_{ij} \qquad (2.16, \text{rptd})$$

This *full* model employs the general mean, μ, and includes parameters α_j to accommodate any influence of the experimental conditions. Essentially, it presumes that subjects' dependent variable scores (data) are best described by the experimental condition means. The full GLM manifests the data description under a nondirectional experimental hypothesis, which may be expressed more formally as

$$\alpha_j \neq 0, \quad \text{for some } j \qquad (2.30)$$

This states that the effect of some experimental conditions does not equal 0. An equivalent expression in terms of the experimental condition means is

$$\mu \neq \mu_j \quad \text{for some } j \qquad (2.31)$$

This states that some of the experimental condition means do not equal the general mean.

It is also possible to describe a reduced model that omits any effect of the experimental conditions. Here, the reduced GLM is described by the equation

$$Y_{ij} = \mu + \varepsilon_{ij} \qquad (2.32)$$

which uses only the general mean of scores (μ) to account for the data. This GLM presumes that subjects' dependent variable scores are best described by the general mean of all scores. In other words, it presumes that the description of subjects' scores would not benefit from taking the effects of the experimental conditions (α_j) into account. The reduced GLM manifests the data description under the null hypothesis. By ignoring any influence of the experimental conditions, the reduced GLM assumes that the experimental conditions do not influence the data. This assumption may be expressed more formally as

$$\alpha_j = 0 \qquad (2.33)$$

which states that all of the experimental conditions have zero effect. An equivalent expression in terms of experimental condition means is

$$\mu = \mu_j \qquad (2.34)$$

This states that the general mean and the experimental condition means are equal. As equation (2.20) defines μ as the mean of the separate experimental condition means, a longer formal version of the null hypothesis expressed in equation (2.34) is

$$\mu_1 = \mu_2 = \mu_3 \qquad (2.35)$$

This is termed an omnibus (or complete or overall) null hypothesis because it states that *all* of the experimental condition means are equal.

If the full GLM, which accommodates the experimental condition effect(s), provides a better description of the data, it should have a smaller error component than the reduced GLM. Moreover, any reduction in the size of the error component caused by including the effects of the experimental conditions will be matched by an equivalent increase in the size of the model component. Therefore, comparing the size of the error components before and after adding the effects of the experimental conditions to the model component provides a method of assessing the consequences of changing the model. Presenting the full and reduced GLM equations together should clarify this point

$$\text{Reduced GLM:} \quad Y_{ij} = \mu + \varepsilon_{ij} \tag{2.32, rptd}$$

$$\text{Full GLM:} \quad Y_{ij} = \mu + \alpha_j + \varepsilon_{ij} \tag{2.16, rptd}$$

Any reduction in the error component of the full GLM can be attributed only to the inclusion of the experimental condition effects, as this is the only difference between the two GLMs.

The reduced GLM defines errors as

$$\varepsilon_{ij} = Y_{ij} - \mu \tag{2.36}$$

Note that the difference between equations (2.22) and (2.35) is simply whether each subjects' dependent variable score (Y_{ij}) is taken to deviate from the pertinent experimental condition mean (μ_j) or the general mean (μ). Of course, as GLM errors sum to 0, interest is in the sum of the squared errors (see Table 2.5). A convenient computational formula for the reduced GLM error sum of squares (SSE) is

$$\text{SSE}_{\text{RGLM}} = \sum_{i=1}^{N} Y_i^2 - \frac{\left(\sum Y_i\right)^2}{N} \tag{2.37}$$

Using the data from Table 2.2 provides

$$\text{SSE}_{\text{RGLM}} = 2108 - \frac{(216)^2}{24} = 164$$

Note that this is equivalent to SS_{total}, described by equation (2.7). The full GLM defines errors as

$$\varepsilon_{ij} = Y_{ij} - (\mu + \alpha_j) \tag{2.38}$$

$$\varepsilon_{ij} = Y_{ij} - u_j \tag{2.23, rptd}$$

A convenient computational formula for the full GLM error term SS is

$$\text{SSE}_{\text{FGLM}} = \sum_{i=1}^{N} Y_{ij}^2 - \frac{\left(\sum Y_{ij}\right)^2}{N_j} \tag{2.39}$$

Using the data from Table 2.3 provides

$$\begin{aligned}\text{SSE}_{\text{FGLM}} &= 304 - \left[\frac{48^2}{8}\right] + 814 - \left[\frac{80^2}{8}\right] + 990 - \left[\frac{88^2}{8}\right] \\ &= 52\end{aligned}$$

As this is equivalent to the $\text{SS}_{\text{within}}$, described by equation (2.9), it should come as no surprise that

$$\begin{aligned}\text{SSE}_{\text{RGLM}} - \text{SSE}_{\text{FGLM}} &= \text{SS}_{\text{total}} - \text{SS}_{\text{within}} = \text{SS}_{\text{between}} \\ &= 164 - 52 = 112\end{aligned} \tag{2.40}$$

In other words, including the experimental condition effects reduces the error component sum of squares by an amount identical to the traditional ANOVA between groups sum of squares. Therefore, the reduction in the error component sum of squares, attributable to the experimental condition effects, is given by

$$\text{SSE}_{\text{RGLM}} - \text{SSE}_{\text{FGLM}} = \sum_{j=1}^{p} N_j(\mu_j - \mu)^2 \tag{2.41}$$

which, of course, is equivalent to equation (2.28) when the sample statistics replace the parameters. Also note that SSE_{RGLM} is equal to what traditional ANOVA labels the Total SS. This emphasizes that the traditional ANOVA Total SS is actually an estimate of the dependent variable scores deviation from the general mean (see Section 2.6).

An F-test of the reduction in the error component attributed to the inclusion of the experimental condition effects is given by

$$F = \frac{(\text{SSE}_{\text{RGLM}} - \text{SSE}_{\text{FGLM}})/(df_{\text{RGLM}} - df_{\text{FGLM}})}{\text{SSE}_{\text{FGLM}}/df_{\text{FGLM}}} \tag{2.42}$$

Therefore

$$F = \frac{164 - 52/23 - 21}{52/21} = \frac{56}{2.476}$$

$$F_{(2,21)} = 22.617$$

A convenient alternative to solving equation (2.40) in a single step is to construct an ANOVA summary table (Table 2.7), similar to Table 2.3. As mentioned in Section 2.6,

Table 2.7 ANOVA Summary Table for Full and Reduced GLMs

Source	SS	df	MS	F	p
Error reduction due to experimental conditions	112.000	2	56.000	22.615	<0.001
FGLM error	52.000	21	2.476		
Total (i.e., SSE_{RGLM})	164.000	23			

if the ANOVA is calculated by hand, or the statistical software used does not output the associated p-values, the table of critical F-values for $\alpha = 0.25$, 0.10, 0.05, and 0.01, provided in Appendix B may be employed.

2.8.3 Regression GLMs

The experimental design GLM equation (2.16) may be compared with an equivalent regression equation

$$Y_i = \beta_0 + \beta_1 X_{i,1} + \beta_2 X_{i,2} + \varepsilon_i \tag{2.43}$$

where Y_i is the dependent variable score for the ith subject, β_0 is a constant, β_1 is the regression coefficient for the first predictor variable X_1, β_2 is the regression coefficient for the second predictor variable X_2, and the random variable ε_i represents error. No i subscript is applied to the regression coefficient parameters, as, in principle, they are common across subjects. Often, however, the subscript i is omitted from the predictor variables because although each subject provides a value for each variable X, this value is common across all of the subjects in an experimental condition. Equation (2.43) describes *multiple* regression, rather than *simple* regression, because $k = 2$ independent or predictor variables are employed, rather than one. Similar to the experimental design GLM, equation (2.43) is a summary of a set of equations, each describing a single dependent variable score.

2.8.4 Schemes for Coding Experimental Conditions

Dummy, effect, and contrast coding schemes are used to represent experimental conditions and other categories of data for GLM analysis. This is done by employing as predictors particular sets of quantitative variables that operate in established formulas to produce "categorical" analyses. Variables used in this manner also may be termed indicator variables.

2.8.4.1 Dummy Coding

The dummy coding scheme uses only 1 and 0 values to denote allocation to experimental conditions. $(p - 1)$ variables are used and one condition (usually last in sequence—180 s), is given 0s across all indicator variables and may be termed the base condition. The other conditions (30 and 60 s) are denoted by 1s rather than 0s on variables X_1 and X_2, respectively. Table 2.8 illustrates the dummy coding of the example data from Table 2.2.

Table 2.9 presents the ANOVA summary table output from statistical software when a regression ANOVA GLM is applied to the data presented in Table 2.2 when dummy, effect, or contrast coding schemes are applied. However, not all regression software provides such ANOVA summary tables and they may need to be constructed from the information provided about the overall regression applied to the data. Most regression software provides the multiple correlation coefficient (R), its square and an adjusted R^2 value. R^2 estimates the proportion of the dependent variable variance that can be attributed to the predictors, but unfortunately this statistic exhibits an

Table 2.8 Dummy Coding Representing Subject Allocation to Experimental Conditions

Conditions	Subjects	X_1	X_2	Y
30 s	s1	1	0	7
	⋮	⋮	⋮	⋮
	s8	1	0	7
60 s	s9	0	1	7
	⋮	⋮	⋮	⋮
	s16	0	1	11
180 s	s17	0	0	8
	⋮	⋮	⋮	⋮
	s24	0	0	12

Table 2.9 ANOVA Summary Table Output From Statistical Software Implementing a Regression ANOVA GLM

Source	SS	df	MS	F	p
Regression	112.000	2	56.000	22.615	<0.001
Residual	52.000	21	2.476		

R: 0.826; R^2: 0.683; adjusted R^2: 0.653.

overestimate bias. The smaller adjusted R^2 attempts to eliminate this bias (see Pedhazur, 1997). Irrespective of the coding scheme employed, the same values are obtained for all of these estimates.

When a regression ANOVA GLM employing dummy coding is applied to the data in Table 2.2, most statistical software provides the estimates presented in Table 2.10.The "Variable" column lists the multiple regression equation variables. The variable labeled "Constant" is the Y-axis intercept β_0 and the variables X_1 and X_2 are the dummy coded predictor variables X_1 and X_2 presented in Table 2.8. The "Coefficient" column provides estimates of the coefficients applicable to the predictor variables (i.e., β_1 and β_2) and the regression intercept β_0. The "Standard Error" column provides estimates of the coefficient standard errors and the "Standard Coefficient" column presents estimates of the standardized regression coefficients. The "t" and "p (two tailed)" columns present t-statistics (some software provides F-statistics) and p-values of tests of the variable coefficients (tests of Coefficient and Standard Coefficient are equivalent), where

$$t = \frac{\text{Coefficient}}{\text{Standard Error}}$$

Table 2.10 Output Pertinent to Multiple Regression Equation for Dummy Coding

Variable	Coefficient	Standard Error	Standard Coefficient	t	p (Two-Tailed)
Constant	11.000	0.556	0.000	19.772	<0.001
X_1	−5.000	0.787	−0.902	−6.355	<0.001
X_2	−1.000	0.787	−0.180	−1.271	0.218

with *dfs* equal to the residual (error) term *dfs*. Predicted scores are given by

$$\hat{Y}_i = \beta_0 + \beta_1 X_{i,1} + \beta_2 X_{i,2} \tag{2.44}$$

Only three experimental conditions are represented by the dummy coding scheme used, so in common with the experimental design GLM, there are only three different predicted scores—the means of the three experimental conditions. Substituting the pertinent X predictor variable dummy codes and the coefficients from Table 2.10 into the system of equations summarized by equation (2.44) provides the means of each of the three experimental conditions

$$\overline{Y}_1 = 11 + (-5)(1) + (-1)(0) = 11 - 5 = 6$$

$$\overline{Y}_2 = 11 + (-5)(0) + (-1)(1) = 11 - 5 = 10$$

$$\overline{Y}_3 = 11 + (-5)(0) + (-1)(0) = 11 - 0 = 11$$

The 180 s experimental condition is the base condition: the condition coded 0 on both dummy predictor variables. However, as this experimental condition is coded 0 on both X_1 and X_2 dummy variables, it follows that

$$\beta_1 X_{i,1} + \beta_2 X_{i,2} = 0$$

and so

$$\overline{Y}_3 = \beta_0$$

In short, the variable labeled "Constant" in Table 2.10 is the Y-axis intercept, β_0, which is equal to the mean of the base condition. Irrespective of balanced (i.e., equal numbers of subjects per condition) or unbalanced data, β_0 equals the mean of the base condition. Nevertheless, the t-test of the "Constant," β_0, which assesses the null hypothesis that the mean of the 180 s condition equals 0, has no corollary in ANOVA.

Equation (2.43) defines the mean of the 30 s experimental condition to be

$$\overline{Y}_1 = \beta_0 + \beta_1 X_{i,1} + \beta_2 X_{i,2}$$

However, as 30 s experimental condition is coded 1 on dummy variable X_1, but is coded 0 on dummy variable X_2, it follows that

$$\beta_2 X_{i,2} = 0$$

Therefore

$$\overline{Y}_1 = \beta_0 + \beta_1 X_{i,1}$$

and so

$$\beta_1 X_{i,1} = \overline{Y}_1 - \beta_0$$

This means that testing the coefficient β_1 assesses the difference between the base condition—the 180 s experimental condition—and the condition coded 1 on predictor variable X_1—the 30 s experimental condition. Table 2.10 presents a t-test assessment of this coefficient, $t(21) = -6.355$, $p < 0.001$.

Equation (2.44) defines the mean of the 60 s experimental condition to be

$$\overline{Y}_2 = \beta_0 + \beta_1 X_{i,1} + \beta_2 X_{i,2}$$

However, as the 60 s experimental condition is coded 1 on dummy variable X_2, but is coded 0 on dummy variable X_1, it follows that

$$\beta X_{i,1} = 0$$

Therefore

$$\overline{Y}_2 = \beta_0 + \beta_2 X_{i,2}$$

and so

$$\beta_1 X_{i,2} = \overline{Y}_2 - \beta_0$$

Consequently, testing the coefficient β_2 assesses the difference between the base condition—the 180 s experimental condition—and the condition coded 1 on predictor variable X_2—the 60 s experimental condition. Table 2.10 presents a t-test assessment of this coefficient, $t(21) = -1.271$, $p = 0.218$.

As described above, dummy coding sets the condition coded 0 on all predictors as a base or reference condition—the experimental condition that is compared to the other experimental conditions. This makes dummy coding ideal when all experimental conditions are to be compared with a control condition. (Of course, any condition can be coded as the base condition so, recoding and reanalyzing the same data with the 30 s or 60 s experimental conditions as the base condition can provide a comparison between the 30 s and 60 s experimental conditions.)

2.8.4.2 Why Only (p – 1) Variables Are Used to Represent All Experimental Conditions?

When a regression equation represents ANOVA, the predictor variables identify allocation to experimental conditions and a parameter is associated with each predictor variable. However, rather than requiring p predictor variables to represent

Table 2.11 Dummy Coding Representing Subject Allocation to Experimental Conditions

Conditions	Subjects	X_1	X_2	X_3	Y
	s1	1	0	0	7
30 s	⋮	⋮	⋮	⋮	⋮
	s8	1	0	0	7
	s9	0	1	0	7
60 s	⋮	⋮	⋮	⋮	⋮
	s16	0	1	0	11
	s17	0	0	1	8
180 s	⋮	⋮	⋮	⋮	⋮
	s24	0	0	1	12

p experimental conditions, the ANOVA regression equation needs only $(p-1)$ predictor variables to represent all of the experimental conditions. This is why there are only two predictors in equation (2.43). Table 2.11 illustrates the dummy coding of the example data from Table 2.2 using $p = 3$ predictor variables, X_1, X_2, and X_3. Allocation to the 30 s experimental condition is denoted by 1s rather than 0s on variable X_1, allocation to the 60 s experimental condition is denoted by 1s rather than 0s on variable X_2, and allocation to the 180 s experimental condition is denoted by 1s rather than 0s on variable X_3. However, closer scrutiny of Table 2.11 reveals that three experimental conditions are represented even if variable X_3 is eliminated. This is because allocation to the 30s and 60 s experimental conditions still is denoted by a 1 on variables X_1 and X_2, respectively, but now only the 180 s experimental condition is denoted by a 0 on X_1*and* a 0 on X_2. Variable X_3 is redundant for the unique specification of the three experimental conditions. Indeed, not only is variable X_3 redundant but it is also necessary to exclude it when regression formulas employ the indicator variables in a quantitative fashion.

The reason why the particular $(p-1)$ predictor variables are used rather than p predictor variables has to do with the linear dependence of predictors. For example, consider the matrix **A**

$$\mathbf{A} = \begin{bmatrix} 1 & 0 & 0 \\ 0 & 1 & 0 \\ 0 & 0 & 1 \end{bmatrix}$$

This matrix contains three rows, with each row corresponding to the coding over the predictors for an experimental condition in Table 2.11. However, in every regression GLM, a variable representing the constant, β_0, also is used as a predictor. As every score is defined with respect to β_0, every row contains a 1 in this predictor column indicating the involvement of β_0 in defining the score. Therefore, the complete model matrix (Kempthorne, 1980) for the regression GLM is

$$\mathbf{B} = \begin{bmatrix} 1 & 1 & 0 & 0 \\ 1 & 0 & 1 & 0 \\ 1 & 0 & 0 & 1 \end{bmatrix}$$

The matrix **B** also can be considered as four (predictor variable) column vectors. Different scalars (s_n) can be associated with each column vector.

$$\begin{matrix} & \mathbf{X}_0 & & \mathbf{X}_1 & & \mathbf{X}_2 & & \mathbf{X}_3 \\ s_0 & \begin{bmatrix} 1 \\ 1 \\ 1 \end{bmatrix} & s_1 & \begin{bmatrix} 1 \\ 0 \\ 0 \end{bmatrix} & s_2 & \begin{bmatrix} 0 \\ 1 \\ 0 \end{bmatrix} & s_3 & \begin{bmatrix} 0 \\ 0 \\ 1 \end{bmatrix} \end{matrix}$$

The column vectors are defined as linearly independent when the equation

$$s_0\mathbf{X}_0 + s_1\mathbf{X}_1 + s_2\mathbf{X}_2 + s_3\mathbf{X}_3 = 0$$

is satisfied *only* when all scalars equal zero. (Linear dependence exists if the equation is satisfied by a set of scalars that are not all equal to zero.) For matrix **B**, a set of scalars, some of which are not 0, can be found to satisfy this equation. For example, the product of the equation below is 0, but all of the scalars in the set applied are non zero.

$$\begin{matrix} & \mathbf{X}_0 & & \mathbf{X}_1 & & \mathbf{X}_2 & & \mathbf{X}_3 & & \mathbf{0} \\ 1 & \begin{bmatrix} 1 \\ 1 \\ 1 \end{bmatrix} & +(-1) & \begin{bmatrix} 1 \\ 0 \\ 0 \end{bmatrix} & +(-1) & \begin{bmatrix} 0 \\ 1 \\ 0 \end{bmatrix} & +(-1) & \begin{bmatrix} 0 \\ 0 \\ 1 \end{bmatrix} & = & \begin{bmatrix} 0 \\ 0 \\ 0 \end{bmatrix} \end{matrix}$$

Therefore, the set of **X** predictor column vectors are not linearly independent. Linear dependence occurs when one column vector (representing a predictor variable) is a linear function of one or more other column vectors (representing other predictor variables) in the matrix. When a matrix of column vectors are linearly dependent, the matrix is described as being singular. In the current example, linear dependence arises because the column vector representing the predictor variable X_3 is included in the set of predictors. With X_3 included, the predictor variable X_0 is a linear function of all the others ($X_0 = X_1 + X_2 + X_3$).

Linear dependency among predictor variables prevents a unique solution to the system of normal simultaneous equations upon which GLM parameter estimation is based. The solution cannot be unique because due to the relations between the predictor variables the same dependent variable variance can be attributed to more than one predictor variable. As there is nothing in the mathematical representation to resolve this ambiguity, the mathematical operations fail. (The nature of this failure is most evident in the matrix algebra representation. The determinant of a matrix decreases as the correlation between predictor variables increases. When predictor variables are correlated perfectly, the determinant of a matrix equals 0, linear dependence exists and the matrix is labeled singular. Singular matrices have no inverse and as they cannot be inverted, the appropriate divisions to obtain estimates cannot be implemented.) However, simply eliminating the redundant predictor variable X_3 results in linear independence among the remaining predictors (X_0, X_1, and X_2) and allows a unique solution to the system of normal simultaneous equations.

2.8.4.3 Effect Coding

Effect coding operates very much similar to dummy coding, but rather than the "last" experimental condition being denoted by all indicator variables taking 0 values, it is denoted by all of these indicator variables taking the value minus 1. Effect coding for the example data in Table 2.2 is presented in Table 2.12.

The ANOVA summary table output from statistical software when a regression ANOVA GLM is applied to the data presented in Table 2.2 using an effect coding scheme is presented in Table 2.9. Most statistical software also provides the estimates presented in Table 2.13. This table compares with Table 2.10, but with estimates based on effect coding.

As with dummy coding, only three experimental conditions are represented by the effect coding scheme, so in common with the experimental design GLM, there are only three different predicted scores—the means of the three experimental conditions. Substituting the pertinent X predictor variable dummy codes and the coefficients from Table 2.12 into the system of equations summarized by equation (2.44) provides the means of each of the three experimental conditions

$$\begin{aligned}
\overline{Y}_1 &= 9 + (-3)(1) + (1)(0) &&= 9 - 3 &&= 6 \\
\overline{Y}_2 &= 9 + (-3)(0) + (1)(1) &&= 9 - 1 &&= 10 \\
\overline{Y}_3 &= 9 + (-3)(-1) + (1)(-1) &&= 9 + 3 - 1 &&= 11
\end{aligned}$$

Table 2.12 Effect Coding Representing Subject Allocation to Experimental Conditions

Conditions	Subjects	X_1	X_2	Y
	s1	1	0	7
30 s	⋮	⋮	⋮	⋮
	s8	1	0	7
	s9	0	1	7
60 s	⋮	⋮	⋮	⋮
	s16	0	1	11
	s17	−1	−1	8
180 s	⋮	⋮	⋮	⋮
	s24	−1	−1	12

Table 2.13 SYSTAT Output Pertinent to Multiple Regression Equation for Effect Coding

Variable	Coefficient	Standard Error	Standard Coefficient	t	p (Two-Tailed)
Constant	9.000	0.321	0.0	28.019	<0.000
X_1	−3.000	0.454	−0.937	−6.604	<0.000
X_2	1.000	0.454	0.312	2.201	0.039

The variable labeled "Constant" in Table 2.13 is the Y-axis intercept, β_0. Similar to the experimental design GLM parameter μ, β_0 with effect coding is the general mean of the experimental condition means. With balanced data, β_0 is also the mean of all the dependent variable scores. However, with unbalanced data, β_0 is not the mean of all dependent variable scores but remains the general mean of the experimental condition means (i.e., the unweighted mean of the experimental condition means). The significance test of "Constant" assesses the null hypothesis that β_0 is equal to 0. (This test also can be regarded as an assessment the variation accommodated by the inclusion of the general mean, β_0 or μ, but this requires a different definition of total variation to that presented here.) Some statistical software (e.g., BMDP) includes this test of the general mean in the ANOVA summary table.

Equation (2.44) defines the mean of the 30 s experimental condition to be

$$\overline{Y}_1 = \beta_0 + \beta_1 X_{i,1} + \beta_2 X_{i,2}$$

However, as the 30 s experimental condition is coded 1 on effect variable X_1, but is coded 0 on effect variable X_2, it follows that

$$\beta_2 X_{i,2} = 0$$

Therefore

$$\overline{Y}_1 = \beta_0 + \beta_1 X_{i,1}$$

and so

$$\beta_1 X_{i,1} = \overline{Y}_1 - \beta_0$$

This means that testing the coefficient β_1 assesses the difference between μ and the condition coded 1 on predictor variable X_1—the 30 s condition. As this difference is the effect of the 30 s experimental condition, testing β_1 tests the effect of the 30 s experimental condition. Table 2.13 presents a t-test assessment of β_1, $t(21) = -6.604$, $p < 0.001$.

Equation (2.44) defines the mean of the 60 s experimental condition to be

$$\overline{Y}_2 = \beta_0 + \beta_1 X_{i,1} + \beta_2 X_{i,2}$$

However, as the 60 s experimental condition is coded 1 on effect variable X_2, but is coded 0 on effect variable X_1, it follows that

$$\beta_1 X_{i,1} = 0$$

Therefore

$$\overline{Y}_2 = \beta_0 + \beta_2 X_{i,2}$$

and so

$$\beta_1 X_{i,2} = \overline{Y}_B - \beta_0$$

Consequently, testing the coefficient β_2 assesses the difference between μ and the condition coded 1 on predictor variable X_2—the 60 s condition. As this difference is the effect of the 60 s experimental condition, testing β_2 tests the effect of the 60 s experimental condition. Table 2.13 presents a t-test assessment of β_1, $t(21) = 2.201$, $p < 0.039$.

Unlike the 30s and 60 s experimental conditions, each of which is coded 1 on an effect predictor, the 180 s experimental condition is coded -1 on both X_1 and X_2 effect variables and so it is not possible to determine the effect of this experimental condition in the manner described for the other experimental conditions. Nevertheless, it was stated in Section 2.8.1 that ANOVA may be defined as the special case of multiple regression where experimental effects sum to zero. With balanced data

$$\sum_{j=1}^{p} \alpha_j = 0 \qquad (2.21, \text{rptd})$$

Given that two of the three effects are known and it is known that all effects sum to zero, it is a relatively simple to determine the effect of the 180 s experimental condition.

$$\text{If} \quad \alpha_1 + \alpha_2 + \alpha_3 = 0, \quad \text{then} \quad \alpha_3 = -\alpha_1 - \alpha_2$$

With α_1 and α_2 identified already as -3 (30 s condition) and 1 (60 s condition), respectively, it follows that

$$\alpha_3 = -\alpha_1 - \alpha_2$$

$$\alpha_3 = -(-3) - 1 = 2$$

and so the effect of the 180 s experimental condition is equal to 2.

Effect coding defines each of the experimental condition means in terms of their difference from μ. Therefore, the regression coefficient estimates equal the respective experimental design GLM α_j parameter estimates (i.e., the effect estimates) in the experimental design GLM. It is this concordance that gives effect coding its name and probably explains why it is the most popular coding scheme.

2.8.5 Coding Scheme Solutions to the Overparameterization Problem

In Section 2.8.1, ANOVA was described as the special case of multiple regression that includes the side condition that the experimental effects sum to zero, that is

$$\sum_{j=1}^{p}(N_i\alpha_j) = 0 \tag{2.22, rptd}$$

In fact, requiring experimental effects to sum to zero is equivalent to eliminating one of the parameters and redefining the condition previously specified by the eliminated parameter in terms of the other conditions. A more formal expression is

$$\text{because} \sum_{j=1}^{p}(N_j\alpha_j) = 0, \quad \alpha_p = -\sum_{j=1}^{p}\alpha_j \tag{2.45}$$

The use of only $p - 1$ predictors, where the "last" experimental condition is defined as the negative of the sum of the remaining conditions (so that experimental effects to sum to 0) is effect coding. Therefore, the implicit consequence of the side condition that effects sum to 0 is made explicit in a regression ANOVA GLM using effect coding. Dummy coding does not result in experimental effects summing to 0, but instead redefines β_0 and the $(p - 1)$ experimental conditions in terms of p, the "last" experimental condition. Therefore, both effect and dummy coding schemes constitute reparameterization solutions to the overparameterization problem.

2.8.6 Cell Mean GLMs

Another solution to the experimental design GLM overparameterization problem is to apply a cell mean GLM. Although this approach is popular with some statisticians, so far it has had relatively little impact in psychology.

Cell mean GLMs describe each dependent variable score as comprising the mean of the experimental condition plus error. The equation for this type of GLM is

$$Y_{ij} = \mu_j + \varepsilon_{ij} \tag{2.46}$$

In contrast to the experimental design GLM, which expresses experimental effects in terms of deviation from the constant μ, the only structure imposed upon the data by the experimental design cell mean model is that of the experimental conditions. (This feature of the experimental design cell mean model becomes more prominent with factorial designs.) As cell mean GLMs do not employ the parameter μ, there are only as many experimental condition means as there are parameters to be estimated.

Apart from solving the problem of overparameterization, the cell mean GLM affords another advantage. When overparameterized experimental design GLMs are used, it is possible to obtain a unique solution to the problem of estimating parameters by reparameterization or estimable function techniques (Searle, 1987). These methods of circumventing the overparameterization problem work well with balanced data, but with unbalanced data, they can result in ambiguous hypothesis tests. In contrast, irrespective of balanced or unbalanced data, when cell mean GLMs are applied there is never any ambiguity about which hypothesis is tested.

2.8.7 Experimental Design Regression and Cell Mean GLMs

Experimental design GLMs offer a number of advantages for the analysis of experimental data beyond those offered by other GLM representations. Experiments almost always are conceived in terms of experimental effects–differences in mean performance across conditions due to the experimental manipulations. The experimental design GLM parameters provide explicit estimates of separate and interactive experimental effects in single and multifactor experiments, and also clearly present the estimates of the multiple error terms arising in related measures experimental designs, so facilitating the accurate assessment of experimental effects. (A more detailed account of the benefits of the experimental design GLM is provided by Maxwell and Delaney, 2004, pp. B23–B25). The merit of experimental design GLMs is supported by the notations employed by the computer programs NAG GLIM, NAG GENSTAT (developed from Wilkinson and Rogers, 1973), SYSTAT, MINITAB, and SAS, as well as by numerous authors (e.g. Kirk, 1995; Howell, 2010; Maxwell and Delaney, 2004; McCullagh and Nelder, 1989; Searle, 1987, 1997; Winer, Brown, and Michels, 1991).

The text also describes how to implement ANOVA using effect coding in regression GLMs. Implementing ANOVAs in such a manner using statistical software packages not only emphasizes the GLM nature of ANOVA, but it also provides a way of accessing the statistical software's regression diagnostics directly. Typically, regression diagnostics will assess GLM assumptions (see Chapter 10) by applying some sophisticated techniques to analyse the regression (i.e., the GLM) residuals – the error term estimates. Nevertheless, even a cursory glance at the descriptions of the regression implementations of the various ANOVA designs illustrates the disparity between the experimental conception and the regression model depiction. As $(p-1)$ predictor variables are required to represent a first factor, $(q-1)$ predictor variables are required to represent a second factor and $(p-1) \times (q-1)$ predictor variables are required to represent their interaction, the simple and explicit match between a GLM parameter estimate and an experimental effect is lost.

In common with regression ANOVA GLMs, cell mean GLMs do not represent experimental main effects and interactions as simply and explicitly as experimental design GLMs. Due to the over parameterization of experimental design GLMs, the main advantage afforded by cell mean GLMs is the elimination of ambiguity over hypothesis tests with unbalanced data. This can be extremely useful, but this advantage is lost with balanced data and here, it is argued that good practice in experimental design should ensure balanced data.

Experimental design GLMs are employed throughout the present text to provide the benefits outline above. However, readers interested in regression GLMs in general and for ANOVA should consult Cohen *et al.* (2003), while readers interested in cell mean GLMs should consult Searle (1987).

CHAPTER 3

Comparing Experimental Condition Means, Multiple Hypothesis Testing, Type 1 Error, and a Basic Data Analysis Strategy

3.1 INTRODUCTION

In Chapter 2, traditional and GLM perspectives on the estimation of experimental effects and error terms for single factor independent measures ANOVA were described and the use of ANOVA summary tables—probably the most useful and effective way of presenting ANOVA results—was illustrated (Tables 2.3, 2.7, and 2.9). Each perspective provided identical estimates of experimental effects and error terms and culminated in an F-test of the omnibus null hypothesis (Section 2.3.2). However, carrying out an omnibus ANOVA usually is only an early step in analyzing experimental data. A significant omnibus F-test informs us that there is a difference among the experimental condition means that is very unlikely to occur ($p \leq 0.05$) if the null hypothesis is true. This is the basis for rejecting the null hypothesis and accepting the experimental hypothesis that the experimental condition means are not equal (i.e., the experimental manipulations exert an effect). However, when there are three or more experimental conditions, accepting the experimental hypothesis on the basis of a significant omnibus F-test says nothing about the location of the difference or differences between the experimental conditions.

The sum of squares due to experimental conditions is the sum of a set of $(p - 1)$ othogonal comparisons between experimental condition means. This key conception of the experimental conditions sum of squares is discussed in Section 3.5 and informs a number of topics presented in this and subsequent chapters. However, the

ANOVA and ANCOVA: A GLM Approach, Second Edition. By Andrew Rutherford.

relationship between the sum of squares due to experimental conditions and the experimental condition means may be expressed more directly.

Equation (2.28) defined the experimental conditions sum of squares as

$$\text{Experimental conditions SS} = \sum_{j=1}^{p} N_j(\mu_j - \mu)^2 \qquad \text{(2.28, rptd)}$$

This describes the experimental conditions sum of squares as the sum of each difference between each experimental condition mean and the general mean, squared and then multiplied by the number of subjects in each experimental condition. As discussed in Section 2.8.2, the full experimental design GLM predicts each subject's dependent variable score to be the experimental condition mean. Therefore, equation (2.28) can be regarded as defining the experimental conditions sum of squares as the sum of *each* subject's (squared) deviation from the general mean. This reflects the extent to which being in a particular experimental condition shifts the subjects' scores away from the general mean they would have obtained if no experimental manipulation had been experienced. However, all of the shifts in subjects' scores due to the experimental conditions experienced are squared and summed and so no information about which experimental conditions cause what direction and size of subject score shifts is retained.

Applying some algebra to equation (2.28) when there are equal numbers of subjects in each condition reveals that the experimental conditions sum of squares also can be defined as

$$\text{Experimental conditions SS} = \frac{N_j}{p} \sum^{up} (\mu_j - \mu_{j'})^2 \qquad (3.1)$$

where *up* is the number of unique pairs of experimental condition mean comparisons, where j and j' represent the particular experimental condition means involved in the unique pair comparisons. The number of unique pairs of comparisons is given by

$$\text{up} = \frac{p(p-1)}{2} \qquad (3.2)$$

For the experimental data presented in Table 2.2, the unique pairs of experimental conditions are presented in Table 3.1.

These unique comparisons are also *pairwise* comparisons—comparisons between the means of two individual experimental conditions (see Section 3.3 for a formal definition). Applying equation (3.1) provides

$$\text{Experimental conditions SS} = \frac{8}{3} \sum (6-11)^2 + (6-10)^2 + (10-11)^2$$

$$\frac{8}{3}(42) = 112$$

Table 3.1 The Unique Pairs of Condition Comparisons in the Study Time Experiment

30 s vs. 60 s	30 s vs. 180 s	60 s vs. 180 s

Therefore, the sum of squares due to experimental conditions is also the *average* of the differences between unique pairs of experimental condition means squared and summed. Again the differences over the unique comparisons of pairs of experimental condition means are squared, summed and averaged, so no information about which unique pair comparisons of experimental condition means cause what direction and size of difference is retained.

The preceding paragraphs emphasize that although there are several conceptions and definitions of the sum of squares due to experimental conditions—all providing equivalent estimates—this estimate always is a composite of differences between experimental condition means. Of course, this is no accident, as the purpose of this ANOVA is to test the omnibus hypothesis that all experimental condition means are equal (see Section 2.8.2). However, it does have the consequence that after a significant omnibus ANOVA, further analyses are required to find out exactly where the difference or differences between the experimental condition means lie. In the following sections, the standard ANOVA approach to categorical differences between experimental condition means is presented. A GLM comparison approach to assessing differences between experimental condition means consistent with the comparison of full and reduced experimental design GLMs described in Section 2.8.2 is available (e.g., Maxwell and Delaney, 2004). However, the standard ANOVA approach also fits well with a GLM approach to ANOVA and is consistent with the form of presentation employed by the majority of articles and books in the multiple hypothesis testing literature. Therefore, the standard approach to assessing the differences between experimental condition means is employed here. Nevertheless, depending on the nature of their experimental hypotheses, some researchers may prefer to examine the linear and curvilinear relations (trends) between the experimental condition means (see Maxwell and Delaney, 2004; Keppel and Wickens, 2004, Kirk, 1995).

3.2 COMPARISONS BETWEEN EXPERIMENTAL CONDITION MEANS

As the conditions in a designed experiment are devised to address particular theoretical or practical research issues, pairwise comparisons between experimental condition means are likely to be of most interest because they specifically address these issues. However, it also may be appropriate to address some issues by comparing the average of a set of experimental condition means with either the mean of an experimental condition or the average of another set of experimental condition means. Such comparisons are termed *nonpairwise* (and sometimes *complex*) comparisons. Although nonpairwise comparisons involve means from more than two experimental conditions, the comparison still is made between two means—between the mean of an experimental condition and the average of other experimental condition means, or between the means of two separate sets of experimental condition means. As only two means are compared in both pairwise and nonpairwise comparisons, the numerator *df* always is equal to 1. Consequently, pairwise and nonpairwise comparisons are termed *single df* comparisons.

Pairwise and nonpairwise comparisons can be illustrated with respect to the study time experiment described in Chapters 1 and 2. Here, it might be predicted that any additional study time increases memory performance, in which case all three experimental condition means should differ and pairwise comparisons between the means would provide appropriate tests of the theoretical predictions. The three experimental conditions mean comparisons of interest are presented in Table 3.1. Bearing in mind that for the 30 s condition $j = 1$, for the 60 s condition $j = 2$, and for the 180 s condition $j = 3$, the three null hypotheses for these comparisons are

$$\mathrm{H}_{01}\colon \mu_1 - \mu_2 = 0$$

$$\mathrm{H}_{02}\colon \mu_1 - \mu_3 = 0$$

$$\mathrm{H}_{03}\colon \mu_2 - \mu_3 = 0$$

However, consider the prediction that the average of the memory performance observed after study times of 30 and 60 s will differ from the memory performance observed after 180 s study time. The null hypothesis for this nonpairwise comparison is

$$\mathrm{H}_{04}\colon \mu_3 - \frac{\mu_1 + \mu_2}{2} = 0$$

3.3 LINEAR CONTRASTS

Comparisons between experimental condition means can be expressed as linear combinations of the condition means with specified coefficients, c_j, where one coefficient does not equal to 0 and all coefficients sum to 0. The general form of a linear contrast is

$$\psi_i = c_1\mu_1 + c_2\mu_2 + \cdots + c_p\mu_p$$

where ψ_i is the ith contrast between the μ_p experimental condition means. With respect to the study time experiment, the null hypotheses, H_{01}, H_{02}, and H_{03} can be expressed, respectively, as the linear contrasts

$$\psi_1 = (-1)\mu_1 + (1)\mu_2 + (0)\mu_3 = 0$$

$$\psi_2 = (-1)\mu_1 + (0)\mu_2 + (1)\mu_3 = 0$$

$$\psi_3 = (0)\mu_1 + (-1)\mu_2 + (1)\mu_3 = 0$$

The linear contrast

$$\psi_4 = (-0.5)\mu_1 + (-0.5)\mu_2 + (1)\mu_3 = 0$$

expresses the nonpairwise comparison null hypothesis H_{04}. Which experimental condition means are assigned the negative and positive coefficients makes no real difference, but as it is easier to work with positive rather than negative numbers, it usually makes sense to try to assign the negative coefficient(s) to the experimental condition(s) with the lower scores.

Unlike the number of pairwise comparisons available in an experiment given by equation (3.2), when $p \geq 3$, an infinite number of nonpairwise comparisons can be expressed (see Kirk, 1995, p.115, for a useful illustration). Relative to the number of comparisons possible in any experiment where $p \geq 3$, only a small number are examined and all should be meaningful in terms of the theoretical or practical issues addressed by the experiment.

3.4 COMPARISON SUM OF SQUARES

To obtain the sum of squares for a linear contrast, the estimated linear contrast, $\widehat{\psi}_i$, is obtained by substituting the population means, μ_j with their estimates, the experimental condition sample means, $\overline{Y}_j$.

$$\widehat{\psi}_i = \sum c_j \overline{Y}_j \tag{3.3}$$

The linear contrast sum of squares is given by

$$SS_{\widehat{\psi}_i} = \frac{N_j \widehat{\psi}_i^2}{\sum c_j^2} \tag{3.4}$$

where N_j is the number of subjects per experimental condition, $\widehat{\psi}_i$ is the contrast or difference between the two experimental conditions being compared and $\sum c_j^2$ is the sum of the squared coefficients. Equations (3.3) and (3.4) are applied to determine the sum of squares for the linear contrasts ψ_1 and ψ_4 in the study time experiment.

$$\psi_1 = (-1)\mu_1 + (1)\mu_2 + (0)\mu_3 = (-1)6 + (1)10 + (0)11 = -6 + 10 + 0 = 4$$

Therefore

$$SS_{\widehat{\psi}_i} = \frac{N_j \widehat{\psi}_i^2}{\sum c_j^2} = \frac{(8)(4)^2}{(-1)^2 + (1)^2 + (0)^2} = \frac{128}{2} = 64$$

One df is associated with $SS_{\widehat{\psi}_i}$, so the mean square for the contrast

$$MS_{\widehat{\psi}_1} = \frac{SS_{\psi_1}}{1} = \frac{64}{1} = 64$$

and so

$$F_{(1,21)} = \frac{MS_{\psi_1}}{MSe} = \frac{64}{2.476} = 25.848$$

$$\begin{aligned}\psi_4 = (-0.5)\mu_1 + (-0.5)\mu_2 + (1)\mu_3 &= (-0.5)6 + (-0.5)10 + (1)11 \\ &= -3 - 5 + 11 = 3\end{aligned}$$

Therefore

$$SS_{\widehat{\psi}_4} = \frac{N_j \widehat{\psi}_i^2}{\sum c_j^2} = \frac{(8)(3)^2}{(-0.5)^2 + (-0.5)^2 + (1)^2} = \frac{72}{1.5} = 48$$

$$MS_{\widehat{\psi}_4} = \frac{SS_{\psi_4}}{1} = \frac{48}{1} = 48$$

and so

$$F_{(1,21)} = \frac{MS_{\psi_1}}{MSe} = \frac{48}{2.476} = 19.386$$

Statistical packages usually have the capability to provide the p-value associated with a F-value with specified numerator and denominator *dfs*. Alternatively, the table of critical F-values presented in Appendix B can be consulted.

3.5 ORTHOGONAL CONTRASTS

The linear independence of regression predictors was discussed in Section 2.8.4.2. Linear independence refers to nonredundant information, but orthogonality refers to nonoverlapping information. Therefore, orthogonality is a special, more restricted case of linear independence (Rodgers, Nicewander, and Toothaker, 1984). Two contrasts are orthogonal when the information used in one contrast is completely distinct from the information used in the other contrast(s). Contrasts are defined as orthogonal if the sum of the products of the coefficients for their respective elements is 0 and not orthogonal if it is anything other than 0. This method simply defines whether or not contrasts are orthogonal—it provides no information on the degree of relation between the contrasts. Given p groups, there are only $(p-1)$ orthogonal contrasts available in any set of contrasts. However, when $p \geq 3$, there are an infinite number of different sets of $(p-1)$ orthogonal contrasts available. Consider the set of contrasts ψ_1 and ψ_2

$$\psi_1 = (-1)\mu_1 + (1)\mu_2 + (0)\mu_3$$

$$\psi_2 = (-1)\mu_1 + (0)\mu_2 + (1)\mu_3$$

The coefficients for these contrasts are

$$\psi_1 = -1 + 1 + 0$$

$$\psi_2 = -1 + 0 + 1$$

the products of the $\psi_1 \times \psi_2$coefficients for their respective elements are

$$\psi_1 \times \psi_2 = (-1)(-1) + (1)(0) + (0)(1)$$

and the sum of these contrasts is

$$\sum \psi_1 \times \psi_2 = 1 + 0 + 0 = 1$$

As the sum is not equal to 0, the two contrasts in this set are not orthogonal. Now consider the linear contrasts

$$\psi_1 = (-1)\mu_1 + (1)\mu_2 + (0)\mu_3$$
$$\psi_4 = (-0.5)\mu_1 + (-0.5)\mu_2 + (1)\mu_3$$

The sum of the products of the $\psi_1 \times \psi_4$ coefficients is

$$\begin{aligned}\sum \psi_1 \times \psi_4 &= (-1)(-0.5) + (1)(-0.5) + (0)(1)\\ &= (0.5) + (-0.5) + (0)\\ &= 0\end{aligned}$$

As the sum is equal to 0, these linear contrasts are orthogonal.

The ψ_4 linear contrast can be assessed in exactly the same fashion as described for the ψ_1 linear contrast in Section 3.4. However, an alternative way to implement orthogonal contrasts is to employ the linear contrast coefficients in the manner of dummy and effect codes. Such an orthogonal coding scheme for the experimental data in Table 2.2 is presented in Table 3.2. Notice that each predictor variable implements a particular linear contrast.

Table 3.2 Orthogonal Coding Representing Subject Allocation to Experimental Conditions

Conditions	Subjects	X_1 (ψ_1)	X_2 (ψ_4)	Y
	s1	−1	−0.5	7
30 s	⋮		⋮	⋮
	s8	−1	−0.5	7
	s9	1	−0.5	7
60 s	⋮		⋮	⋮
	s16	1	−0.5	11
	s17	0	1	8
180 s	⋮	⋮	⋮	⋮
	s24	0	1	12

Table 3.3 Output Pertinent to Multiple Regression Equation for Orthogonal Coding

Variable	Coefficient	Standard Error	Standard Coefficient	t	p (Two-Tailed)
Constant	9.000	0.321	0.000	28.019	<0.000
X_1	2.000	0.393	0.625	5.084	<0.000
X_2	2.000	0.454	0.541	4.403	<0.000

Irrespective of the set of orthogonal contrasts used, when a regression ANOVA GLM employing orthogonal coding is applied to the data in Table 2.2, the omnibus ANOVA always is equivalent to that obtained by all other approaches (see Tables 2.3, 2.7 and 2.9). Most statistical software packages also provide the estimates presented in Table 3.3.

As with dummy and effect coding, only three experimental conditions are represented by the orthogonal coding scheme, so only three different scores are predicted—the means of the three experimental conditions. Substituting the pertinent predictor variable orthogonal codes and regression coefficients from Table 3.3 into equation (2.44) provides the means of each of the three experimental conditions

$$\overline{Y}_1 = 9 + (2)(-1) + (2)(-0.5) = 9 - 2 - 1 = 6$$

$$\overline{Y}_2 = 9 + (2)(1) + (2)(-0.5) = 9 + 2 - 1 = 10$$

$$\overline{Y}_3 = 9 + (2)(0) + (2)(1) = 9 + 0 + 2 = 11$$

The variable labeled Constant, the Y-axis intercept, β_0, reflects the general mean of the dependent variable scores. Therefore, only with balanced data does β_0 under orthogonal coding provide an estimate of the experimental design GLM parameter μ—the general mean of the experimental condition means. As with effect coding, the significance test of this variable assesses the null hypothesis that β_0 equals 0. With balanced data, this test is equivalent to a test of β_0 with effect coding (see Section 2.8.4.3 for further discussion of this test), but this test has no ANOVA corollary when data are unbalanced.

Under orthogonal coding, the tests of the coefficients β_1 and β_2 have simple interpretations: each assesses the linear contrast coded by the predictor linear contrast coefficients. As indicated in Table 3.2, predictor X_1 codes the linear contrast $\widehat{\psi}_1$ and predictor X_2 codes the linear contrast $\widehat{\psi}_4$. Therefore, tests of the β_1 and β_2 regression coefficients provide direct tests of the $\widehat{\psi}_1$ and $\widehat{\psi}_4$ linear contrasts. Both t-tests of these coefficients indicate significant linear contrasts. Indeed, as

$$t^2_{(df)} = F_{(1,df)} \qquad (2.1, \text{rptd})$$

it is a simple matter to show that the t-values associated with the β_1 and β_2 regression coefficients for predictors X_1 and X_2, respectively, are equivalent to the F-values associated with the linear contrasts $\widehat{\psi}_1$ and $\widehat{\psi}_4$ presented at the end of Section 3.4

$$\text{For } \beta_1\text{: } t = (-5.084)^2 = 25.848 = F_{\widehat{\psi}_1}$$

$$\text{For } \beta_2\text{: } t = (4.403)^2 = 19.386 = F_{\widehat{\psi}_4}$$

The total sum of squares for these two orthogonal contrasts is equal to the study time experiment omnibus ANOVA sum of squares ($\text{SS}_{\widehat{\psi}_i} = 64 + \text{SS}_{\widehat{\psi}_4} = 48 =$ Experimental Conditions SS $= 112$). This demonstrates that the omnibus ANOVA experimental conditions sum of squares is equal to the sum of squares for the contrasts in a complete orthogonal set. This equality is reflected in the experimental conditions *dfs*—the set of $(p - 1)$ orthogonal contrasts. The equality is a consequence of the fact that all comparisons between p means can be constructed from $(p - 1)$ orthogonal contrasts and the ANOVA test of the omnibus null hypothesis is equivalent to a simultaneous test of the null hypothesis that all possible contrasts among the p means are equal to 0.

If the two orthogonal contrasts $\widehat{\psi}_1$ and $\widehat{\psi}_4$ sum to the experimental conditions sum of squares, then any other contrast will reuse variance already attributed to the $\widehat{\psi}_1$ and $\widehat{\psi}_4$ contrasts. Does this mean that only orthogonal contrasts should be examined? The answer to this question is emphatically no. Although there should be economical use of nonorthogonal contrasts, it is agreed that research issues and not the orthogonality of comparisons should determine which hypotheses are tested (e.g., Howell, 2010; Keppel and Wickens, 2004; Kirk, 1995; Maxwell and Delaney, 2004; Myers, Well, and Lorch, 2010; Winer, Brown, and Michels, 1991). So why do statistics texts devote so much space to orthogonal contrasts? Well, the distinction between orthogonal and nonorthogonal contrasts is important for understanding the nature of the variance attributed to experimental conditions in ANOVA, as well as understanding how this variance is used to assess different hypotheses. Orthogonal comparisons employ non-overlapping information, but across nonorthogonal comparisons, information used in one comparison is used again in another comparison. There are only $(p-1)$ ways in which the experimental conditions can differ and so, nonorthogonal comparisons simply express the orthogonal comparisons in an alternative fashion. Kirk (1995, p. 118) provides an excellent illustration of this. Consider the three comparisons

$$\psi_{01} : \mu_1 - \mu_2$$

$$\psi_{02} : \mu_1 - \mu_3$$

$$\psi_{03} : \mu_2 - \mu_3$$

However, note that the comparison ψ_{03} also can be obtained from the comparisons ψ_{01} and ψ_{02}

$$\psi_{03} = (\mu_1 - \mu_3) - (\mu_1 - \mu_2) = \mu_2 - \mu_3$$

$$\psi_{03} = \quad \psi_{02} \quad - \quad \psi_{01} \quad = \mu_2 - \mu_3$$

This demonstrates that although the way in which comparison ψ_{03} is expressed suggests new information is involved, in fact, exactly the same information is involved in comparison ψ_{03} as is involved in comparisons ψ_{01} and ψ_{02}.

Orthogonal contrasts maintain in data analytic terms the conceptual and logical distinctions between hypotheses and provide an unambiguous breakdown of the variance attributed to the experimental conditions in ANOVA. This simplifies the relationship between hypotheses and data, and facilitates interpretation of the hypotheses. However, another reason for considering the orthogonality of contrasts is its relevance for multiple hypothesis testing described below.

3.6 TESTING MULTIPLE HYPOTHESES

Different hypotheses are tested when different experimental condition means are compared. So, when multiple comparisons are carried out, multiple hypotheses are tested. Research in this area is driven by three awkward facts—as the number of hypothesis tests increases so does the Type 1 error rate, but controlling the Type 1 error rate reduces analysis power and controlling the Type 1 error rate too strictly can reduce analysis power substantially. Multiple hypothesis testing research might be characterized as the study of these problems and how to deal with them.

Fisher (1935b) was one of the first to present multiple hypothesis testing procedures (Fisher's LSD and the Bonferroni procedure), but multiple hypothesis testing research appeared in fits and starts until the 1980s (Harter, 1980), when it began to receive greater sustained attention from statisticians in experimental, clinical, and epidemiological research (e.g., Shaffer, 1995). More recently, most research on this topic has cohered around the *Closure Principle* (Marcus, Peritz, and Gabriel, 1976). A family of hypotheses is closed if the family includes all the hypotheses that intersect with these hypotheses (e.g., Westfall et al., 1999). For example, consider the omnibus hypothesis, $A = B = C$. A number of hypotheses are implied or intersect with this omnibus null hypothesis. These include the three pairwise hypotheses: $A = B$, $A = C$, and $B = C$. A closed procedure applied to these three pairwise comparison tests would allow rejection of one or more of the three pairwise tests only if the omnibus hypothesis also was rejected.

Developing an understanding of the multiple hypothesis testing literature is difficult for a number of reasons. First, multiple hypothesis testing research provides a large and complex literature and too much still is confined to mathematical and statistical journals. More accessible accounts of key developments in the multiple hypothesis testing literature would be extremely beneficial for those researchers expected to apply the contemporary techniques, but who possess expertise in the application areas rather than in mathematical statistics. Ironically, researchers working in clinical and epidemiological research opposed to controlling Type 1 error when multiple hypotheses are tested (e.g., Aickin, 1999; Perneger, 1998; Rothman, 1990; Saville, 1990; Savitz and Olshan, 1995, 1998) have succeeded in making their arguments accessible to a less mathematically sophisticated audience. This achievement seems to be under appreciated by some of those advocating Type 1 error rate control (e.g., Bender and Lange, 2001). Another impediment to developing an understanding of this literature is a lack of consistent terminology for more than 30 different multiple hypothesis testing procedures (Kirk, 1995), as well as inconsistent and sometimes contradictory recommendations about which multiple hypothesis testing procedures should be applied in which circumstances (Keppel and Wickens,

2004; O'Neil and Wetherill, 1971; Pedhazur, 1997; Preece, 1982; Toothaker, 1991). More recently, genetics and brain imaging research have joined those areas whose needs drive multiple hypothesis testing research, with the result that the large, complex, and demanding literature on multiple hypothesis testing is beginning to separate according to application area (e.g., Dudoit, Shaffer, and Boldrick, 2003; Farcomeni, 2008; Turk et al., 2008).

The focus here is the approach to multiple hypothesis testing in experimental psychology (e.g., Keppel and Wickens, 2004; Keselman, Holland, and Cribbie, 2005; Kirk, 1994; Shaffer, 1995). Although multiple hypothesis testing in experimental psychology exhibits some coherence, there is still variation in approaches and recommendations. Both aspects should be apparent in the following sections. First, Type 1 and Type 2 error rates, different approaches to these error rates, and the nature of the Type 1 error rate inflation due to multiple hypothesis testing are described, as are different conceptions of Type 1 error rate and current definitions of what constitutes a family of hypotheses. This is followed by discussion of logical and empirical relations among hypotheses, and the place of planned and unplanned comparisons. Subsequently, there is consideration of which of the many different multiple hypothesis test procedures available are most appropriate and powerful, and finally, a general strategy for analyzing experimental data is outlined.

3.6.1 Type 1 and Type 2 Errors

A Type 1 error occurs when a true null hypothesis is rejected. For a Type 1 error to be committed, two events need to occur together—the null hypothesis must be true (i.e., the effect does not exist in the population) and the null hypothesis must be rejected. A Type 1 error cannot occur if the null hypothesis is false (i.e., the effect exists in the population). As described in Section 2.3, an F-statistic (representing the effect) is calculated from the data. Comparing this F-statistic with the appropriate F-distribution provides the probability of obtaining an F-value equal to or greater than that calculated when the null hypothesis is true. This provides the probability of observing the effect observed when the null hypothesis is true. When the probability of observing an effect when the null hypothesis is true is sufficiently low, the null hypothesis can be rejected and the experimental hypothesis accepted. The probability point at which it is deemed reasonable to reject the null hypothesis is given by the significance level, α. Therefore, when $\alpha = 0.05$, the probability of rejecting a null hypothesis when the null hypothesis is true $= 0.05$. In other words, when the null hypothesis is true, the α and the Type 1 error rate are different descriptions of the same criterion point (see Figure 2.1).

A Type 2 error occurs when a false null hypothesis is accepted. For a Type 2 error to be committed, two events need to occur together—the null hypothesis must be false (i.e., the effect exists in the population) and the null hypothesis must be retained. A Type 2 error cannot occur if the null hypothesis is true (i.e., the effect does not exist in the population).

Scientists generally regard Type 1 errors as more serious than Type 2 errors. This is because a Type 1 error identifies a false effect that can misdirect theory development and empirical effort, and requires empirical and theoretical effort to remedy. In contrast, when a Type 2 error is made, although a real effect is missed, no misdirection occurs and further experimentation is very likely to reveal the effect. The replication of study findings is a key requirement in science and so further experimentation on a topic always is likely, especially if the issue is relevant to theory development or has important practical relevance (e.g., Bakan, 1966; Clark, 1976).

The convention in most scientific disciplines is to set $\alpha = 0.05$. In the classic statistical conception, where one test is applied to experimental data, this has the consequence of setting the probability of a Type 1 error $= 0.05$. The contemporary setting of Type 2 error rates owes most to Cohen (1969, 1988, 1992a,b), who argued that researchers should invest as much time and effort ensuring the study is able to detect the effect under examination as they spend ensuring they will not erroneously report a false effect. In other words, it also makes sense to try and minimize the likelihood of a Type 2 error. Cohen defines power as the probability of correctly rejecting a false null hypothesis when an experimental hypothesis is true

$$\text{Power} = (1 - \beta) \tag{3.5}$$

where β is the Type 2 error rate. Cohen (e.g., 1988) recommends high analysis power and his examples indicate a power level of 0.8 is appropriate. When this level of power is achieved, equation (3.5) reveals the probability of a Type 2 error $= 0.2$. Setting Type 1 error rate $= 0.05$ and $\beta = 0.2$ confirms that Type 1 errors are regarded as more serious than Type 2 errors.

Another influence on the setting of Type 1 and 2 error rates is the exploratory or confirmatory nature of the study and its data analysis. Jöreskog (1969) and Tukey (1977) were among the first to distinguish between exploratory and confirmatory data analysis. Tukey demonstrated how a variety of procedures, particularly graphical procedures, could be used in an exploratory fashion to facilitate and improve confirmatory analyses. Since then the concept of exploratory data analysis has developed and now the exploratory label can extend to the nature of a study. Essentially, confirmatory study and data analysis is about confirming (or rejecting) hypotheses derived from theory, as is emphasized in this text, whereas exploratory study and data analysis is about exploring the data to outline interesting relationships. Exploratory and confirmatory studies may employ the same statistical procedures. However, in confirmatory studies, the emphasis is on minimizing Type 1 error, while ensuring Type 2 error is held at a level low enough to ensure sufficient power, whereas exploratory studies emphasize the minimization of Type 2 rather than Type 1 errors. Consequently, in exploratory studies the Type 1 error rate usually is substantially above the established discipline level and so these exploratory study results cannot be used to assess theory. Nevertheless, exploratory studies can provide useful insights and some indication as to the value of investigating the issues further by conducting confirmatory studies and analyses (e.g., Jaeger and Halliday, 1998).

The focus here is on Type 1 errors due to rejecting null hypotheses about the equivalence of means, but it is worth mentioning that Type 1 errors can be made in many other situations. For example, if a null hypothesis concerning a linear relationship between two quantitative variables is rejected, then a Type 1 error is possible. Indeed, whenever inferences are generalized from samples to populations, Type 1 errors are possible and just as likely when mean differences are assessed using confidence intervals (e.g., Feise, 2002; Hochberg and Tamhane, 1987), or when graphical procedures or any other informal procedures are applied to data and inferences are drawn (e.g., see Section 10.5).

3.6.2 Type 1 Error Rate Inflation with Multiple Hypothesis Testing

When one statistical test is applied to a data set, under the null hypothesis, the chosen significance level (α) determines the Type 1 error rate. For example, if $\alpha = 0.05$, then the Type 1 error rate $= 0.05$ and the probability of avoiding a Type 1 error $=$ $(1 - 0.05)$. However, if three tests are carried out, then avoiding a Type 1 error requires that each of the three tests avoids a Type 1 error. If the three tests are independent (i.e., orthogonal), then the probability of avoiding a Type 1 error across all of the tests is equal to the joint probability of all three (Type 1 error avoided) events occurring together. Therefore

$$\text{Pr(No Type 1 errors)} = (1 - 0.05)\,(1 - 0.05)\,(1 - 0.05) = 0.857$$

In more general terms

$$\text{Pr(No Type 1 errors)} = (1 - \alpha)^c \tag{3.6}$$

where α is the significance level and c is the number of independent tests or comparisons over which the Type 1 error rate applies. Equation (3.6) gives the probability of no (i.e., zero) Type 1 errors. The probability of at least one Type 1 error occurring when a number of hypotheses are tested is given by

$$\text{Pr(At least one Type 1 error)} = 1 - (1 - \alpha)^c \tag{3.7}$$

Therefore, when $\alpha = 0.05$ and $c = 3$, the probability of at least one Type 1 error occurring is

$$\text{Pr(At least one Type 1 error)} = 1 - (0.857) = 0.143$$

There is considerable difference between the probability of a Type 1 error of a single test (0.050) and the probability of at least one Type 1 error when $c = 3$ tests are applied (0.143). The cause of this Type 1 error rate inflation when multiple hypotheses are tested is the accumulation of the Type 1 error rates of each of the c individual hypothesis tests. Equation (3.7) describes the manner of accumulation of these individual hypotheses Type 1 errors.

The account of Type 1 error rate presented above assumes orthogonal tests of hypotheses. However, when tests are not orthogonal, equation (3.6) underestimates the probability of avoiding a Type 1 error and equation (3.7) overestimates the probability of at least one Type 1 error occurring. Unfortunately, when hypothesis tests are related, the Type 1 error rate also depends on the correlation structures of the tests. As no simple formulas exist to estimate Type 1 error rates in these circumstances, nearly all discussions assume independent tests of hypotheses and accept that the Type 1 error rate will be lower in circumstances where tests of hypotheses are related.

3.6.3 Type 1 Error Rate Control and Analysis Power

Equation (3.7) describes how Type 1 error rate accumulates as the number of hypothesis tests increase with result that the probability of falsely rejecting any of these null hypotheses is greater than that set by the conventional Type 1 error rate. Researchers usually want to maintain the conventional low levels of α and Type 1 error rate. To control the increase in Type 1 error rate due to multiple hypothesis testing (i.e., retain the conventional 0.05 levels) requires the probability point at which a true null hypothesis is rejected to be made more stringent. Of course, as this probability point is also that at which it is deemed reasonable to reject the null hypothesis (i.e., α), lowering the probability of a Type 1 error makes rejecting the null hypothesis less likely (i.e., the α p-value is reduced) and so the power of the statistical test is diminished.

One of the first methods proposed to control the Type 1 error rate inflation due to multiple hypothesis testing was the Bonferroni adjustment. The Bonferroni adjustment is also known as Dunn's procedure. Although the inequality was presented by Boole (1854, see Seneta, 1993), Bonferroni (1936) is credited with providing the mathematical proof, but, after Fisher (1935b), Dunn (1961) was the first to apply the inequality to multiple hypothesis testing. The inequality

$$1 - (1 - \alpha)^c \leq c(\alpha) \tag{3.8}$$

when α is between 0 and 1, states that the probability of independent events occurring together cannot be greater than the sum of their individual probabilities. Therefore, the sum of the individual event probabilities sets an upper limit on the probability of these joint events. Comparing equations (3.7) and (3.8) reveals that the probability of independent joint events, $1 - (1 - \alpha)^c$, equals the probability of at least one Type 1 error occurring. In short, the probability of rejecting at least one of c null hypotheses when all null hypotheses are true cannot be greater than $c(\alpha)$. Therefore, when c hypotheses are tested at α, setting what will be called a nominal α maintains the actual α. The Bonferroni adjustment describes how to obtain the Bonferroni nominal α ($B_n\alpha$)

$$B_n\alpha = \frac{\alpha}{c} \tag{3.9}$$

Testing and rejecting each of the set of c null hypotheses only when the probability of obtaining an effect $\leq B_n\alpha$ ensures that the probability of a Type 1 error cannot be greater than α. For example, applying equation (3.9) when three hypotheses are tested at $\alpha = 0.05$, provides

$$B_{\mathrm{n}}\alpha = \frac{\alpha}{c} = \frac{0.05}{3} = 0.01667$$

Therefore, when three hypotheses are tested and the Bonferroni adjustment is applied, setting nominal $\alpha = 0.01667$ for each of the three tests provides a set of three tests over which the actual α and Type 1 error rate $= 0.05$.

Subsequently, Sidak (1967) suggested using the specific relationship described in equation (3.8) to provide a more exact and slightly less conservative nominal α, which will be termed $S_{\mathrm{n}}\alpha$. (Dunn did not address the Sidak inequality, but due to the link with the Bonferroni inequality, the Sidak adjustment sometimes is known as the Dunn–Sidak adjustment.)

$$S_{\mathrm{n}}\alpha = 1 - (1-\alpha)^{1/c} = 1 - \sqrt[c]{(1-\alpha)} \tag{3.10}$$

Applying equation (3.10) when three hypotheses are to be tested at an actual $\alpha = 0.05$, provides

$$S_{\mathrm{n}}\alpha = 1 - \sqrt[3]{(1-0.05)} = 1 - 0.98305 = 0.01695$$

Therefore, when three hypotheses are tested and the Sidak adjustment is applied, setting a nominal significance level of 0.01695 for each of the three tests provides a set of three tests over which the actual α and Type 1 error rate $= 0.05$. In this situation, the Sidak adjustment provides a slight advantage. As $S_{\mathrm{n}}\alpha > B_{\mathrm{n}}\alpha$, the $S_{\mathrm{n}}\alpha$ sets a less stringent criterion for significance and so, more hypotheses will be rejected using the Sidak adjustment than the Bonferroni adjustment.

Although the difference between Bonferroni and Sidak adjustments increase as c increases, to detect these differences requires the probabilities to be calculated beyond three decimal places. Given this relatively slight difference, there is much to be said for the simplicity of the Bonferroni adjustment.

The Bonferroni adjustment also offers another advantage. The Bonferroni nominal αs sum to the actual α and Type 1 error rate. Therefore, provided the nominal αs and Type 1 error rates per hypothesis also sum to the actual Type 1 error rate desired, different nominal αs can be allocated to each hypothesis depending on the power desired to assess each hypothesis. For example, if one hypothesis had much more theoretical relevance than the other two (which were equally relevant), greater power could be achieved with respect to rejecting this hypothesis with $B_{\mathrm{n}}\alpha = 0.03$. The other two theoretically less important hypotheses could be assessed with $B_{\mathrm{n}}\alpha = 0.01$. Nevertheless, as the sum of the three $B_{\mathrm{n}}\alpha$s $(0.03 + 0.01 + 0.01) = 0.05$, the actual Type 1 error rate is maintained at ≤ 0.05.

It should be borne in mind that as the Bonferroni and Sidak adjustments are based on inequalities, the actual Type 1 error rates could be much lower than those estimated. Bonferroni and Sidak adjustments also assume the c hypothesis tests are orthogonal, so their adjustments will be too conservative when hypothesis tests are related. Bonferroni and Sidak adjustments are calculated and then applied to all hypothesis tests, so they are termed simultaneous or single-step methods. Nevertheless, most of the criticism directed at multiple hypothesis testing Type 1 error rate control is really criticism of the conservative nature (i.e., low power) of Bonferroni and Sidak adjustments (e.g., Bender and Lange, 1999; Perneger, 1998). However, more powerful methods for testing multiple hypothesis have been developed and some are presented later in this chapter.

3.6.4 Different Conceptions of Type 1 Error Rate

One of the main questions in multiple hypothesis testing is over which hypothesis tests should Type 1 error be controlled? Three approaches to organizing Type 1 error control over hypotheses have been proposed. These approaches differ in the way they organize hypotheses into the groups over which Type 1 error control is exerted. At the conservative extreme is *experimentwise* Type 1 error rate and at the liberal extreme is *testwise* or *per comparison* Type 1 error rate. Between these two extremes lies the notion of *familywise* Type 1 error rate.

Consider the experiment described in Chapter 2 with three experimental conditions. A single factor ANOVA applied to the data from this experiment provides a significant F-test and so, the omnibus null hypothesis of equal means is rejected. The next question is which experimental condition means differ? Applying equation (3.2) reveals that with three experimental conditions, there are three unique pairs or pairwise comparisons

$$\begin{aligned} up &= \frac{p(p-1)}{2} \\ &= \frac{3(3-1)}{2} = \frac{6}{2} = 3 \end{aligned} \tag{3.2, rptd}$$

The three unique pairwise comparisons are (1) 30 s versus 60 s, (2) 30 s versus 180 s, and (3) 60 s versus 180 s. We shall assume that the experimenter has interest only in the three hypotheses tested by these three pairwise comparisons.

3.6.4.1 Testwise Type 1 Error Rate

Testwise Type 1 error rate is defined individually for each hypothesis test. This means that $c = 1$, as the Type 1 error rate is controlled over each individual hypothesis test. Applying equation (3.7) when $\alpha = 0.05$ provides

$$\begin{aligned} \text{Pr(At least one Type 1 error)} &= 1 - (1-\alpha)^c \\ & 1 - (1 - 0.05)^1 \\ & 0.05 \end{aligned}$$

Under the testwise conception, each hypothesis is tested alone, at the conventional alpha and Type 1 error rate levels, usually with both alpha and Type 1 error rate = 0.05, irrespective of the number of hypotheses actually tested. Essentially, the testwise conception ignores the effect of testing other hypotheses on the Type 1 error rate. Statistical tables and the p-values typically reported by statistical software are based on the assumption that a single hypothesis test is applied to the data. For convenience, this can be called the classic statistical conception. Employing testwise Type 1 error rate applies the classical statistical conception to each test when multiple hypotheses are tested. Unfortunately, it is very likely that testwise Type 1 error rate is adopted unknowingly by many users in multiple hypothesis testing situations due to an unwarranted belief in the relevancy and accuracy of statistical software.

3.6.4.2 Familywise Type 1 Error Rate

Familywise Type 1 error rate is defined over specified groups or families of hypotheses. The notion of a family of hypotheses was introduced by Tukey (1953, and until Braun, 1994, this was an unpublished, but much distributed manuscript). Under this conception, control of Type 1 error rate is exerted across each family of hypotheses. For example, in the study experiment described above, the experiment may have been conducted to test a main hypothesis manifest in the comparison of the 30 s versus 180 s experimental conditions. Subsequently, ancillary hypothesis are tested by the 30 s versus 60 s and the 60 s versus 180 s experimental condition comparisons. This theoretical perspective provides two families of hypotheses. The first family comprises the single hypothesis tested by the planned comparison of the 30 s versus 180 s experimental conditions. The second family comprises the two unplanned hypothesis tests assessed by comparing the 30 s versus 60 s and the 60 s versus 180 s experimental conditions. Applying equation (3.3) separately to each family provides the Type 1 error rate per family. For the first family, $c = 1$, as the Type 1 error rate is controlled over the single planned hypothesis test. Applying equation (3.7) provides

$$\begin{aligned} \Pr(\text{At least one Type 1 error}) &= 1 - (1 - \alpha)^c \\ & 1 - (1 - 0.05)^1 \\ & 0.05 \end{aligned}$$

This is identical to the testwise situation where Type 1 error rate is controlled separately over single hypotheses. For the second family of hypotheses, $c = 2$, as the Type 1 error rate is controlled over the two unplanned hypothesis tests. Applying equation (3.7) provides

$$\begin{aligned} \Pr(\text{At least one Type 1 error}) &= 1 - (1 - \alpha)^c \\ & 1 - (1 - 0.05)^2 \\ & 0.098 \end{aligned}$$

3.6.4.3 Experimentwise Type 1 Error Rate

Experimentwise Type 1 error rate is defined and controlled over all of the hypotheses tested in an experiment. In the hypothetical experiment, a total of three hypotheses were tested, so $c = 3$. Applying equation (3.3) provides

$$
\begin{aligned}
\Pr(\text{At least one Type 1 error}) &= 1 - (1 - \alpha)^c \\
& 1 - (1 - 0.05)^3 \\
& 0.143
\end{aligned}
$$

The testwise, familywise, and experimentwise conceptions Type 1 error rate control demonstrate that Type 1 error rate increases with the number of hypotheses over which the error rate is defined. In consequence, testwise Type 1 error rate control, operating over individual tests of hypotheses, always provides the smallest Type 1 error rate. Testwise Type 1 error rate is equal to the α level chosen (usually 0.05). Familywise Type 1 error rate always will be greater than the testwise Type 1 error rate—provided the family contains more than one hypothesis. Provided a family of hypotheses contains fewer than all of the hypotheses tested over the whole experiment, the familywise Type 1 error rate always will be less than the experimentwise Type 1 error rate. The experimentwise Type 1 error rate is always greater than testwise and familywise Type 1 error rates. The only exception to this is when the set of hypotheses tested over the whole experiment is defined as the hypotheses family (see below). In these circumstances, experimentwise and familywise Type 1 error rates will be equivalent.

3.6.4.4 False Discovery Rate

The false discovery rate (FDR) is a relatively new and different conception of error rate proposed by Benjamani and Hochberg (1995, 2000). FDR is the ratio of the expected number of erroneous rejections to the total number of null hypothesis rejections—it is the expected proportion of falsely rejected null hypotheses (i.e., false discoveries: Type 1 errors). However, FDR control is less stringent than familywise Type 1 error rate control and Benjamani and Hochberg have suggested FDR may be most appropriate for situations where rejecting a few true null hypotheses would be tolerable, as in exploratory research. In line with these ideas, genetics and brain imaging research, where hundreds to thousands of multiple comparisons are applied, already are making great use of FDR (e.g., Dudoit, Shaffer, and Boldrick, 2003; Farcomeni, 2008). However, as psychological research generally and experimental research in particular does not need to cope with so many multiple comparisons, the case for applying FDR is weaker. Indeed, Keselman, Holland, and Cribbie (2005) state that researchers employing FDR should be able to justify why it provides a more appropriate form of error control than familywise Type 1 error rate control (see Maxwell and Delaney, 2004, for further discussion of FDR).

3.6.5 Identifying the "Family" in Familywise Type 1 Error Rate Control

There is general agreement that Type 1 error rate should be controlled at the level of the hypotheses family. Familywise Type 1 error rate control provides the best balance between the costs associated with Type 1 and Type 2 errors. However, what constitutes a family is a key issue for familywise Type 1 error rate and its control.

Tukey (1953) introduced the notion of families of hypotheses, but still seemed to regard all hypotheses tested in an experiment as constituting a single family and a number of researchers have maintained this view (e.g., Ryan, 1959, 1960). In the first text on multiple comparison procedures, Miller (1966, p. 34) states, "The *natural family ... in the majority of instances ...* is the *individual experiment* of a *single researcher*" (his italics). Nevertheless, distinguishing between experimentwise and familywise error rate seems rather pointless if the family is defined as comprising all of the hypotheses tested in an experiment (i.e., when experimentwise error rate = familywise error rate). Both Miller (1966, 1981) and Ludbrook (1998) regard the *individual experiment* family as including all of the hypotheses assessed by a global statistical procedure, such as ANOVA. In experimental research, this would mean that all of the hypotheses addressed by the ANOVA of all the experimental data would constitute the experiment family. Miller and Ludbrook's definition can be implemented easily when an ANOVA is applied to the data from single factor experiments. However, the approach has the potential to create problems for theory assessment. For example, consider a single ANOVA that addresses a fairly large number of hypotheses. All of these hypotheses would be defined as a single family, even although subsets of hypotheses may address different theoretical issues. One of these theoretical issues also could be addressed specifically with a regression analysis. The single or small number of hypotheses addressed by the regression would define a different family of hypotheses. In such circumstances, the most frequent outcome would be rejection of the regression null hypotheses and acceptance of the null hypotheses tested by the ANOVA, due to lower analysis power for the larger hypothesis family. Defining hypothesis families by statistical procedures without regard for the theoretical issues the study was designed and conducted to address is very likely to create problems for the theoretical interpretation of the analysis. It also ignores the matter of analyses communalities, for example, as regression and ANOVA are different instances of the GLM, are the regression and ANOVA hypotheses really assessed by different global statistical procedures? Moreover, when factorial experimental designs are used, the recommendation conflicts with the established ANOVA convention of treating each main and interaction effect as separate hypothesis families (demonstrated by the use of unadjusted p-values for each omnibus F-test in factorial ANOVA). Therefore, defining families in terms of the set of hypotheses assessed by different global statistical procedures is an unattractive analysis strategy because it is likely to create theoretical anomalies and it contradicts standard ANOVA practice.

In only the second text on multiple comparison procedures, Hochberg and Tamhane (1987, p. 5) define the hypothesis family as, "Any collection of inferences for which it is meaningful to take into account some combined measure of errors." However, "If these inferences are unrelated in terms of their content or intended use (although they may be statistically dependent), then they should be treated separately and not jointly." (Hochberg and Tamhane, 1987, p. 6). Therefore, theoretical or practical issues appear to be the basis for a "meaningful" collection of inferences. Fortunately, descriptions of what constitutes a family have become clearer. Shaffer (1995) identifies the family in terms of analysis purpose. Westfall et al. (1999, pp. 10 and 16) provide more detail and recommend that hypothesis families should be as small as possible and should form a natural and coherent unit. Keppel and Wickens (2004) describe a hypothesis family as a set of theoretically (or practically) related hypotheses. These descriptions make it clear that a hypothesis family should cohere around a theoretical or practical issue.

One of the difficulties in multiple hypothesis testing is obtaining accurate Type 1 error rate estimates when hypotheses are related. Defining hypotheses families as sets of related hypotheses when most available calculations assume independent tests of hypotheses might seem to be a recipe for substantial hypothesis family Type 1 error rate overestimation and an associated loss of analysis power. However, in univariate experimental studies, the theoretical relations between hypotheses are of a different kind to the logical and empirical relations affecting Type 1 error rates. The nature of the logical and empirical relations affecting Type 1 error rate estimates and adjustments are discussed briefly below.

3.6.6 Logical and Empirical Relations

Multiple hypotheses can be related logically, while the hypothesis tests can be related empirically. Addressing the logical relations between hypotheses can reduce the number of null hypotheses over which Type 1 error rate is controlled and this can have substantial consequences for the power of all multiple hypothesis testing methods. Indeed, if hypotheses are logically related, accommodating these relations and the consequences for the number of possibly true null hypotheses can provide an increase in power greater than is provided by other techniques. Empirical relations between tests of experimental hypotheses were accommodated by several of the classic multiple comparison procedures developed in the 1950s. However, rather than addressing the issue of empirically related hypotheses directly, like Bonferroni and Sidak adjustments, many recent multiple hypothesis tests simply accept that their Type 1 error rate adjustments, which assume independent hypotheses, will be conservative when related hypotheses are tested.

3.6.6.1 Logical Relations

Shaffer (1986) described the nature and consequences of logically related hypotheses (LRH). For example, consider again the hypothetical study time experiment with the three experimental conditions, 30s, 60s and 180 s. Here the omnibus hypothesis

$$\mu_1 = \mu_2 = \mu_3 \qquad (2.35, \text{rptd})$$

states that the means of all the experimental condition are equal. The omnibus null hypothesis may be expanded to specify the pairwise hypotheses

$$\mu_1 = \mu_2$$
$$\mu_1 = \mu_3$$
$$\mu_2 = \mu_3$$

The omnibus null hypothesis is rejected when a significant omnibus F-test is obtained. Indeed, when the omnibus null hypothesis is false, logic dictates that at most, only one of the three pairwise null hypotheses can be true. This is because

(1a) If $\mu_1 \neq \mu_2$, then: $\mu_1 = \mu_3$ is possible, but entails, $\mu_2 \neq \mu_3$
or (1b) If $\mu_1 \neq \mu_2$, then: $\mu_2 = \mu_3$ is possible, but entails, $\mu_1 \neq \mu_3$

(2a) If $\mu_1 \neq \mu_3$, then: $\mu_1 = \mu_2$ is possible, but entails, $\mu_2 \neq \mu_3$
or (2b) If $\mu_1 \neq \mu_3$, then: $\mu_2 = \mu_3$ is possible, but entails, $\mu_1 \neq \mu_2$

(3a) If $\mu_2 \neq \mu_3$, then: $\mu_2 = \mu_1$ is possible, but entails, $\mu_1 \neq \mu_3$
or (3b) If $\mu_2 \neq \mu_3$, then: $\mu_1 = \mu_3$ is possible, but entails, $\mu_2 \neq \mu_1$

In each of the three situations where one pairwise null hypothesis is rejected, the logical consequence is that at least one other null hypothesis also must be rejected. Therefore, when there are three means and one pairwise null hypothesis is rejected, the maximum number of pairwise null hypotheses that can be true = 1.

However, consider situations (1a) and (1b) again. (1a) shows that when $\mu_1 = \mu_2$ is rejected, then $\mu_1 = \mu_3$ is possible, while (1b) shows that when $\mu_1 = \mu_2$ is rejected, then $\mu_2 = \mu_3$ is possible. Although the logical consequence of rejecting one pairwise null hypothesis in this situation is only one other null hypothesis can be true, it is still not known which null hypothesis is true. Moreover, assume situation (1a) is true: $\mu_1 \neq \mu_2$ and $\mu_2 \neq \mu_3$, but $\mu_1 = \mu_2$. The first hypothesis test carried out demonstrates $\mu_1 \neq \mu_2$ and the second hypothesis test carried out demonstrates $\mu_2 \neq \mu_3$. Unfortunately in situation (1a), $\mu_2 \neq \mu_3$ was entailed by $\mu_1 \neq \mu_2$ and so $\mu_1 = \mu_3$ still may be true. In short, when there are three pairwise null hypotheses, rejecting one pairwise null hypothesis means that at most only one pairwise null hypothesis can be true, but it is not known which of the two remaining null hypotheses is false. Consequently, after the first pairwise null hypothesis is tested, only one null hypothesis can be true, but even after the second null hypothesis has been rejected, the same pairwise null hypothesis still could be true (see Table 3.7).

Due to LRH, rejection of the omnibus hypothesis limits the number of possibly true null hypotheses. Moreover, the rejection of other specific pairwise null hypotheses further limits the number of possibly true null hypotheses. As Type 1 error rate control refers to the probability of rejecting true null hypotheses, only the number of possibly true null hypotheses are of concern—there is no need to exert Type 1 error control over false null hypotheses (see Section 3.6.1). Therefore, Type 1 error rate control must accommodate only the number of possibly true null hypotheses. By focusing on only the number of possibly true null hypotheses, the estimate of Type 1 error rate is reduced and so analysis power is increased. When one statistical test is used to determine the circumstances for the next statistical test, it can be termed a stepwise procedure.

Shaffer (1986) presented two stepwise procedures to handle LRH. The more complex and more powerful of these procedures is known as S2 in the multiple hypothesis testing literature. Each time a null hypothesis (or a set of null hypotheses) is rejected, the S2 procedure employs the specific knowledge of which hypothesis or hypotheses are rejected to determine which and how many null hypotheses still could be true. In contrast, the less complex, but less powerful, S1 procedure does not employ specific knowledge of which hypotheses are rejected to determine which and how many null hypotheses possibly still are true. Instead, at each stage the S1 procedure determines the maximum number of null hypotheses that possibly could be true given that any one or more unspecified null hypotheses have been rejected (see Sections 3.8 and 3.8.1.2, and Table 3.7).

Of course, Shaffer's (1986) account of LRH is not limited to pairwise comparisons between experimental condition means in single factor experiments. Shaffer's LRH account also applies to main effect and interaction analyses in factorial experimental designs whenever pairwise comparisons are involved. Unfortunately, psychology researchers generally have made very little use of the S1 and S2 procedures (and another approach presented by Shaffer, 1979) to increase the power of range tests by incorporating LRH.

3.6.6.2 Empirical Relations

Empirical relations between multiple hypothesis tests can arise from two sources: Correlations between dependent variables and correlations due to multiple comparisons that employ the same treatment group or groups (Westfall and Young, 1993). However, as the vast majority of experimental studies apply univariate, rather than multivariate, analyses, even when there is more than one dependent variable (e.g., hit and false alarm rates), multiple comparisons involving the same experimental conditions are the main source of empirical relations in the analysis of experimental data. For example, with regard to the study time experiment described in Chapter 2 and this chapter, the null hypotheses H_{01} and H_{02}

$$H_{01}: -\mu_1 + \mu_2 = 0$$

$$H_{02}: -\mu_1 + \mu_3 = 0$$

are assessed by the linear contrasts ψ_1 and ψ_2 (see Sections 3.2 and 3.3)

$$\psi_1 = (-1)\mu_1 + (1)\mu_2 + (0)\mu_3$$

$$\psi_2 = (-1)\mu_1 + (0)\mu_2 + (1)\mu_3$$

Table 3.4 Orthogonal Coding Representing Subject Allocation to Experimental Conditions

Conditions	Subjects	X_1 (ψ_1)	X_2 (ψ_2)	X_4 (ψ_4)
	s1	−1	1	−0.5
30 s	⋮		⋮	⋮
	s8	−1	1	−0.5
	s9	1	0	−0.5
60 s	⋮		⋮	⋮
	s16	1	0	−0.5
	s17	0	−1	1
180 s	⋮	⋮	⋮	⋮
	s24	0	−1	1

The 30 s experimental condition mean is employed in both linear contrasts, but each contrast compares it with a different experimental condition mean (the 60 s mean for ψ_1 and the 180 s mean for ψ_2). Therefore, the correlation between these linear contrasts is 0.5. The veracity of this correlation can be established by determining the correlation between the linear contrast coefficients as described in Section 3.5. Table 3.4 presents the coding scheme for the ψ_1, ψ_2, and ψ_4 linear contrasts of the experimental data presented in Table 2.2. The correlation between the X_1 and X_2 variables is 0.5, confirming the figure above. Variable X_4 represents the contrast ψ_4, which is orthogonal to the ψ_1 contrast. This is confirmed by the zero correlation between the X_1 and X_4 variables. However, these correlation coefficients are obtained only when there are equal numbers of subjects per condition. When there are unequal numbers of subjects per condition (i.e., unbalanced data), the relative numbers per condition influence the degree of relation between conditions and more complicated statistical procedures are required to obtain parameter estimates. Particularly problematic are those estimates needed for specific comparisons between experimental conditions. In such circumstances, it should not be assumed that the often opaque strategies adopted by statistical software packages will be appropriate in all situations (Searle, 1987, and see Section 2.1 for the benefits of balanced data).

The example above illustrates the two independent correlations between three linear contrasts. However, even in this relatively simple situation, the correlation between all three of the linear contrasts is not addressed (although the multiple correlation coefficient offers one way of measuring this dependence) nor is the way in which the correlation estimates affect the Type 1 error rate estimates. As the complexity of the correlation structure increases with an increase in the number of experimental conditions and contrasts it can be appreciated why independent hypothesis tests are assumed and Type 1 error rate overestimates are accepted in the knowledge that they set conservative limits.

3.7 PLANNED AND UNPLANNED COMPARISONS

The classic statistical conception assumes a confirmatory approach with a separate experiment designed and conducted to test each hypothesis (e.g., Hochberg and Tamhane, 1987). In such circumstances, testwise, familywise, and experimentwise Type 1 error rate and α are equivalent and usually $= 0.05$. Nevertheless, conducting a large experiment designed to test several hypotheses may provide a better scientific strategy (Fisher, 1926) and also provides greater statistical and economic efficiency (Hochberg and Tamhane, 1987; Westfall and Young, 1993). It is in this context that Shaffer (1995) states that any single experiment may have several purposes, with the consequence that the data set may be analyzed under different hypothesis family configurations.

In psychology and most other research areas, the classic and multipurpose perspectives on hypothesis testing have been reconciled. Consistent with the classic statistical conception, the experiment will have been designed to test a hypothesis and there is no reason to preclude testing this hypothesis according to the classic statistical conception just because the experiment also is designed to test other hypotheses. However, an experiment cannot have been designed and conducted to test unplanned comparisons, so it is appropriate for unplanned comparisons to be Type 1 error rate adjusted.

The high regard researchers have for *a priori* predictions receives strong support from the philosophy of science (for lucid introductions to Popper, Kuhn, and Lakatos, see Dienes, 2008). For Popper and Lakatos, novel theoretical predictions are extremely important and valued. A prediction is novel with respect to a theory if the prediction information did not contribute to the construction of the theory. Novel predictions are necessary for a progressive program of scientific research and their corroboration provides conjectural signs of theoretical truth.

Philosophy of science perspectives award a special status to theoretical predictions—predictions derived from theory that has not already incorporated the prediction information. The classic statistical conception also awards a special status to predictions that experiments are designed and conducted to test (e.g., O'Neil and Wetherill, 1971). When planned comparisons manifest these features, it is appropriate that their hypothesis tests are not Type 1 error rate adjusted. Nevertheless, researchers tend to follow a *convention* that places a limit on the number of unadjusted planned comparisons that can be conducted. Usually, the number of *dfs* associated with the particular factor under examination defines this limit on the number of unadjusted planned comparisons. This limit also equals the number of orthogonal contrasts available with respect to the factor. Kirk (1995) recommends testwise Type 1 error rate is applied only to orthogonal planned comparisons. Howell (2010) comments that over time statisticians have become less concerned about contrasts being orthogonal. This may explain in part why most researchers limit the number of unadjusted planned comparisons, but do not require these comparisons to be orthogonal (e.g., Gamst, Meyers, and Guarino, 2008; Howell, 2010; Keppel and Wickens, 2004; Pedhazur, 1997; Tabachnik and Fidell, 2007; Winer, Brown, and Michels, 1991).

However, there are alternative recommendations. Myers, Well and Lorch (2010) suggest that all planned hypotheses should be included in a single family and, consistent with the ANOVA convention, the use of a common error term should define the other hypothesis families. Maxwell and Delaney (2004) and advocate a more conservative strategy that ignores the recent recommendations on how a family of hypotheses should be defined. They state that (irrespective of their nature) all planned hypothesis tests constitute a single family over which Type 1 error rate should be controlled. This position is consistent with the views of Ryan (1959), who takes the most conservative position with respect to planned comparisons. He considers those hypotheses experiments have been designed and conducted to test to be no different to those hypotheses developed and tested after the completion of the experiment and review of the experimental data. Consequently, Ryan advocates that all comparisons should be treated alike and advocates the application of experimentwise Type 1 error control.

In a single factor experiment, the *dfs* associated with the factor and so the maximum number of unadjusted planned comparisons is given by $(p-1)$. For example, if a single factor experiment has three levels $(p=3)$, no more than two (i.e., $3-1$) unadjusted planned comparisons should be conducted. In a two-factor experiment, q defines the number of levels of the second factor and $(q-1)$ defines the *dfs* associated with the second factor and the limit on the number of unadjusted planned comparisons applied to the second factor. Similarly, $(p-1)(q-1)$ defines the *dfs* associated with the factor interaction and the limit on the number of unadjusted planned comparisons applied to the interaction. A two-factor experiment with three levels of Factor A $(p=3)$ and four levels of Factor B $(q=4)$ is conducted. With respect to the three experimental conditions defined by Factor A, no more than two (i.e., $3-1$) planned comparisons should be conducted. With respect to the four experimental conditions defined by Factor B, no more than three (i.e., $4-1$) planned comparisons should be conducted. And with respect to the 12 experimental conditions defined by the interaction between Factors A and B, no more than six [i.e., $(3-1)(4-1)$] planned comparisons should be conducted. Therefore, over the whole experiment, a total of 11 (i.e., $2+3+6$) planned comparisons could be conducted. Each planned comparison may be considered as the sole member of a hypotheses family.

There is an implicit assumption that all planned comparisons involve pairwise comparisons. This is a reasonable expectation because the experimental manipulation of key theoretical factors usually is conceived and designed in terms of differences between experimental conditions. It is rare for an experiment to be conceived and designed with a plan to address theoretical issues using nonpairwise comparisons and similarly, even unplanned comparisons are more likely to be contemplated in terms of pairwise comparisons, at least initially. As theoretical interest is most likely confined to a subset of pairwise comparisons, usually the limits on the number of unadjusted planned comparisons just described are more than adequate for a full analysis of the experimental data.

3.7.1 Direct Assessment of Planned Comparisons

It is perfectly legitimate to carry out planned comparisons directly, irrespective of the ANOVA omnibus *F*-test outcome. When such direct comparisons are conducted, a choice needs to be made about the error term(s) to be used in the comparisons. This choice is between the omnibus ANOVA MSe and separate error terms per comparison based on only that data in the two experimental conditions being compared. The omnibus ANOVA MSe is a weighted pooled error variance estimate (see Section 2.6) – all of the experimental data is involved in providing the omnibus ANOVA MSe error variance estimate, with each of the separate group variance estimates weighted by the group sample size (i.e., $N_j - 1$). Statistical estimates improve as a function of the sample size upon which they are based, so, when the GLM assumptions are tenable, particularly the independent measures variance homogeneity assumption (see Chapter 10), the omnibus ANOVA MSe should provide the best estimate of error variance. Moreover, being based on all of the experimental data also means that the omnibus ANOVA MSe estimate is associated with the greatest number of *dfs* available with the experimental data. Consequently, assuming all else is equal, use of the omnibus MSe provides the most powerful multiple comparison tests (see Chapter 4). When the GLM assumptions are not tenable, particularly the independent measures variance homogeneity assumption, using separate error terms based on only the two experimental conditions being compared is a valid alternative. Indeed, when related measures designs are applied, due to concerns about the nature of the omnibus ANOVA MSe error, the recommendation is always to employ error terms based on only the data involved in the pairwise comparisons. Concerns about sphericity are eliminated in related designs when only two sets of data are compared – when pairwise (or nonpairwise) comparisons are assessed, there will be only a single difference between subjects' scores, so the sphericity assumption, which is about equality of variances of differences, cannot apply (see Chapter 10). However, in independent measures designs, the two experimental conditions being compared must exhibit homogeneous variances.

3.7.2 Contradictory Results with ANOVA Omnibus *F*-Tests and Direct Planned Comparisons

It is possible for a directly assessed planned pairwise comparison to be significant when the omnibus *F*-test is not significant. This is one reason why planned pairwise comparisons should be assessed directly. One reason for these different outcomes is that omnibus *F*-tests and pairwise (and nonpairwise) comparisons address different questions. For example, with respect to the hypothetical study time experiment, the omnibus ANOVA tests the omnibus null hypothesis

$$\mu_1 = \mu_2 = \mu_3 \qquad (2.35, \text{rptd})$$

whereas the specific pairwise comparisons test the following null hypotheses

$$\mu_1 = \mu_2$$

$$\mu_1 = \mu_3$$

$$\mu_2 = \mu_3$$

Under the null hypotheses, the probability of observing a difference over three experimental conditions need not equal the probability of observing the same difference over only two experimental conditions. An appropriate test of the omnibus null hypothesis is provided by the omnibus F-test. However, the F-test numerator is the experimental conditions sum of squares – the sum of $(p - 1)$ orthogonal contrasts – divided by $(p - 1)$ *dfs*. In other words, the omnibus F-test numerator estimates the average influence of the $(p - 1)$ orthogonal contrasts and it is this averaging process that can diminish the influence of specific pairwise comparisons, which would be significant if considered alone.

It also is possible for the omnibus F-test to be significant, but no pairwise comparison to be significant. This is a direct consequence of the omnibus F-test indicating that at least one of the $(p - 1)$ orthogonal contrasts is significantly different. However, there is no guarantee that the significant contrast(s) in the set of $(p - 1)$ orthogonal contrasts will include a pairwise comparison. It is quite feasible that the significant contrast(s) will be one (or more) of the nonpairwise comparisons. (This issue arises in the context of Shaffers R test described in Section 3.8.1.2.)

3.8 A BASIC DATA ANALYSIS STRATEGY

Experiments are designed and conducted for a whole variety of theoretical and practical purposes, and these purposes determine which data analyses are most appropriate and direct their application. As varied experimental purposes will require varied analyses of experimental data, no single data analysis strategy will be appropriate for all experimental situations. Nevertheless, the basic data analysis strategy presented should provide some useful guidance, especially for new researchers, and it also may be helpful in providing an initial analysis to consider and from which to develop alternatives.

The strategy presented below is in the form of three stages, but depending on the research purpose some stages may be more relevant than others and some may be omitted all together. A number of useful analysis techniques researchers may wish to employ are not included (e.g., trend analysis) and some researchers may prefer to apply valid alternative techniques.

One important omission from the data analysis strategy presented below is any account of exploratory data analysis. Exploratory data analysis allows the researcher to develop an appreciation of their data, particularly by examining data distributions. The original account of exploratory analysis is provided by Tukey (1977), but strongly recommended are concise and more recent introductions provided by Kirk (1995) and Howell (2010).

3.8.1 ANOVA First?

Many statistical texts present the omnibus ANOVA as the first step in data analysis, with further analysis contingent upon the outcome of the omnibus ANOVA F-test—further analysis occurs only when the F-test is significant. However, several of the classic multiple comparison procedures (e.g., Dunn, 1961; Dunnett, 1955; Scheffe, 1953; Tukey, 1953) assess hypotheses more specific than the omnibus null hypothesis and were designed to be applied directly to data irrespective of the outcome of the omnibus ANOVA F-test. When these multiple comparison procedures are applied as designed, appropriate Type 1 error rates are observed. Frequently, however, these procedures are applied only after a significant omnibus F-test and when this is done, the probability of a Type 1 error occurring with respect to at least one of the hypotheses tested by the pairwise comparisons is reduced. The Type 1 error rate diminishes because the significant omnibus ANOVA F-test α level sets an upper limit on the pairwise comparison Type 1 error rate (e.g., Bernhardson, 1975; Wilcox,1987, 2003). Consequently, rather than testing the pairwise comparison hypotheses at α, a lower more stringent criterion is applied, so lowering comparison power—the likelihood of detecting a significant difference. The current data analysis strategy advocates applying planned comparisons to data directly, irrespective of the outcome of the omnibus ANOVA F-test. As the plan to apply these comparisons is drawn before the data is collected, the ANOVA outcome has no bearing on the planned comparisons. Therefore, applying an omnibus ANOVA before the planned comparisons is quite acceptable and offers practical benefits such as the provision of an omnibus error term in independent measures designs. (A different strategy often is applied in related measures designs due to concerns about the consequences of omnibus MSe assumption violations for pairwise comparisons—see Section 6.7)

The omnibus ANOVA outcome determines the need to analyze the data further only when there are no planned comparisons to assess. There has been such a focus here on designing and conducting experiments to test theory or examine practical issues that it may seem odd for an experiment to be designed and conducted without a planned comparison in mind. Nevertheless, there may be circumstances when theory or practical issues do not drive the experiment, or when theory or practical issues make such weak predictions that they fail to discriminate between experimental condition outcomes. Alternatively, a researcher may decide to follow Ryan's conservative recommendations. In such situations, data analysis would start at *Stage 2*.

3.8.2 Strong and Weak Type 1 Error Control

Under the complete null hypothesis, all null hypotheses are assumed true, whereas under partial null hypotheses, only a subset of the null hypotheses are assumed true. Weak Type 1 error rate control refers to Type 1 error rate control only under the complete null hypothesis. Strong Type 1 error rate control refers to Type 1 error rate control under partial null hypotheses (Hochberg and Tamhane, 1987). Fisher's (1935) LSD test provides an example of weak Type 1 error rate control, as it controls Type 1 error rate appropriately only under the complete null hypothesis. Therefore, when all null hypotheses are true, Fisher's LSD test will reject true null hypotheses only 5% of the time, but when some null hypotheses are false, Fisher's LSD test will reject true null hypotheses more than 5% of the time. Fisher's LSD test is applied only after a significant omnibus F-test, but this means that Fisher's LSD test is applied only under partial null hypotheses and so can apply only weak Type 1 error rate control. The Newman–Keuls procedure (Keuls, 1952; Newman, 1939) and Duncan's (1955) multiple range test also apply only weak Type 1 error control. All of the techniques recommended below apply the required strong Type 1 error rate control (Shaffer, 1995).

Hayter (1986) modified Fisher's LSD test to provide strong Type 1 error rate control for pairwise comparisons. Ramsey (1993) and Seaman, Levin, and Serlin (1991) report that over multiple comparisons, the Fisher–Hayter test is nearly as powerful as the REGW and Peritz tests. (The REGW and Peritz tests are two of the most powerful multiple comparison procedures, but neither is presented here because their computational demands make them difficult and laborious to apply—for an example of the REGW test, see Kirk, 1995.) The Fisher–Hayter test is recommended highly (Keppel and Wickens, 2004; Seaman, Levin, and Serlin, 1991) because of its Type 1 error rate control, its power, its ability to accommodate unbalanced data, and its ease of application. Both Keppel and Wickens (2004) and Kirk (1995) provide excellent descriptions of its application.

3.8.3 Stepwise Tests

Fisher's LSD and the Fisher-Hayter test sometimes are described as stepwise procedures—step 1 is the omnibus F-test and step 2 is the application of the Fisher LSD or Fisher-Hayter test. However, stepwise tests of multiple hypotheses really began with Holm's (1979) stepdown test.

Stepdown procedures order the pairwise comparison p-values from smallest to largest (from most likely to least likely to be significant). Next, the smallest comparison p-value is assessed. If this smallest comparison p-value is significant, then its corresponding null hypothesis is rejected and the second smallest comparison p-value is assessed. If this second smallest comparison p-value is significant, then its corresponding null hypothesis is rejected and the third smallest comparison p-value is assessed and so on. The stepdown procedure continues until a comparison p-value is assessed as not significant and its corresponding null hypothesis is accepted. At this point, the procedure terminates and all of the null hypotheses corresponding with the remaining pairwise comparisons are accepted.

Stepup procedures also begin by ordering the pairwise comparison p-values from smallest to largest. However, it is the largest comparison p-value that is assessed first. If this largest comparison p-value is significant, then its corresponding null hypothesis and all subsequent null hypotheses are rejected. However, if this comparison p-value is not significant, then its corresponding null hypothesis is accepted and the second largest comparison p-value is assessed. If this second largest comparison p-value is significant, then its corresponding null hypothesis and all subsequent null hypotheses are rejected. However, if this second largest comparison p-value is not significant, then its corresponding null hypothesis is accepted and the third largest comparison p-value is assessed and so on.

One attractive feature of most stepdown and stepup procedures is they require only classic p-values. This means that most stepdown and stepup procedures can be applied not only across different tests, but also across all types of test that provide classic p-values (e.g., Ramsey, 2002). Although it is more useful and now more common for stepdown and stepup procedures to be described with respect to the comparison

p-values (as above), the stepdown and stepup labels refer to the comparison test statistics. As larger test statistics are associated with smaller p-values, this explains what are now counter intuitive labels for these procedures.

3.8.4 Test Power

Stepwise procedures tend to be more powerful than single step procedures. However, Kirk (1995) notes that as the number of pairwise comparisons diminish, so too does the power of the Fisher-Hayter test—with only four comparisons, the Bonferroni and Sidak single step procedures and Holm's (1979) stepwise procedure are more powerful than the Fisher-Hayter test. So, clearly, not all stepwise procedures are more powerful than all single step procedures in all circumstances.

Due to the different orders in which stepdown and stepup procedures are applied, stepup procedures are likely to assess and possibly reject hypotheses that stepdown procedures do not assess, with the result that stepup procedures tend to be more powerful than stepdown procedures. However, again, not all stepdown procedures are less powerful than all stepup procedures (e.g., Lehman, Romano, and Shaffer, 2005).

The three most powerful stepwise procedures are Rom's (1990) and Hommel's (1988) procedures, closely followed by Hochberg's (1988) procedure (Dunnett and Tamhane, 1993; Shaffer, 1995). However, Hommel's procedure is difficult to understand and apply (Shaffer, 1995), while Rom's (1990) procedure is actually Hochberg's procedure employing more accurate and less stringent (i.e., higher) critical p-value levels.

Hochberg and Rom (1995) describe how to increase test power by employing Shaffer's LRH modifications with their stepwise procedures. Therefore, due to its power and relative ease of use, Rom's test, implemented with Shaffer's (1986) simpler S1 procedure, is the multiple hypothesis test procedure recommended here. After discussion with Dror Rom and Juliet Shaffer, a novel and simple implementation of Rom's test with Shaffer's S1 procedure was described by Dror Rom, which he named as Shaffer's R test (personal communications, February, 2010). This test is presented in Section 3.8.1.2.

Hochberg's and Rom's multiple hypothesis tests are based on Simes' (1986) inequality and the closure principle. Standard ANOVA, with normally distributed data and nondirectional hypothesis tests (see Section 2.1 and Chapter 10), typically provides positively dependent test statistics. For positively dependent and independent test statistics, the validity of multiple hypothesis test Type 1 error rates based on Simes' inequality is established (e.g., Benjamini and Yekutieli, 2001; Chang, Rom, and Sarkar, 1996; Hochberg and Rom, 1995; Samuel-Cahn, 1996; Sarkar, 1998; Sarkar, 2002; Sarker and Chang, 1997; Simes, 1986). ANOVA with normally distributed data and directional hypothesis tests may provide negatively dependent test statistics. Hochberg and Rom (1995) observed that negatively dependent test statistics can increase the multiple hypothesis test Type 1 error rates based on Simes' inequality, but they also demonstrated the upper bound on this increase differed minimally from the true level. Rodland (2006) also reported a strong bound on the average deviation between Simes corrected and true Type 1 error rates with negatively dependent test statistics. Rodland also confirmed that the instances where Simes' inequality fails more dramatically with negatively dependent test statistics are limited to highly artificial (pathological) examples at specific significance levels (Block, Savits and Wang, 2008; Hommel, 1983). In short, multiple hypothesis tests based on the Simes' inequality provide accurate control of the Type 1 error rate, particularly with standard ANOVA.

3.9 THREE BASIC STAGES OF DATA ANALYSIS

3.9.1 Stage 1

The first act at Stage 1 is to apply the omnibus ANOVA and then assess the GLM assumptions. The omnibus ANOVA MSe provides the best estimate of error variation and provides the most powerful tests when these assumptions are tenable. The second act at Stage 1 is to conduct the planned comparisons using the omnibus ANOVA MSe—if supported by the results of step 1 (see Chapter 10 for GLM assumptions). Planned comparisons need no Type 1 error rate adjustment, but attention needs to be paid to the limit on the number of unadjusted planned comparisons that can be applied (see Section 3.7).

3.9.2 Stage 2

At this point attention turns to interpreting the results of the omnibus ANOVA and to any specific and theoretically relevant unplanned comparisons. (Any hypotheses tested already at *Stage 1* should be omitted from *Stage 2* analyses.) Investigation of unplanned comparisons is warranted by a significant omnibus F-test. However, if any of the planned comparisons assessed in *Stage 1* are significant, then the omnibus F-test may be significant due to these differences alone. Therefore, it is important to employ comparison methods that exert *strong* control over the Type 1 error rate (see Section 3.7). Although it is perfectly legitimate to apply nonpairwise comparisons, for the reasons described in Section 3.7, it is most likely that pairwise comparisons will be of greatest interest. Once all hypotheses to be assessed have been specified, hypothesis families need to be constructed (see Section 3.6.5) and the appropriate familywise Type 1 error rate control applied to assess these unplanned hypotheses tests.

Consider a hypothetical experiment with five experimental conditions. According to equation (3.2), 5 experimental conditions provide 10 unique pairwise comparisons. As each of these pairwise comparisons is assessed by an F-test (see Sections 3.3 and 3.4), 10 classic test p-values are obtained. In order to focus on the application of the pairwise comparison tests, it will be assumed that no hypotheses were assessed with planned comparisons at *Stage 1* and all 10 of these unplanned pairwise comparisons constitute a single family of comparisons.

3.9.2.1 Rom's Test

In the standard application of Rom's (1990) test to such data, the 10 classic p-values are ordered from smallest to largest, as presented in Table 3.5. Next, the largest

Table 3.5 Information Relevant to Hochberg's (1988) and Rom's (1990) Tests

Classic F-Test p-Values $p_{(1,\ldots,m)}$	Rank i, Where $i = 1, \ldots, m$ i	Adjustment Factor $m - i + 1$	Hochberg's Adjusted α $\alpha/(m - i + 1)$	Rom's Adjusted α $\alpha'_{(i)}$
0.0003	1	10	0.0050	0.0051
0.0036	2	9	0.0056	0.0057
0.0062	3	8	0.0063	0.0064
0.0081	4	7	0.0071	0.0073
0.0124	5	6	0.0083	0.0085
0.0222	6	5	0.0100	0.0102
0.0363	7	4	0.0125	0.0127
0.0593	8	3	0.0167	0.0169
0.0651	9	2	0.0250	0.0250
0.1582	10	1	0.0500	0.0500

All adjustments assume $\alpha = 0.05$.

p-value, $p_{(m)}$, is compared with Rom's adjusted critical value, $\alpha'_{(m)}$. If $p_{(m)} \leq \alpha'_{(m)}$, then the null hypothesis, H_m, and all the other null hypotheses, which have lower p-values, are rejected. As can be seen from Table 3.5, the classic test p-value for the null H_m of 0.1582 is greater than Rom's α'_m of 0.0500 and so the null H_m is retained. Attention now turns to the hypothesis associated with the second largest classic p-value, $H_{(m-1)}$. If $p_{(m)} \leq \alpha'_{(m-1)}$, then the $(m - 1)$ null hypothesis and all other null hypotheses with lower classic p-values are rejected. However, again it can be seen that the classic p-value for the null $H_{(m-1)}$ of 0.0651 is greater than the α' of 0.0250. Therefore, the $(m - 1)$ null hypothesis is retained and attention turns to $H_{(m-2)}$ and so on in this fashion until all m null hypotheses have been retained or until the first null hypothesis is rejected. When this first null hypothesis is rejected, it and all other hypotheses lower in rank (i.e., with lower classic p-values) are rejected also. With regard to the hypothetical experimental data presented in Table 3.5, the first null hypothesis with a p-value $\leq \alpha'_{(m)}$ is the null with $i = 3$. Therefore, H_3 (classic F-test p-value $= 0.0062$) is rejected, as are all other null hypotheses with lower classic p-values. This means that the three null hypotheses designated H_3, H_2, and H_1 are rejected (see Olejnik et al., 1997, for another worked example).

3.9.2.2 Shaffer's R Test

When an ANOVA is applied to the five experimental conditions and a significant omnibus F-test is observed, it implies that one or more of the null hypotheses assessed are false. In Table 3.6, column one presents the number of conditions in an experiment, column two presents the initial number of pairwise null hypotheses available in experiments with 3–8 conditions and subsequent columns present the number of possibly true null hypotheses remaining after one pairwise null hypothesis has been rejected, after two pairwise null hypotheses have been rejected, after three pairwise null hypotheses have been rejected and so on. For example, when there are four experimental conditions, six pairwise comparisons are possible and so at the

Table 3.6 Number of Experimental Conditions, Number of Pairwise Comparison Null Hypotheses Available Initially, and the Number of Possibly True Null Hypotheses After Each Rejection of One More Pairwise Null Hypothesis

Experimental	Initial	Number of Pairwise Null Hypotheses Rejected																											
Conditions	Nulls	1	2	3	4	5	6	7	8	9	10	11	12	13	14	15	16	17	18	19	20	21	22	23	24	25	26	27	28
3	3	1	1	0																									
4	6	3	3	3	2	1	0																						
5	10	6	6	6	6	4	4	3	2	1	0																		
6	15	10	10	10	10	10	7	7	7	6	4	4	3	2	1	0													
7	21	15	15	15	15	15	15	11	11	11	11	10	9	7	7	6	5	4	3	2	1	0							
8	28	21	21	21	21	21	21	21	16	16	16	16	16	15	13	13	12	11	10	9	8	7	6	5	4	3	2	1	0

outset, there are six null experimental hypotheses. When any one of these null hypotheses is rejected, logic dictates that a maximum of only three null hypotheses possibly can be true. When a second null hypotheses is rejected, logic dictates that the maximum number of possibly true null hypotheses is still three (see Section 3.6.6.1) and even when three null hypotheses have been rejected, the maximum number of possibly true null hypotheses is still three. However, when four pairwise null hypotheses have been rejected, the maximum number of possibly true null hypotheses drops to two and when five pairwise null hypotheses have been rejected, only one pairwise null hypothesis can be true and clearly, when all six null hypotheses have been rejected then zero null hypotheses can be true.

Shaffer's R test applies a short cut method (to reduce the number of hypotheses tests required to comply with the closure principle) to implement a hierarchical stepdown procedure with a stepup test at each step. At each step (i.e., stepdown), the maximum number of possibly true null hypotheses is determined using Shaffer's S1 procedure (see Table 3.6) and Rom's test is applied to assess the significance of the pairwise comparisons.

Shaffer's R test could be applied directly to the pairwise comparisons between the experimental condition means. However, the strategy advocated here assumes Shaffer's R test is applied only after a significant ANOVA F-test is obtained. Of course, applying Shaffer's R test directly to a set of experimental condition means and applying an F-test followed by Shaffer's R test to the same set of experimental condition means are mutually exclusive approaches-applying both approaches invalidates both approaches.

When the strategy advocated here is applied, it must be remembered a significant ANOVA F-test indicates that at least one of the set of orthogonal contrasts assessed is significant. (The F-test numerator *dfs* define the number of orthogonal contrasts assessed.) However, as all sets of orthogonal contrasts include a nonpairwise comparison, it is possible that the significant F-test is attributable only to a significant nonpairwise comparison. In such circumstances, no pairwise comparison is significant and no follow up test should identify a significant pairwise comparison. This possibility is accommodated by Shaffer's R test at Step 1.

3.9.2.3 Applying Shaffer's R Test After a Significant F-Test

Consider the hypothetical experiment with five experimental conditions. Apply an ANOVA to the experimental data. If the omnibus F-test is significant, conduct all of the pairwise comparisons and order these on the basis of their associated classic p-values from smallest (ranked 1) to largest (ranked m) as presented in Table 3.5.

Step 1: The significant F-test indicates a reduction in the number of possibly true hypotheses, but it does not identify which (if any) of the pairwise null hypotheses should be rejected. Therefore, six pairwise null hypotheses are possibly true and it is necessary to identify which pairwise null hypotheses are false. This is done by testing the null hypothesis corresponding with the $p_{(1)}$-value plus the null hypotheses corresponding with the five largest $p_{(i)}$-values. Rom's test applied to these ordered p-values is presented in Table 3.7. The first

Table 3.7 Step 1 of Shaffer's *R* Test

Ordered p-Values	Rank i	Rom's $\alpha'_{(i)}$ for Six Null Hypothesis Tests (see Table 3.5)	Reject?
0.0003	1	0.0085	Yes
0.0222	6	0.0102	No
0.0363	7	0.0127	No
0.0593	8	0.0169	No
0.0651	9	0.0250	No
0.1582	10	0.0500	No

hypothesis to be rejected is that corresponding with $p_{(1)}$. This is because $p_{(1)} = 0.0003 \leq$ Rom's α'_6 (for six null hypothesis tests) $= 0.0085$. As neither of the other two null hypotheses' $p_{(i)}$-values are less than or equal to their respective critical $\alpha'_{(i)}$ values (the Rom adjusted significance levels), no other hypotheses are rejected at this step.

When Shaffer's R test is applied following a significant omnibus F-test, the first significant pairwise comparison not only indicates that the specific experimental condition means differ significantly, but also that a pairwise difference contributed to the significant omnibus F-test—it was not due to only a significant nonpairwise difference (see Section 3.7.2). Therefore, as Table 3.6 shows, having rejected only one pairwise null hypothesis, the first of a series of six null hypotheses are possibly true and a total of five pairwise null hypotheses must be rejected before the number of possibly true null hypotheses drops to a maximum of four pairwise null hypotheses possibly being true.

Step 2: Although a pairwise null hypothesis has been rejected, Table 3.6 still shows a maximum of six null hypotheses may be true. Therefore, Rom's test is applied again to assess six null hypotheses by testing the null hypothesis corresponding with the $p_{(2)}$-value plus the null hypotheses corresponding with the five largest $p_{(i)}$-values (see Table 3.8). The first and only hypothesis to be rejected is that corresponding with $p_{(2)}$, as $p_{(2)} = 0.0036 \leq$ Rom's $\alpha'_6 = 0.0085$.

Table 3.8 Step 2 of Shaffer's *R* Test

Ordered p-Values	Rank i	Rom's $\alpha'_{(i)}$ for Six Null Hypothesis Tests (see Table 3.6)	Reject?
0.0036	2	0.0085	Yes
0.0222	6	0.0102	No
0.0363	7	0.0127	No
0.0593	8	0.0169	No
0.0651	9	0.0250	No
0.1582	10	0.0500	No

Table 3.9 Step 3 of Shaffer's *R* Test

Ordered p-Values	Rank i	Rom's $\alpha'_{(i)}$ for Six Null Hypothesis Tests (see Table 3.5)	Reject?
0.0062	3	0.0085	Yes
0.0222	6	0.0102	No
0.0363	7	0.0127	No
0.0593	8	0.0169	No
0.0651	9	0.0250	No
0.1582	10	0.0500	No

Step 3: Two pairwise null hypotheses now have been rejected, but Table 3.6 still shows a maximum of six null hypotheses may be true. Rom's test is applied again to assess six null hypotheses by testing the null hypothesis corresponding with the $p_{(3)}$-value plus the null hypotheses corresponding with the five largest $p_{(i)}$-values (see Table 3.9). The first and only hypothesis to be rejected is that corresponding with $p_{(3)}$, as $p_{(3)} = 0.0062 \leq$ Rom's $\alpha'_6 = 0.0085$.

Step 4: Three pairwise null hypotheses have been rejected, but Table 3.6 continues to show that a maximum of six null hypotheses still may be true. Rom's test is applied again to assess six null hypotheses by testing the null hypothesis corresponding with the $p_{(4)}$-value plus the five null hypotheses corresponding with the five largest $p_{(i)}$-values (see Table 3.10).The first and only hypothesis to be rejected is that corresponding with $p_{(4)}$, as $p_{(4)} = 0.0081 \leq$ Rom's $\alpha'_6 = 0.0085$.

Step 5: Four null hypotheses now have been rejected, but Table 3.6 still shows a maximum of six null hypotheses may be true. Rom's test is applied again to assess six null hypotheses by testing the null hypothesis corresponding with the $p_{(5)}$-value plus the null hypotheses corresponding with the five largest

Table 3.10 Step 4 of Shaffer's *R* Test

Ordered p-Values	Rank i	Rom's $\alpha'_{(i)}$ for Six Null Hypothesis Tests (see Table 3.5)	Reject?
0.0081	4	0.0085	Yes
0.0222	6	0.0102	No
0.0363	7	0.0127	No
0.0593	8	0.0169	No
0.0651	9	0.0250	No
0.1582	10	0.0500	No

Table 3.11 Step 5 of Shaffer's *R* Test

Ordered p-Values	Rank i	Rom's $\alpha'_{(i)}$ for Six Null Hypothesis Tests (see Table 3.5)	Reject?
0.0124	5	0.0085	No
0.0222	6	0.0102	No
0.0363	7	0.0127	No
0.0593	8	0.0169	No
0.0651	9	0.0250	No
0.1582	10	0.0500	No

$p_{(i)}$-values (see Table 3.11). Table 3.11 presents all of the remaining null hypotheses and their corresponding p-values. As none of these ordered p-values are lower than the pertinent Rom $\alpha'_{(i)}$, no hypotheses are rejected. As no null hypotheses are rejected, there can be no reduction in the number of possibly true null hypotheses and no change to Rom's $\alpha'_{(i)}$, the testing procedure terminates at this point.

If a pairwise null hypothesis (corresponding with the $p(5)$-value) had been rejected at Step 5, then at Step 6, only four null hypotheses possibly could be true and so, Rom's test would be applied to assess four null hypotheses by testing the null hypotheses corresponding to the $p(4)$-value plus the three largest p(i)-values. In the example just presented, only one pairwise null hypothesis was rejected at each step, but rejecting more than one pairwise null hypothesis per step is a possibility. However, in these circumstances, all hypotheses with a rank higher than the lowest rank hypothesis rejected also are rejected and the following step simply employs Rom's test to assess the number of pairwise null hypotheses still possibly true given the full number of pairwise null hypotheses rejected at that point.

In the example presented, the final outcome of applying Shaffer's *R* test is to reject the four null hypotheses, H_1, H_2, H_3, and H_4, corresponding with the four $p_{(1\text{–}4)}$-values. When compared with Rom's conventional procedure (one of the most powerful multiple hypothesis testing procedures), Shaffer's *R* test demonstrates its greater power by rejecting an additional hypothesis.

3.9.3 Stage 3

After specific unplanned comparisons have been assessed, a researcher may want to examine the experimental data even further. The extensive comparisons likely to be undertaken at this stage might include comparing a control condition to all other experimental conditions, conducting all possible pairwise comparisons between experimental conditions and conducting all possible pairwise and nonpairwise comparisons between experimental conditions. Each of these three situations is discussed below.

When a control condition is compared to all other experimental conditions, due to the control condition mean being involved in every comparison, each comparison has a 0.5 correlation with every other comparison (see Section 3.6.6.2). Dunnett (1955) described a single-step test that accommodates this degree of correlation, while controlling the experimentwise Type 1 error rate (see Kirk, 1995 for an excellent description of its application). Dunnett's test often is described as being appropriate for experiments designed and conducted to compare a control with all other experimental conditions. However, if each of these comparisons genuinely was motivated theoretically, it would be more appropriate to consider them as planned comparisons and to test the hypotheses as described at *Stage 1*.

When all possible pairwise comparisons between experimental conditions are to be assessed, Rom's test and Shaffer's R test may be applied. In fact, due to the assumptions that no planned hypothesis tests were carried out and all of the pairwise hypotheses constituted a single family of hypotheses, these tests already have been illustrated for all possible pairwise comparisons in *Stage 2* above. This also reveals that while Stage 1 and Stage 2 differ conceptually and in terms of the control of Type 1 error rate, the only differences between Stage 2 and 3 are conceptual, as both stages apply the same Type 1 error rate control.

The single-step Tukey highest significant difference (HSD) or wholly significant difference (WSD) test (Tukey, 1953) is appropriate for all possible pairwise comparisons between experimental conditions. Although two-tailed tests are the norm, this is all Tukey's HSD test can apply. The Tukey–Kramer test (Kramer, 1956; Tukey, 1953) accommodates unbalanced data, but still requires normally distributed data and equal variances. In common with the Fisher–Hayter test, as the number of pairwise comparisons diminishes, so too does the power of Tukey's HSD relative to the Bonferroni and Sidak adjustments. However, as the more powerful and more easily applied Fisher–Hayter test is applicable to all pairwise comparisons between experimental conditions, it is recommended in preference to Tukey's HSD.

When all possible pairwise comparisons between experimental conditions are to be assessed, Scheffe's test (Scheffe, 1953) applies appropriate experimentwise Type 1 error rate control. Scheffe's single step test shares all of ANOVA GLM assumptions, but it is also one of the least powerful multiple comparison procedures.

Examining for differences between experimental conditions at *Stage 3*, and sometimes at *Stage 2*, is called "data snooping." The pejorative label is applied because this approach to testing hypotheses compares poorly with planned comparison hypothesis tests (see Westfall and Young, 1993, Chapter 1, for further discussion). However, rather than struggling to retain, or even force, a confirmatory data analytic approach, an alternative would be to abandon confirmatory data analysis and replace it with an exploratory data analytic approach. Instead of trying to control Type 1 error rates, conventional analyses or analyses employing FDR control could be applied to explore possible relationships. Although an exploratory approach may indicate relations worth investigating in a confirmatory fashion, it is important to appreciate and acknowledge that these relationships do not meet the conventional standards for statistical significance.

3.10 THE ROLE OF THE OMNIBUS *F*-TEST

Given that specific hypotheses manifest in planned comparisons can be tested directly, some authors have questioned the worth of any ANOVA omnibus *F*-test (e.g., Howell, 2010; O'Brien, 1983; Rosnow and Rosenthal, 1989; Wilcox, 1987). Although habitual application of ANOVA and further analyses only after a significant *F*-test is a poor research strategy, when researchers genuinely want to know if there any differences between all of the experimental condition means, then the ANOVA omnibus *F*-test is appropriate. However, if the experimenter only wants to know about specific differences between particular experimental condition means, then carrying out planned or unplanned comparisons directly is the appropriate strategy.

The GLM perspective provides another reason to continue applying ANOVA omnibus *F*-tests. An overall *F*-test reveals if the complete GLM significantly predicts the dependent variable scores. In factorial studies, the omnibus *F*-tests (for main effects and interactions) reveal whether particular components of the GLM make significant contributions to the prediction. If the overall *F*-test and the omnibus *F*-tests in factorial studies indicate significant prediction, further analyses are carried out to identify how the particular component elements manifest the prediction. Testing an omnibus null hypothesis also fits very well with stepwise multiple comparison procedures. Applying an ANOVA omnibus *F*-test can provide additional power for some stepwise multiple comparisons (Seaman, Levin, and Serlin, 1991), it may be required if specific closed procedures are implemented and its calculation conveniently provides the ANOVA MSe. In short, the GLM perspective provides for the use of omnibus *F*-tests and multiple comparisons as part of a coherent data analysis strategy that can accommodate ANOVA, ANCOVA, and regression. However, if the aim is only to test specific differences for significance, then conducting planned and/or unplanned comparisons directly may be appropriate. This sort of strategy may seem simple and easy at first, but its soon can develop into a piecemeal approach, which can be the source of confusion and mistakes in the control of Type 1 error. As the coherent rationale underlying the application of planned and/or unplanned comparisons directly to experimental data is achieved by implicit reference to the GLM, it seems sensible to make this explicit from the start.

CHAPTER 4

Measures of Effect Size and Strength of Association, Power, and Sample Size

4.1 INTRODUCTION

The main approach presented in this text is known as null hypothesis testing. Fisher (1925, 1926, 1935) developed and presented an approach for assessing the null hypothesis, but in psychology and other disciplines, what came to be known as null hypothesis testing was a combination of Fisher's approach and a much more regimented scheme for deciding between null and alternate hypotheses presented by Neyman and Pearson (1928, 1933). Although never free of criticism (e.g., Berkson, 1938), null hypothesis testing soon became the dominant data analytic approach in psychology and other disciplines. Nevertheless, recent criticism of null hypothesis testing led to serious discussion and debate about statistical reporting, and the formation of an APA Task Force on statistical methods (Wilkinson and Task Force on Statistical Inference, 1999). There now appears to be a general consensus that while null hypothesis testing should be retained, it should be supplemented with further information to provide greater appreciation of the effects examined. In particular, information on effect sizes should be reported and rather than simply stating the α level at which the decision to accept or reject the null hypothesis is made, the exact p-value associated with the statistical test should be reported. The provision of mean square error values also can be very useful information. In many ways the consensus has moved statistical reporting away from the barren application of Neyman and Pearson's decision strategy in the direction of the reasoned and evidenced argument approach to null hypothesis assessment advocated by Fisher. (Lehman, 1993, provides an account of Fisher's and Neyman and Pearson's statistical approaches, as do Maxwell and Delaney, 2004, who also provide an excellent summary of the recent debate about null hypothesis testing.)

ANOVA and ANCOVA: A GLM Approach, Second Edition. By Andrew Rutherford.

One of the main concerns and criticisms of null hypothesis testing is that the test outcome is influenced greatly by the size of the sample. In fact, as Rosenthal (1987) describes, test statistics such as F-values are related to the sample size and the size of the effect in the following way

$$\text{Test statistic} = \text{Sample size} \times \text{Effect size} \tag{4.1}$$

Equation (4.1) reveals that test statistics can be increased by increasing the sample size. In other words, the same effect size could be declared significant in one experiment, but not significant in another experiment, simply because the former experiment had recruited more subjects. It also reveals that even very small effect sizes may be declared significant if the sample size is large enough. As increasing the sample size improves the estimates of the experimental condition means and so the ability to distinguish between these means, detecting a significant influence of a smaller size of effect with a larger sample is entirely appropriate. Nevertheless, the facility to compare estimates of effect sizes that are free of the influence of sample size would be very useful for both theoretical and practical purposes. In the following sections, different conceptions of effect size and different measures of these conceptions are presented and discussed. Subsequently, analysis power is considered and its relationship with effect size and sample size is described, before a way of deciding upon the number of subjects to include in an experiment is presented.

4.2 EFFECT SIZE AS A STANDARDIZED MEAN DIFFERENCE

Cohen (1969, 1988) defined effect size as the difference between the population means of two experimental conditions divided by the standard deviation of this difference

$$d = \frac{(\mu_1 - \mu_2)}{\sigma_\varepsilon} \tag{4.2}$$

$(\mu_1 - \mu_2)$ represents the difference between the two population means and σ_ε is the standard deviation of the difference between these two population means (often δ-delta is employed rather than d to designate the effect size parameter). However, different experiments conducted in different laboratories may employ different dependent variable measures. If the effect sizes observed in these different experiments are to be compared, the different effect sizes need to be expressed on a common scale. Cohen's estimate of effect size provides such a common scale by expressing the difference between the two population means in standard deviation units (cf. z scores). When the effect size is estimated from sample data

$$\hat{d} = \frac{(\overline{Y}_1 - \overline{Y}_2)}{\sqrt{\text{MSe}}} \tag{4.3}$$

where $(\overline{Y}_1 - \overline{Y}_2)$ represents the difference between the two experimental condition means and MSe is the mean square error from the omnibus analysis of variance

(ANOVA). d is employed when $p = 2$, but when $p > 2$, f is employed. f is defined with respect to the population means of the experimental conditions

$$f = \sqrt{\frac{\sum (\mu_j - \mu)^2 / p}{\sigma_\varepsilon}} \tag{4.4}$$

where μ_j represents each of the different experimental condition population means, μ is the general population mean, p is the number of experimental conditions, and σ_ε is the standard deviation of the differences between the experimental condition population means (Cohen,1969, 1988). When the effect size is estimated from sample data

$$\hat{f} = \sqrt{\frac{\sum (\overline{Y}_j - \overline{Y}_G)^2 / p}{\text{MSe}}} \tag{4.5}$$

where $(\overline{Y}_j - \overline{Y}_G)$ is the difference between each of the experimental condition means and the general mean, and MSe is the mean square error from the omnibus ANOVA. Equations (4.2)–(4.5) reveal that the effect size increases as the difference between the means increases, but diminishes as the standard deviation of the differences between the experimental condition means increases.

Unfortunately, however, sampling error causes $\hat{d}$ and $\hat{f}$ to overestimate the population effect sizes. Why $\hat{d}$ and $\hat{f}$ provide overestimates of population effect sizes may be appreciated by considering the situation where no effect exists in the population. In such circumstances, although μ_1 and μ_2 will be equal, it is very unlikely that the sample mean estimates, $\overline{Y}_1$ and $\overline{Y}_2$ will be exactly equal, but, nevertheless, the formulas for $\hat{d}$ and $\hat{f}$ simply attribute any difference between means to the effect size estimate. As this also occurs when a real effect exists in the population, so $\hat{d}$ and $\hat{f}$ continue to overestimate real population effect sizes.

The extent of the effect size overestimation is related inversely to the size of the sample—the overestimate decreases as the sample size increases. Therefore, any adjustment to rectify the overestimate has to include the sample size (i.e., N). An adjusted $\hat{f}$ can be defined as

$$\text{Adjusted } \hat{f} = \sqrt{\frac{(p-1)}{N} \frac{(\text{MS}_{\text{effect}} - \text{MSe})}{\text{MSe}}} = \sqrt{\frac{(p-1)}{N}(F-1)} \tag{4.6}$$

Also note that when $p = 2$, the adjusted $\hat{f}$ will be comparable to an adjusted $\hat{d}$. Tables 2.6 and 2.7 provide the information required to calculate $\hat{f}$ and adjusted $\hat{f}$

$$\hat{f} = \sqrt{\frac{14/3}{2.476}} = 1.373$$

while

$$\text{Adjusted}\,\widehat{f} = \sqrt{\frac{(3-1)}{24}(22.615-1)} = 1.339$$

4.3 EFFECT SIZE AS STRENGTH OF ASSOCIATION (SOA)

Another way of conceiving of effect size is in terms of the strength of association (SOA) between the predictor variable and the dependent variable. As this description suggests, SOA has its origin in correlational rather than experimental research and is defined with respect to variances, rather than in terms of differences between experimental condition means and their standard deviations.

Many statistical packages provide R^2 as a standard part of the regression output and R^2 is a frequently employed index of SOA (e.g., see Table 2.9). In regression terms, R^2 is the square of the (multiple) correlation between the predictor variables (i.e., the experimental conditions in ANOVA) and the dependent variable, and can be interpreted as the proportion of dependent variable variance accommodated by, or attributable to, the experimental manipulation. In the experimental design GLM context, R^2 may be defined as

$$R^2 = \frac{\text{SSE}_{\text{RGLM}} - \text{SSE}_{\text{FGLM}}}{\text{SSE}_{\text{RGLM}}} \tag{4.7}$$

Applying the estimates calculated in Section 2.8.2 provides

$$R^2 = \frac{164-52}{164} = \frac{112}{164} = 0.683$$

Therefore, it may be said that 68.3% of the variance in the dependent variable is attributable to the experimental manipulation.

R^2 is equivalent to another SOA measure, η^2, described originally by Pearson (1905). Unfortunately, however, like $\widehat{d}$ and $\widehat{f}$, R^2 and η^2 overestimate the relationship between the experimental manipulation(s) and the dependent variable, with the R^2 overestimate related inversely to the sample size. Wherry (1931) described an adjusted R^2 that can be defined as

$$\text{Adjusted}\,R^2 = 1 - \left[\frac{N-1}{N-p}(1-R^2)\right] \tag{4.8}$$

Adjusted R^2 is presented as part of Table 2.9. Independently, Kelley (1935) presented another SOA measure, ε^2 as an unbiased alternative to η^2. In fact, ε^2 is equivalent to adjusted R^2 (Cohen and Cohen, 1975). (Current practice is to use lowercase

Greek letters to denote population parameters and to denote their estimates from sample data by placing a "hat" on these Greek letters. However, this was not standard practice when Pearson and Kelley published their work—indeed, Kelley denoted the population correlation ratio as, $\tilde{\eta}$, and its estimate squared as, ε^2. Therefore, to avoid the suggestion that η^2 or ε^2 are estimates of population parameters, no "hats" are employed. However, this does require appreciation that η^2 and ε^2 are unusual in denoting sample-based estimates of SOA. See Richardson, 1996, for further discussion.)

The final SOA measure to be considered is ω^2, which Hays (1963) defined as

$$\omega^2 = \frac{\sigma^2_{\text{total}} - \sigma^2_e}{\sigma^2_{\text{total}}} \tag{4.9}$$

where σ^2_{total} is the total variance (i.e., the reduced GLM error sum of squares (SS), or the sum of the traditional ANOVA effect and error SS) and σ^2_{e} is the full GLM error SS (i.e., the traditional ANOVA error SS). As $(\sigma^2_{\text{total}} - \sigma^2_{\text{e}}) = \sigma^2_{\text{expt. effect}}$ (i.e., the experimental effect). Therefore, ω^2 is the proportion of the total variance attributed to the experimental effect. (Following Keppel and Wickens, 2004, this estimate of effect size can be termed as the complete ω^2 SOA for the omnibus effect.)

When estimated from sample data, $\hat{\omega}^2$ can be defined by

$$\hat{\omega}^2 = \frac{(\text{SSE}_{\text{RGLM}} - \text{SSE}_{\text{FGLM}}) - (p-1)(\text{SSE}_{\text{FGLM}}/df_{\text{FGLM}})}{\text{SSE}_{\text{RGLM}} + (\text{SSE}_{\text{FGLM}}/df_{\text{FGLM}})} \tag{4.10}$$

Equation (4.10) also can be expressed in more traditional ANOVA summary table terms

$$\hat{\omega}^2 = \frac{\text{Experimental conditions SS} - (p-1)\text{MSe}}{\text{Total SS} + \text{MSe}} \tag{4.11}$$

or simply in terms of the number of experimental conditions, the F-statistic and the number of subjects per condition (in a balanced design)

$$\hat{\omega}^2 = \frac{(p-1)(F-1)}{(p-1)(F-1) + pN_j} \tag{4.12}$$

Equations (4.10)–(4.12) show that the $\hat{\omega}^2$ estimate addresses the overestimate bias by including sample size information. This is expressed as *dfs* in equation (4.10) and as N_j in equation (4.12), while the use of MSe in equation (4.11) incorporates the influence of *dfs*. Equation (4.12) also addresses the overestimation through the use of the $(F-1)$ components. As $F = 1$ when there are no experimental effects (i.e., under the null hypothesis), subtracting this value from each F removes the influence of chance differences between the condition means. Applying equation (4.12) to the data presented in Table 2.7 provides

$$\widehat{\omega}^2 = \frac{(3-1)(22.615-1)}{(3-1)(22.615-1)+3(8)} = \frac{43.230}{67.230} = 0.64$$

This reveals that the experimental effect explains 64% of the population variance. Applied to the same data, adjusted $R^2 = 0.65$ (see Section 2.6.3), which is 0.01 greater than the $\widehat{\omega}^2$ estimate. This difference is consistent with Maxwell and Delaney's (2004) claim (based on Maxwell, Camp, and Arvey, 1981, review of Carroll and Nordholm, 1975) that usually adjusted R^2 is no more than 0.02 greater than $\widehat{\omega}^2$.

4.3.1 SOA for Specific Comparisons

The type of ω^2 estimate considered so far refers to the overall or omnibus effect. However, there is likely to be interest in the SOA for particular comparisons between experimental conditions. For example, in the hypothetical study time experiment, there is specific interest in the SOA between the 30 and 180 s conditions. Similar to the omnibus ω^2, it is possible to obtain such an ω^2 estimate simply by substituting the $(\sigma^2_{\text{total}} - \sigma^2_{\text{e}})$ numerator in equation (4.9) with the specific comparison SS. Keppel and Wickens (2004) label this type of ω^2 estimate as the complete ω^2. This complete ω^2 can be defined as

$$\omega^2_{\psi_C} = \frac{\sigma^2_{\psi}}{\omega^2_{\text{total}}} \tag{4.13}$$

The complete ω^2 is estimated easily from

$$\widehat{\omega}^2_{\psi_C} = \frac{F_{\psi} - 1}{(p-1)(F_O - 1) + pN_j} \tag{4.14}$$

where F_{ψ} is the specific comparison F-value and F_O is the omnibus F-value—indeed, the whole denominator corresponds with the omnibus F-test.

Another type of ω^2 estimate, which Keppel and Wickens term the partial ω^2 (also see Grissom and Kim, 2005), can be defined as

$$\omega^2_{<\psi>} = \frac{\sigma^2_{\psi}}{\sigma^2_{\psi} + \sigma^2_{\text{e}}} \tag{4.15}$$

where (following Keppel and Wickens, 2004), the partial nature of the effect estimate is denoted by placing the effect in angled brackets. The partial $\omega^2_{<\psi>}$ is the specific comparison variance expressed as a proportion of the specific comparison variance plus the error variance, which in single factor experiments is the full experimental design GLM error variance. This can be estimated by

$$\hat{\omega}^2_{<\psi>} = \frac{F_\psi - 1}{F_\psi - 1 + 2N_j} \tag{4.16}$$

4.4 SMALL, MEDIUM, AND LARGE EFFECT SIZES

A simple relationship exists between effect size indexed by $\hat{f}$ and the strength of association defined by $\hat{\omega}^2$

$$\hat{f} = \sqrt{\frac{\hat{\omega}^2}{1 - \hat{\omega}^2}} \tag{4.17}$$

and

$$\hat{\omega} = \frac{\hat{f}^2}{(\hat{f}^2 + 1)} \tag{4.18}$$

For guidance only, Cohen (1969, 1988) suggested *d* and *f* values that would indicate small, medium, and large effect sizes. These are presented in Table 4.1, along with *f* values converted to $\hat{\omega}^2$ using equation (4.17). Cohen's guidelines suggest a large effect size is observed in the hypothetical experiment on study time.

4.5 EFFECT SIZE IN RELATED MEASURES DESIGNS

When related measures experimental designs are applied, a complicating issue arises with respect to the estimation of effect size. As described in Chapter 6, one of the main benefits of related measures experimental designs is reduced experimental design GLM error term. This reduction is achieved by accommodating the covariation between the dependent variable scores from the same or similar subjects rather than letting it contribute to the experimental design GLM error term.

Table 4.1 Cohen's Small, Medium, and Large Effect Sizes for d, f, and $\hat{\omega}^2$

Size	d	f	ω^2
Small	0.2	0.10	0.01
Medium	0.5	0.25	0.06
Large	0.8	0.40	0.14

Equations (4.3), (4.5), and (4.6), which provide $\widehat{d}$, $\widehat{f}$, and adjusted $\widehat{f}$ and equation (4.11), which provides $\widehat{\omega}^2$, reveal that the experimental design GLM MSe appears as part of the denominator to estimate all of the effect sizes. Therefore, if related measures designs provide reduced MSe terms compared to equivalent independent measures designs, it follows that the same difference between means, or the same SOA will be estimated as greater with a related measures experimental design than with an independent measures experimental design.

As the purpose of effect size estimates is to enable comparisons across studies free of the influence of sample size, it is far from ideal that effect size estimates can depend on whether related or independent measures designs are applied. However, as mentioned above, related measures designs differ from independent measures designs because subjects provide scores under more than one experimental condition and the covariation between these scores can be estimated and attributed to a random factor, usually labeled "subjects," so removing it from the experimental design GLM error term. In other words, related measures designs provide specific additional information that is used to reduce the error term. Therefore, effect size comparability across related and independent measures designs could be achieved simply by ignoring the additional information provided by related measures designs and estimating related measures design effect sizes as if they had been obtained with equivalent independent measures designs. The equivalent independent measures design is identical to the related measures design, but omits the "subjects" random factor. This approach to effect size estimation in related measures designs is described further in Chapter 6.

4.6 OVERVIEW OF STANDARDIZED MEAN DIFFERENCE AND SOA MEASURES OF EFFECT SIZE

Although sample size information has to be accommodated in effect size estimates to offset the overestimation bias, the major benefit provided by effect size measures is they cannot be inflated by increasing the size of the sample. However, in common with other statistics and aspects of null hypothesis testing, the validity of all effect size estimates is influenced by the extent to which the experimental data complies with GLM assumptions (see Chapter 10).

Standardizing mean difference provides a potentially boundless value indicative of effect size. In contrast, SOA measures provide a value between 0 and 1 that expresses effect size in terms of the proportion of variance attributable to the conditions. Therefore, by default, SOA measures also provide the proportion of variance not attributable to the experimental conditions—if 0.25 (i.e., 25%) of the variance is due to experimental conditions, then 0.75 (i.e., 75%) of the variance is not due to experimental conditions. This aspect of SOA measures can provide a greater appreciation of the effect size by highlighting how much of the performance variation is and is not due to the experimental manipulation. A comparison of the equivalent SOA and standardized mean difference measures presented in Table 4.1 shows that SOA measures can be very low. As performance also is affected by multiple genetic and experiential factors, it might be expected that the proportion of all performance variance attributable uniquely to some experimental conditions may be relatively low. Nevertheless, it is important that less experienced researchers do not undervalue low SOA measures of effect and, equally, that genuinely low SOA measures alert researchers to the possibility of additional causal factors or a lack of experimental control, or both (see Grisom and Kim, 2005, for further discussion).

Most authors recommend the use of $\widehat{\omega}^2$ to measure effect size (e.g., Keppel and Wickens, 2004; Maxwell and Delaney, 2004). Although R^2 and η^2 provide a valid description of the effect size observed in the sample data, they are inflated and poor estimates of the effect size in the population. $\widehat{\omega}^2$ minimizes the overestimation bias of the population effect size better than other effect size estimates and ω^2 estimates have been specified for most ANOVA designs. However, it should be appreciated that $\widehat{\omega}^2$ is not entirely bias free - like all of the effect size measures considered its overestimate bias increases as the sample size decreases.

Nevertheless, despite the many recommendations to use $\widehat{\omega}^2$ effect size measures, other estimates continue to be used. For example, η^2 is observed frequently in journal articles, and $\widehat{d}$ and $\widehat{f}$ are used by many statistical packages for power analysis and sample size calculation (e.g., G*Power, nQuery Advisor). For this reason, the next section considers the use of power analysis to determine sample size with respect to $\widehat{\omega}^2$ and f.

4.7 POWER

As mentioned in the previous chapter, the credit for drawing attention to the important issue of power is due to Jacob Cohen (1969, 1988). Although Cohen's work focused primarily on having sufficient power to detect the effects of interest in psychological research, his work also influenced research in many other disciplines. Cohen defined power as the probability of correctly rejecting a false null hypothesis when an experimental hypothesis is true

$$\text{Power} = (1-\beta) \tag{3.5, rptd}$$

where β is the Type 2 error rate (i.e., the probability of accepting a false null hypothesis, see Section 3.6.1). Alternatively, power is the probability of detecting a true effect. As described in Chapter 3, Cohen (1988) recommends power of at least 0.8. When this is achieved, equation (3.5) reveals $\beta = 0.2$.

4.7.1 Influences on Power

The sampling distribution of F under the null hypothesis was discussed in Section 2.3. This is the sampling distribution of F used to assess the tenability of the null hypothesis. When the null hypothesis is true, the sampling distribution of F has a *central distribution*, which depends on only two parameters: the F-value numerator and denominator *dfs* (see Figure 4.1). However, when the null hypothesis is false, F

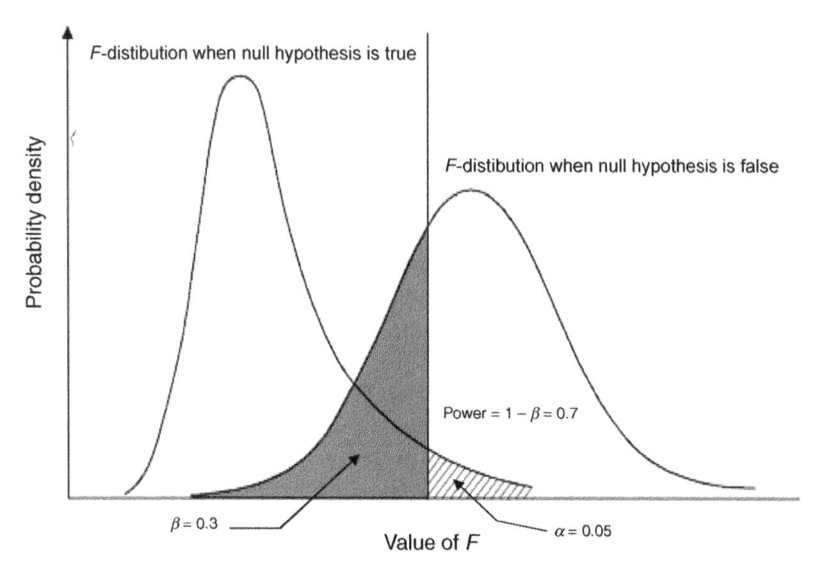

Figure 4.1 Sampling distributions of F when the null hypothesis is true and when it is false.

has a *noncentral distribution*. It is the noncentral distribution that is used to determine the power of a test (see Figure 4.1). The noncentral distribution depends on three parameters: the F-value numerator and denominator *dfs*, plus the noncentrality parameter λ. The noncentrality parameter is defined as

$$\lambda = \frac{\sum_{j=1}^{p} \alpha_j^2}{\sigma_{\mathrm{e}}^2 / N_j} \tag{4.19}$$

where $\sum_{j=1}^{p} \alpha_j^2$ is the sum of the squared experimental effects, σ_{e}^2 is the variance associated with these effects and N_j is the sample size per condition. Applying equation (4.19) to the data presented in Tables 2.2, 2.6, and 2.7 provides

$$\widehat{\lambda} = \frac{14}{2.476/8} = 45.234$$

Horton (1978) describes how λ may be estimated from

$$\widehat{\lambda} = \text{Experimental effect SS/MSe} \tag{4.20}$$

Applying equation (4.20) to the data in Table 2.7 provides

$$\widehat{\lambda} = 112/2.476 = 45.234$$

As equations (4.19) and (4.20) show, λ reflects the ratio of the sum of the squared experimental effects to the (mean square) error associated with this effect. In short, λ is another expression of effect size. Assuming the F-value numerator and denominator *dfs* do not change, any increase in λ will shift the noncentral distribution in a positive direction (see Figure 4.1). In fact, the power to detect an effect can be defined by the proportion of the noncentral F distribution that lies above the critical (central F) value used to define a significant effect. Therefore, it follows that power is determined by λ (i.e., effect size) and the noncentral F-distribution numerator and denominator *dfs*. The final determinant of power is the level of significance adopted. A more stringent level of significance reduces the likelihood that an effect will be detected, so reducing power.

Effect size increases with greater differences between the experimental condition means, or lower error variance, or both. Although the differences between the experimental condition means may be increased by selecting extreme factor levels and error variance may be constrained by the implementation of appropriate experimental controls, effect size really is set by nature and not the experimenter. Acceptable significance level (and so Type 1 error rate) is set by strong discipline conventions, while the numerator *dfs* (specifically, the number of experimental conditions) is determined by the experimental design appropriate to investigate the theoretical or practical issue. Therefore, the most easily manipulated experimental feature affecting analysis power is the denominator *dfs*, which is determined by the sample size. Consequently, most attempts to increase analysis power involve increasing the size of the sample.

Power refers to the ability of a statistical analysis to detect significant effects. However, because all of the information needed to assess power is determined by the nature of the experiment or study conducted, many researchers refer to experiment or study power. When the most powerful test appropriate for the data is applied, analysis and experiment or study power will be at a maximum and any of these labels will be acceptable. However, if the most powerful test is not applied to the data, a discrepancy can exist between the analysis power achieved and the analysis power possible given the nature of the study conducted. In such circumstances, it might be useful to distinguish between analysis power and study power, with the latter referring to the power achievable if the most powerful test is applied to the data.

4.7.2 Uses of Power Analysis

Murphy and Myors (2004) describe four uses of power analysis. First, power analysis can be applied to determine the sample size required to achieve a specific power of analysis. Second, power analysis can be applied to determine the power level of a planned or a completed study. Third, power analysis can be applied to determine the size of effect that a study would declare significant. The fourth and final use of power analysis is to determine an appropriate significance level for a study. However, only the two most important uses of power analysis will be considered here: employing power analysis to determine the sample size required to achieve a specific power and employing power analysis to determine the power level of a planned or completed study. An excellent overview of sample size planning is provided by Maxwell, Kelly, and Rausch (2008), while readers interested in the other uses of power analysis should consult Murphy and Myors (2004).

4.7.3 Determining the Sample Size Needed to Detect the Omnibus Effect

Power analysis can be employed to determine the sample size required to achieve a specific level of power to ensure that the study to be conducted will be able to detect the effect or effects of interest. Cohen (1962) noted that the low level of power apparent in many published studies across a range of research areas made the detection of even medium-sized effects unlikely. Even experienced and statistically sophisticated researchers can underestimate how many subjects are required for an experiment to achieve a set level of power (Keppel and Wickens, 2004). Unfortunately, recent surveys have indicated that despite the emergence of a considerable literature on power analysis and the issue of underpowered studies, the problem of underpowered studies persists, creating difficulty for the coherent development of psychological theory (see Maxwell, 2004, for review and discussion). Therefore, employing power analysis to determine the sample size required to achieve a specific level of power is by far the most important use of power analysis (e.g., Keppel and Wickens, 2004; Maxwell and Delaney, 2004).

Four pieces of information are required to determine the sample size needed to obtain a specific power. They are

- The significance level (or Type 1 error rate)
- The power required
- The numerator *dfs*
- The effect size

Acceptable significance levels are set by discipline conventions. In psychology, usually α is set at 0.05, although 0.01 may be used in some situations. Here, the usual $\alpha = 0.05$ is employed. Again, the convention in psychology is to aim for power ≥ 0.8. The numerator *dfs* is set by the number of experimental conditions. For the hypothetical single independent measures factor experiment presented in Chapter 2, numerator $dfs = (p - 1) = (3 - 1) = 2$.

The final, but possibly the most difficult piece of information required is the effect size. In an ideal world, researchers simply would apply their research knowledge to describe the effect size to be detected. However, even researchers quite familiar with a research topic and area can find it difficult to predict effect sizes, especially if the purpose of the study is to examine some novel influence. Nevertheless, if a similar study has been conducted already then this data may be useful for deriving an effect size. Alternatively, Keppel and Wickens (2004) suggest researchers to obtain an effect size by considering what minimum differences between means would be of interest. However, the overestimation bias of effect size measures needs to be considered when differences between sample data means provide the effect size estimates. In such circumstances, $\widehat{\omega}^2$ or adjusted f effect size estimates should be employed.

When no similar or sufficiently similar studies exist and researchers are unsure what minimum differences between means would be of interest, then Cohen's effect size guidelines can be useful. Nevertheless, researchers using Cohen's guidelines still need to decide whether large, medium, or small effects are to be detected and these categories may depend upon the research topic, the research area, or both. Here, a medium effect size is to be detected, corresponding to $\widehat{\omega}^2 = 0.06$ or $f = 0.25$.

Probably the easiest way to determine the sample size required to achieve a specific level of power is to use power analysis statistical software. Many statistical packages are now available to conduct power analysis and sample size calculations. Statistical software developed specifically for power analysis and sample size calculation is available commercially (e.g., nQuery Advisor) and as freeware (e.g., G*Power 3, Faul et al., 2007), while some of the larger commercially available statistical software packages (e.g., GenStat, SYSTAT) also include the facility to conduct power analysis and sample size calculation. If you have access to any of these statistical packages, it is recommended they are used, as these programs will provide the most accurate results.

Those without access to power analysis statistical software still can conduct power and sample size calculation in the "old-fashioned" way, using power charts. (Power charts are presented in Appendix C.) The power charts plot power $(1 - \beta)$ against the effect size parameter, ϕ, at $\alpha = 0.05$ and at $\alpha = 0.01$, for a variety of different denominator *dfs*. ϕ is related to λ as described below

$$\phi = \sqrt{\frac{\lambda}{p}} \tag{4.21}$$

The use of power charts is illustrated below for $\widehat{\omega}^2$ and f. The same iterative procedure is employed size irrespective of whether the $\widehat{\omega}^2$ or f effect size estimates are used. The only difference is whether equation (4.22) or (4.23) is applied

$$\phi = \sqrt{\frac{\omega^2}{1 - \omega^2}}\sqrt{N_j} \tag{4.22}$$

and

$$\phi = f\sqrt{N_j} \tag{4.23}$$

With $\widehat{\omega}^2 = 0.06$, the first calculation estimates ϕ, $N_j = 20$. This provides

$$\phi = \sqrt{\frac{\omega^2}{1-\omega^2}}\sqrt{N_j} \qquad (4.22, \text{rptd})$$

$$\phi = \sqrt{\frac{0.06}{1-0.06}}\sqrt{20}$$

$$\phi = 0.25(4.47)$$

$$\phi = 1.12$$

Examination of the power function chart for numerator *dfs* $(\nu_1) = 2$, $\alpha = 0.05$, denominator *dfs* $(\nu_2) = (p \text{ x } N_j - 3 = 3 \text{ x } 20\ 3)\ 57$, and $\phi = 1.12$, reveals power $= 0.37$. To increase power, the second calculation increases N_j to 50. This provides

$$\phi = \sqrt{\frac{0.06}{1-0.06}}\sqrt{50}$$

$$\phi = 0.25(7.07)$$

$$\phi = 1.77$$

Examination of the same power function chart, but now with denominator *dfs* $(\nu_2) = (p \times N_j - p = 3 \times 50 - 3)$ 147, $N = 150$, and $\phi = 1.77$, reveals power ~ 0.8. (In fact, the more accurate G*Power 3 estimate reveals that to obtain power $= 0.8$, *dfs* $(\nu_2) = 156$, $N = 159$. Nevertheless, the power charts allow derivation of quite an accurate estimate of the sample size required.)

To obtain sample size estimates using *f*, similar procedures are implemented, but rather than using equation (4.22), equation (4.23) is employed

$$\phi = f\sqrt{N_j} \qquad (4.23, \text{rptd})$$

Applying equation (4.23), with $f = 0.25$ and $N_j = 50$, equation (4.23) provides

$$\phi = 0.25\sqrt{50}$$

$$\phi = (0.25)(7.07)$$

$$\phi = 1.77$$

The equivalence of the $\widehat{\omega}^2$ and f calculations above can be appreciated by considering equations (4.17) and (4.22). This reveals that to calculate ϕ, $\widehat{\omega}^2$ is converted to *f*.

4.7.4 Determining the Sample Size Needed to Detect Specific Effects

The analyses considered so far have focused on determining the sample size required to achieve a specific level of power to allow rejection of the omnibus null hypothesis. However, as discussed in Section 3.2, the omnibus null hypothesis is rarely the hypothesis in which there is real interest. Usually, the real interest is in the hypotheses manifest in specific pairwise comparisons between the means of specific experimental conditions. In factorial experiments, there is likely to be great interest in whether the interaction effects are consistent with theoretical predictions (see Chapter 5).

In Section 3.6.4, it was suggested that in the hypothetical experiment, the comparison of most interest was the 30 s versus 180 s experimental conditions. (For simplicity, it will be assumed that this is the only pairwise comparison of interest in this experiment.) As this is a planned comparison (i.e., the experiment was designed with the intention of comparing performance in these experimental conditions), it follows that the sample size chosen for the experiment should take into account the level of power required for this comparison. As specific pairwise comparisons employ only a subset of the data involved in assessing the omnibus null hypothesis, determining the sample size needed to achieve a set level of power for pairwise comparisons is most likely to provide greater power for the omnibus null hypothesis assessment.

The key piece of information required to determine the sample size to enable a pairwise comparison to operate at the set power level is the partial ω^2 (see equation (4.15) or (4.16)) or the equivalent *f* measure. Once the size of the (pairwise) effect to be detected is expressed as a partial ω^2 or *f*, the procedure for determining the required sample size continues as was described for the omnibus effect.

The hypothetical experiment presented in Chapter 2 employs three conditions and it may be determined that for the 30 s versus 180 s pairwise comparison to operate with a power of 0.8 (when numerator *dfs* $= 1$, $\alpha = 0.05$), a sample size of $N_j = 20$ is required. Therefore, there needs to be 20 subjects in the 30 s experimental condition and 20 subjects in the 180 s experimental condition. It was established earlier that to detect the anticipated omnibus effect with power $\sim$0.8, required a sample size where $N_j = 15$. Therefore, one possibility would be to conduct an experiment with the number of subjects per condition as shown in Table 4.2.

If the experiment was run with the 55 subjects shown in Table 4.2, rather than the 45 (i.e., 3×15) subjects required to detect the anticipated omnibus effect (with power ~0.8), then, the power of the analysis to detect the anticipated omnibus effect would be >0.8, while the power to detect the effect of the 30 s versus 180 s pairwise comparison would = 0.8. As the purpose of power analysis is to ensure that sufficient power is available to detect effects, having more than the conventional power requirement of

Table 4.2 Possible Numbers of Subjects per Experimental Condition

Experimental condition	30 s	60 s	180 s
Number of subjects	20	15	20

0.8 to detect the omnibus effect is not a problem. However, in Section 2.1, allocating equal numbers of subjects to experimental conditions to obtain a balanced experimental design was advocated as a good design practice. The example above shows that applying the power analysis results above could lead to an unbalanced experimental design, but this could be resolved by employing 20 subjects in all conditions.

In contrast to the view that good design practice involves balanced experimental designs, McClelland (1997) argues that psychologists should optimize their experimental designs to increase the power of the important experimental comparisons by varying the number of subjects allocated to the different experimental conditions. To make his case, McClelland addresses the reasons for favoring balanced data.

The ease of calculation and the interpretation of parameter estimates with balanced data are dismissed by McClelland as insufficient to justify balanced data. McClelland claims that the widespread use of computer-based statistical calculation has made ease of calculation with balanced data irrelevant. However, while the ease of the statistical calculation may no longer be the issue it once was, the same cannot be said about the issue of statistical interpretation with unbalanced data. There are a number of different ways to implement ANOVA. With balanced data in factorial experiments, factors and their interactions are orthogonal and so, the same variance estimates are obtained irrespective of the order in which the variance is attributed. However, with unbalanced data, factors and their interactions are not orthogonal and so, appropriate analysis techniques must be employed to obtain accurate estimates of the variance due to the factors and their interactions. The overparameterization problem solved by cell mean models discussed in Chapter 2 is also relevant. Essentially, with unbalanced data, reparameterization and estimable function techniques can provide parameter estimates that are ambiguous and so provide ambiguous hypothesis tests, and this problem is compounded by the opacity of much statistical software (Searle, 1987). Therefore, the use of statistical software to ease calculation with unbalanced data can exacerbate the more serious problem of understanding what the statistics mean. McClelland also argues that rather than simply relying on balanced data to make ANOVA robust with respect to violations of distribution normality and variance homogeneity assumptions, the tenability of these assumptions should be assessed empirically and then, if necessary, remedied by data transformation, or the adoption of modern robust comparison methods. Unfortunately, however, the situation regarding statistical assumptions is not so simple and clear cut. To begin with, some authors now advise against assumption tests and instead advocate that the experimental design should minimize or offset the consequences of assumption failures (see Chapter 10). From this perspective, balanced experimental designs would be a standard component of any such design. Moreover, McClelland seems over reliant on data transformation and modern robust comparison methods. Certain assumption violations simply cannot be remedied by data transformation and even when data transformation does remedy the assumption violation(s), issues can arise as to the interpretation of transformed data analyses depending on the nature of the transformation applied. Similarly, the adoption of modern robust comparison methods may not be the panacea suggested—not all ANOVA techniques have an equivalent robust comparison method and not all robust comparison methods are considered equally valid.

Optimizing experimental designs by allocating different numbers of subjects to different experimental conditions to increase the power of the comparisons can be a very useful approach, but it is not without drawbacks. ANOVA is not robust to violations of the normality and homogeneity assumptions with unbalanced data. Therefore, if such assumption violations are detected with unbalanced data, a researcher already has abandoned one of their key strategies for dealing with such a situation and is reliant entirely on the success of the available data transformation or robust comparison method strategies to deal with the problems identified. Moreover, although the general availability of statistical software has eliminated concerns about calculation difficulty and error, the accurate statistical interpretation of results obtained with unbalanced data remains problematic. As accuracy is paramount, it may be better for less sophisticated or less confident data analysts to err on the side of inefficient, but equally powerful, balanced data designs, than risk misinterpreting the results of optimally designed experiments.

4.7.5 Determining the Power Level of a Planned or Completed Study

Although the best practice is to employ power analysis to plan and design a study, it also may be applied to determine the power of a study to detect the effects of interest. This might be done as a check before the study is conducted. Alternatively, when a study has been conducted, but no significant effect was detected, a power analysis can be applied to ensure that the study had sufficient power to detect the effect(s) of interest.

In any of these situations, study power can be assessed by applying equation (4.22) or (4.23), depending on the measure of effect size employed. For example, assuming it is planned to conduct a study to detect a large effect size, $\widehat{\omega}^2 = 0.14$, over 4 experimental conditions, with 10 subjects per condition and the significance level set at the conventional 0.05. Applying equation (4.22), provides

$$\phi = \sqrt{\frac{0.14}{1 - 0.14}}\sqrt{10}$$

$$\phi = (0.40)(3.16)$$

$$\phi = 1.26$$

Examination of the power function chart for numerator *dfs* $(\nu_1) = 3$, $\alpha = 0.05$, denominator *dfs* $(\nu_2) = (p \times N_j - p = 4 \times 10 - 4)$ 36, $N = 40$, and $\phi = 1.26$, reveals power $= 0.5$. As this value falls some way below the conventionally required power of 0.8, it is necessary to increase the sample size to obtain the required power. In fact, even when a large effect is to be detected with power $= 0.8$, in an experiment with numerator *dfs* $(\nu_1) = 3$ and $\alpha = 0.05$, $N_j = 19$. Therefore, the total sample size required $= (4 \times 19) = 76$.

When power analysis is applied to determine the study sample size needed to achieve a specific level of power to detect the effect or effects of interest, essentially, a prediction is made with respect to the effect size anticipated. Likewise, when power analysis is applied to check a study has sufficient power to detect an anticipated effect size after the study has been planned and designed, but before the study is conducted, the anticipated effect size is also a predicted effect size. However, when a study has been conducted without detecting a significant effect and a power analysis is applied to ensure that the study had sufficient power to detect the effect(s) of interest, perhaps it is less obvious that the anticipated effect size again is a predicted effect size. In short, all power analyses should employ effect size measures estimated before the study is conducted, or at least independent of the actual observed effect size(s).

4.7.6 The Fallacy of Observed Power

Despite the statement above that all power analyses should employ effect sizes anticipated or predicted before the study is conducted, some statistical packages (e.g., IBM SPSS) provide what is termed, observed power. Observed power employs the sample data to provide direct estimates of the parameters required for the power analysis and so, it is supposed to describe the power of the actual analysis conducted. This means that observed power is based on the untenable assumption that the observed sample means are equivalent to the population means. However, as discussed in Sections 4.2 and 4.3, it is known that sampling error is responsible for the sample data overestimating the population effect size. Nevertheless, sometimes it is argued—if observed power is high but no effect is detected, then the failure to detect the effect cannot be attributed to low power and so, some sort of support for the null hypothesis is provided. However, there is an inverse relationship between power and the p-value associated with any effect—as power increases, (the size of the test statistic increases and) the p-value decreases. Therefore, not only does observed power provide no new information but, by definition, the power of a test that declares an effect not to be significant cannot be high. Consequently, there is general agreement that the notion of observed power is meaningless and should be avoided, and that the appropriate role for power analysis is in planning and designing an experiment or other type of study (Hoenig and Heisey, 2001; Keppel and Wickens, 2004; Maxwell and Delaney, 2004).

CHAPTER 5

GLM Approaches to Independent Measures Factorial Designs

5.1 FACTORIAL DESIGNS

Factorial designs are the most common type of design applied in psychological research. While single factor experiments manipulate a single variable, factorial experiments manipulate two or more variables (i.e., factors) at the same time. As naturally occurring circumstances involve an interplay between a multitude of variables, there is a sense in which factorial designs offer closer approximations to reality than single factor studies. Of course, this line of reasoning leads to the conclusion that the most ecologically valid approach is to observe reality. However, as reality involves this interplay between a multitude of variables, usually it is far too difficult to determine causality simply by observing by observation. An experimental approach tackles the issue of causality in a different way. Rather than deal with the full complexity of reality, experiments simplify by focusing on only those aspects of reality thought relevant to the causal relations under investigation, which are expressed and measured as factors and dependent variables. The experimental factors and dependent variables must be relevant to what occurs in the real world, but maintaining relevance does not necessarily mean replicating real world occurrences. When experiments test psychological theory, factors and dependent variables must be theoretically relevant, but a psychological process may be assessed pertinently by a dependent variable obtained from an artificial task. Therefore, an experiment may have low ecological validity, but still have high external validity—its findings generalize appropriately to the theoretical issues under examination (Brewer, 2000; Shadish, Cook, and Campbell, 2002). While all experiments need to be externally valid, the particular purpose of the experimental investigation determines whether ecological validity also is a concern.

ANOVA and ANCOVA: A GLM Approach, Second Edition. By Andrew Rutherford.

Although the experimental conditions under which performance is observed are defined by two or more factors, factorial designs allow the effects attributable to the separate factors to be estimated. The separate factor effect estimates are termed main effects and compare with the estimates of the effects of experimental conditions in single factor studies. Nevertheless, the unique feature of factorial designs is the ability to observe the way in which the manipulated factors combine to affect behavior. The pattern of performance observed over the levels of a single factor may change substantially when combined with the levels of another factor. The influence of the combination of factors is called an interaction effect and reflects the variation in performance scores resulting specifically from the combination of factors. In other words, an interaction effect is in addition to any factor main effects. Indeed, in many factorial experiments whether there are factor interactions may be of more interest than whether there are main effects.

5.2 FACTOR MAIN EFFECTS AND FACTOR INTERACTIONS

The nature of main effects and factor interactions probably is explained best by way of an example. Consider an extension to the hypothetical experiment examining the influence of study time on recall. A researcher may be interested to know the consequences for recall of different encoding strategies when the same periods of study time are available. To examine this issue, the study time periods could be crossed with two forms of encoding instruction. Half of the subjects would be instructed to "memorize the words," just as before, while the other subjects would be instructed to use the story and imagery mnemonics (i.e., construct a story from the stimulus words and imagine the story events in their mind's eye). As before, the recall period lasts for 4 minutes and begins immediately after the study period ends. The data obtained from this hypothetical independent two-factor (2 × 3) design is presented in Table 5.1, along with some useful summary statistics.

Table 5.1 Experimental Data and Summary Statistics

Encoding Instructions	a1 Memorize			a2 Story and Imagery		
Study Time	b1 30 s	b2 60 s	b3 180 s	b1 30 s	b2 60 s	b3 180 s
	7	7	8	16	16	24
	3	11	14	7	10	29
	6	9	10	11	13	10
	6	11	11	9	10	22
	5	10	12	10	10	25
	8	10	10	11	14	28
	6	11	11	8	11	22
	7	11	12	8	12	24
$\sum Y$	48	80	88	80	96	184
$(\sum Y)^2$	304	814	990	856	1,186	4,470
$\sum Y^2$	2,304	6,400	7,744	6,400	9,216	33,856

Table 5.2 Means and Marginal Means for the Experimental Data in Table 5.1

	Study Time			
Encoding Instructions	b1 30 s	b2 60 s	b3 180 s	Marginal Means
a1 Memorize words	6	10	11	9
a2 Story and imagery mnemonics	10	12	23	15
Marginal means	8	11	17	12

Two conventions are relevant here. The first is that the factor with the fewest levels is labeled A, the factor with the next fewest levels is labeled B, and so on. The second is upper case (capital) letters are used to denote factors and lower case letters denote factor levels. Therefore, Factor A represents the two levels of encoding, with a1 and a2 representing the memorize, and story construction and imagery conditions, respectively, while Factor B represents the three levels of study time, with b1, b2, and b3 representing the 30, 60, and 180 s conditions, respectively.

An alternative presentation of the six experimental conditions comprising the two-factor design is presented in Table 5.2. Here, the mean memory performance observed in each of the experimental conditions is presented, as are marginal means (so termed because they appear in the table margins), which provide performance estimates under the levels of one factor, averaged over the influence of the other factor. (The mean in the bottom right corner is the average of all the averages.) Six experimental conditions result from crossing a two-level independent factor with a three-level independent factor. This type of factorial design can be called a fully crossed factorial design to distinguish it from other multifactor designs that do not fully cross all factor levels (e.g., see Kirk, 1995 on nested designs). However, fully crossed factorial designs are far more frequently employed than nested factor designs. Consequently, in common with linearity as applied to GLMs, nested factorial designs typically are described specifically as nested designs and any design described as factorial can be assumed to be fully crossed, unless specified otherwise.

The equation

$$Y_{ijk} = \mu + \alpha_j + \beta_k + (\alpha\beta)_{jk} + \varepsilon_{ijk} \tag{5.1}$$

describes the experimental design GLM for the independent measures, two-factor ANOVA applicable to the data presented in Table 5.1. Y_{ijk} is the ith subject's dependent variable score in the experimental condition defined by the jth level of Factor A, where $j = 1, \ldots, p$, and the kth level of Factor B, where $k = 1, \ldots, q$. As in the single factor design, the parameter μ is the general mean of the experimental condition population means. The parameter α_j is the effect of the j Factor A levels and the parameter β_k is the effect of the k Factor B levels. The effect of the interaction between Factors A and B over the j and k levels is represented by the parameter $(\alpha\beta)_{jk}$. Finally, the random variable, ε_{ijk}, is the error term, which reflects variation due to any uncontrolled source. Again equation 5.1 summarizes a set of equations, each of which describes the constitution of a single dependent variable score.

The general mean parameter, μ, is defined as

$$\mu = \frac{\sum_{j=1}^{p} \sum_{k=1}^{p} \mu_{jk}}{pq} \tag{5.2}$$

The Factor A effect parameters, α_j, are defined as

$$\alpha_j = \mu_j - \mu \tag{5.3}$$

where μ_j is the marginal mean for Factor A, level j, and μ is the general mean as defined. Therefore, the effect of the j levels of Factor A is given by the difference between the j Factor A marginal means and the general mean. The (marginal) mean for the jth level of Factor A is defined as

$$\mu_j = \frac{\sum_{k=1}^{q} \mu_{jk}}{q} \tag{5.4}$$

where q is the number of levels of Factor B. Therefore

$$\mu_1 = \frac{6 + 10 + 11}{3} = 9$$

$$\mu_2 = \frac{10 + 12 + 23}{3} = 15$$

and so

$$\alpha_1 = 9 - 12 = -3$$

$$\alpha_2 = 15 - 12 = 3$$

This shows that overall, the effect of the Factor A Level 1 manipulation (the *memorize* instruction) is to reduce memory performance by three words, whereas the Factor A Level 2 manipulation (the *story and imagery* mnemonics) is to increase memory performance by three words.

The Factor B effect parameters, β_k, are defined as

$$\beta_k = \mu_k - \mu \tag{5.5}$$

where μ_k is the marginal mean for Factor B, Level k. Therefore, the effect of the k levels of Factor B is given by the difference between the k Factor B marginal means and the general mean. The (marginal) mean for the *j*th level of Factor B is defined as

$$\mu_k = \frac{\sum_{j=1}^{p} \mu_{jk}}{p} \tag{5.6}$$

where p is the number of levels of Factor A. Therefore

$$\mu_1 = \frac{6 + 10}{2} = 8$$
$$\mu_2 = \frac{10 + 12}{2} = 11$$
$$\mu_3 = \frac{11 + 23}{2} = 17$$

and so

$$\beta_1 = 8 - 12 = -4$$
$$\beta_2 = 11 - 12 = -1$$
$$\beta_3 = 17 - 12 = 5$$

This shows that overall, the effect of the Factor B Level 1 manipulation (30 s study time) is to reduce memory performance by four words, the Factor B Level 2 manipulation (60 s) reduces memory performance by one word, while the Factor B Level 3 manipulation (180 s) increases memory performance by five words.

Comparisons between marginal means compare with the effect of experimental conditions in the single independent factor design presented earlier. However, while main effects in factorial designs bear comparison with experimental condition effects in single factor designs, across the two experiments, subjects' performance is very unlikely to have been observed under identical circumstances. In single factor experiments, there should be no systematic variation between experimental conditions other than the experimental manipulations defining the levels of the single factor and this will be reflected in the experimental condition means. However, the factorial design counterparts of the single factor experimental condition means—the marginal means—are estimated by averaging across any influence of the other factor and so, they incorporate any influence of this factor. Therefore, factorial experiment marginal means are likely to differ from comparable single factor experimental condition means (e.g., compare the Factor B marginal means in Table 5.2 with the original "memorize the words" single factor condition means—Level a1 in Table 5.2). Nevertheless, because the factors are crossed, the two marginal means for Factor A average over exactly the same levels of Factor B and so, the only difference between the two marginal means for Factor A is in terms of the distinction between the Factor A levels (encoding instructions).

Similarly, the three marginal means for Factor B are averaged over the same levels of Factor A and so, the only difference between the three marginal means for Factor B is in terms of the distinction between the Factor B levels (study time). Consequently, the averaging procedures result in orthogonal comparisons between the levels of Factor A and the levels of Factor B and the interaction between the Factor A and the Factor B levels.

The Factor A and Factor B interaction effect parameters, $(\alpha\beta)_{jk}$, are defined as

$$(\alpha\beta)_{jk} = \mu_{jk} - (\mu + \alpha_j + \beta_k) \tag{5.7}$$

where μ_{jk} represents the separate experimental condition means. Therefore, each interaction effect is the extent to which each separate experimental condition mean diverges from the additive pattern of main effects. Hopefully, this gives some substance to the earlier claim that the interaction effects were over and above any factor main effects. For the current example

$$\begin{aligned}
(\alpha\beta)_{11} &= \mu_{11} - (\mu + \alpha_1 + \beta_1) = 6 - (12 - 3 - 4) = 1 \\
(\alpha\beta)_{12} &= \mu_{12} - (\mu + \alpha_1 + \beta_2) = 10 - (12 - 3 - 1) = 2 \\
(\alpha\beta)_{13} &= \mu_{13} - (\mu + \alpha_1 + \beta_3) = 11 - (12 - 3 + 5) = -3 \\
(\alpha\beta)_{21} &= \mu_{21} - (\mu + \alpha_2 + \beta_1) = 10 - (12 + 3 - 4) = -1 \\
(\alpha\beta)_{22} &= \mu_{22} - (\mu + \alpha_2 + \beta_2) = 12 - (12 + 3 - 1) = -2 \\
(\alpha\beta)_{23} &= \mu_{23} - (\mu + \alpha_2 + \beta_3) = 23 - (12 + 3 + 5) = 3
\end{aligned}$$

The effect of the interaction is to increase or to decrease subjects' memory performance in each of the six experimental conditions by the number of words shown. For instance, memory performance in the experimental condition in which subjects were instructed to *memorize* the words and had 180 s to do so (condition a1b3) was poorer by 3 words than would be expected if Factors A and B had exerted only additive effects. The interaction effect indicates that the particular combination of these two-factor levels affects memory performance in a manner different to the aggregate of the separate effects of the two factors.

Based on the model component of the two-factor GLM equation, predicted scores are given by

$$\hat{Y}_{ijk} = \mu + \alpha_j + \beta_k + (\alpha\beta)_{jk} \tag{5.8}$$

and so, the last parameters, the error terms, which represent the discrepancy between the actual scores observed and the scores, predicted by the two-factor GLM, are defined as

$$\hat{\varepsilon}_{ijk} = Y_{ijk} - \hat{Y}_{ijk} \tag{5.9}$$

The experimental design GLM underlying the two factor independent measures ANOVA has been described and its parameters defined. Attention now turns to how well the GLMs incorporating some or all of these parameters accommodate the experimental data. Two strategies for carrying out ANOVA by comparing GLMs will be considered. The first is a simple extension of the comparison between full and restricted GLMs for factorial experimental designs, while the second concords with incremental linear modeling. Subsequently, the use of some computer software packages for implementing these strategies is discussed.

In factorial designs with balanced data, all factor main effects and interaction effects are orthogonal. (This may be checked by examining the zero correlation between the variables that effect code the factors and their interactions in Table 5.4 when equal numbers of condition instances, cf. subjects, are employed. NB. correlations between these variables are caused by unbalanced data.) As multicolinearity problems (see Section 1.4) do not arise in factorial designs with balanced data, no matter the order in which the sum of squares estimates for the factor main effects, interaction effects, and error are calculated, the same values always are obtained.

5.2.1 Estimating Effects by Comparing Full and Reduced Experimental Design GLMs

The major issue for a GLM comparison approach is what are the pertinent GLMs to compare? The equation

$$Y_{ijk} = \mu + \alpha_j + \beta_k + (\alpha\beta)_{jk} + \varepsilon_{ijk} \qquad (5.1, \text{rptd})$$

describes the full experimental design GLM underlying the independent measures, two-factor ANOVA. The hypotheses concerning the main effect of Factor A, the main effect of Factor B, and the effect of the interaction between Factor A and Factor B are assessed by constructing three reduced GLMs, which manifest data descriptions under the three different null hypotheses. Subsequently, the error components of these three reduced GLMs are compared with the error component of the full model above.

The Factor A null hypothesis states that the Factor A manipulation does not affect the data. Consequently, the reduced experimental design GLM does not need to accommodate an influence of Factor A. However, any influence of Factor B does need to be accommodated, as does any interactive influence of the Factors A and B. Therefore, the reduced GLM that omits the influence of Factor A, which is used to assess the influence of Factor A is

$$Y_{ijk} = \mu + \beta_k + (\alpha\beta)_{jk} + \varepsilon_{ijk} \qquad (5.10)$$

The null hypothesis that the levels of Factor A do not influence the data may be expressed more formally as

$$\alpha_j = 0 \tag{5.11}$$

Applying the same rationale, the reduced GLM for assessing the effect of Factor B is

$$Y_{ijk} = \mu + \alpha_j + (\alpha\beta)_{jk} + \varepsilon_{ijk} \tag{5.12}$$

The null hypothesis that the levels of Factor B do not influence the data may be expressed more formally as

$$\beta_k = 0 \tag{5.13}$$

Finally, the reduced GLM for assessing the effect of the interaction between Factors A and B is

$$Y_{ijk} = \mu + \alpha_j + \beta_k + \varepsilon_{ijk} \tag{5.14}$$

and the null hypothesis that the interaction between the levels of Factors A and B do not influence the data may be expressed more formally as

$$(\alpha\beta)_{jk} = 0 \tag{5.15}$$

Note that the null hypotheses are expressed in terms of zero effects and not in terms of the equivalence of the general mean, the marginal means, and the experimental condition means. This is because the marginal and experimental condition means may vary from the general mean as a consequence of one effect when another effect is assessed as being equal to zero.

The next step is to calculate the error sums of squares for the two factor independent measures ANOVA, full and reduced GLMs. For both full and reduced GLMs, the error SS can be defined as

$$\text{SSE} = \sum_{i=1}^{N} (\varepsilon_{ijk})^2 = \sum_{j=1}^{p} \sum_{k=1}^{q} \sum_{i=1}^{N} (Y_{ijk} - \widehat{Y})^2 \tag{5.16}$$

where $\widehat{Y}$ is the predicted scores from either the full or the reduced GLM. For the full GLM, the estimate of $\widehat{Y}$ is provided by $\overline{Y}_{jk}$ and for the Factor A reduced GLM, the estimate of $\widehat{Y}$ is provided by$(\overline{Y}_{jk} - \alpha_j)$. Therefore

$$\text{SSE}_{\text{ARGLM}} = \sum_{j=1}^{p} \sum_{k=1}^{q} \sum_{i=1}^{N} (Y_{ijk} - \overline{Y}_{jk} - \alpha_j)^2 \tag{5.17}$$

Contained within the brackets of equation (5.17) are the full GLM error $(Y_{ijk} - \overline{Y}_{jk})$ and the effect of Factor A (α_j). The effect of Factor A has been defined as

$$\alpha_j = \mu_j - \mu \tag{5.3, rptd}$$

and is estimated by

$$\widehat{\alpha}_j = \overline{Y}_j - \overline{Y}_{\mathrm{G}} \tag{5.18}$$

where $\overline{Y}_j$ represents the Factor A marginal means and $\overline{Y}_{\mathrm{G}}$ is the general mean. Substituting these terms and applying some algebra reveals

$$\mathrm{SSE}_{\mathrm{ARGLM}} = \mathrm{SSE}_{\mathrm{FGLM}} + N_{jk}q\sum_{j=1}^{p}(\overline{Y}_j - \overline{Y}_{\mathrm{G}})^2 \tag{5.19}$$

where N_{jk} is the number of subjects in each experimental condition. It follows from equation (5.19) that

$$\mathrm{SSE}_{\mathrm{ARGLM}} - \mathrm{SSE}_{\mathrm{FGLM}} = N_{jk}q\sum_{j=1}^{p}(\overline{Y}_j - \overline{Y}_{\mathrm{G}})^2 \tag{5.20}$$

Equation (5.20) specifies the reduction in the GLM error term when the effect of Factor A is accommodated in comparison with not accommodating only the Factor A effect in the reduced GLM, and is equal to the main effect sum of squares for Factor A (SS_{A}). A similar logic reveals the main effect sum of squares for Factor B (SS_{B}) as

$$\mathrm{SSE}_{\mathrm{BRGLM}} - \mathrm{SSE}_{\mathrm{FGLM}} = N_{jk}p\sum_{k=1}^{q}(\overline{Y}_k - \overline{Y}_{\mathrm{G}})^2 \tag{5.21}$$

where $\overline{Y}_k$ represents the Factor B marginal means. Finally, the sum of squares for the interaction between the levels of Factors A and B ($\mathrm{SS}_{\mathrm{AB}}$) is given by

$$\mathrm{SSE}_{\mathrm{ABRGLM}} - SSE_{\mathrm{FGLM}} = N_{jk}\sum_{j=1}^{p}\sum_{k=1}^{q}(\overline{Y}_{jk} - \overline{Y}_j - \overline{Y}_k + \overline{Y}_G)^2 \tag{5.22}$$

Applying these sums of squares formulas to the example memory experiment data provides

$$\begin{aligned}\mathrm{SSE}_{\mathrm{ARGLM}} - SSE_{\mathrm{FGLM}} &= 8(3)[(9-12)^2 + (15-12)^2]\\ &= 24[18]\\ &= 432\end{aligned}$$

$$\begin{aligned}
SSE_{BRGLM} - SSE_{FGLM} &= 8(2)[(8-12)^2 + (11-12)^2 + (17-12)^2] \\
&= 16[42] \\
&= 672
\end{aligned}$$

$$\begin{aligned}
SSE_{ABRGLM} - SSE_{FGLM} &= 8[(6-9-8+12)^2 + (10-9-11+12)^2 \\
&\quad + (11-9-17+12)^2 + (10-15-8+12)^2 \\
&\quad + (12-15-11+12)^2 + (23-15-17+12)^2] \\
&= 8[28] \\
&= 224
\end{aligned}$$

In addition to the SS for main and interaction effects, the associated *dfs* are required. Previously, *dfs* were described as the number of scores employed in constructing the estimate that genuinely was free to vary. Equivalently, the *dfs* may be defined in accordance with the model comparison approach. The *dfs* for a GLM equals the number of scores minus the number of independent parameters employed in the model. And just as main and interactive effects are defined as the difference between reduced and full GLM errors, the main and the interactive effect *dfs* can be defined as the difference between the reduced and the full GLM *dfs*.

The ANOVA solution to the overparameterization problem for experimental design GLMs is to constrain effects to sum to zero (see Section 2.8.5). Therefore, μ constitutes one parameter, there are $(p-1)$ parameters required to distinguish the levels of Factor A, $(q-1)$ parameters are required for Factor B, and $(p-1)(q-1)$ parameters are required for the interaction between Factors A and B. For the independent (2×3) factors experimental design GLM, a total of six independent parameters are employed. Consequently, for the full independent (2×3) factor experimental design GLM applied to the memory experiment data, there are

$$(N-6) = (48-6) = 42\ dfs$$

For the Factor A reduced GLM, the $(p-1)$ parameters distinguishing the Factor A levels are omitted, leaving, $1 + (q-1) + (p-1)(q-1) = 1 + (3-1) + (2-1)(3-1) = 5$. Therefore, for the Factor A reduced GLM, there are

$$48 - 5 = 43\ dfs$$

As the Factor A reduced GLM has 43 *dfs* and the full independent (2×3) factor experimental design GLM has only 42 *dfs*, it follows that the main effect of Factor A has 1 *df*. For the Factor B reduced GLM, the $(q-1)$ parameters distinguishing the Factor B levels are omitted, leaving, $1 + (p-1) + (p-1)(q-1) = 1 + (2-1) + (2-1)(3-1) = 4$. Therefore, for the Factor B reduced GLM, there are

$$48 - 4 = 44\ dfs$$

Table 5.3 ANOVA Summary Table

Source	SS	*df*	MS	*F*	*p*
Encoding instructions (A)	432.000	1	432.000	47.747	>0.001
Study time (B)	672.000	2	336.000	37.137	>0.001
Encoding instructions × Study time (A × B)	224.000	2	112.000	12.379	>0.001
Error (A × B × S)	380.000	42	9.048		
Total	1708.000	47			

As the Factor B reduced GLM has 44 *dfs* and the full experimental design GLM has 42 *dfs*, it follows that the main effect of Factor B has 2 *dfs*. For the AB Factors interaction reduced GLM, the $(p-1)(q-1)$ parameters distinguishing the separate experimental conditions are omitted, leaving, $1+(p-1)+(q-1)=1+(2-1)+(3-1)=4$. Therefore, for the AB Factors interaction reduced GLM, there are

$$48-4=44\, dfs$$

As the AB Factors interaction reduced GLM has 44 *dfs* and the full experimental design GLM has 42 *dfs*, again it follows that the AB interaction effect has 2 *dfs*.

Armed with sums of squares and degrees of freedom for the two main effects and the interaction effect, the ANOVA summary table can be constructed.

The last column in the ANOVA summary table (Table 5.3) provides the probability of observing these F-values under the null hypothesis. (The tabled critical F-values presented in Appendix B may be used to determine significance if hand calculation is employed or the statistical software employed does not output the required p-values.) As all of the probabilities are less than 0.05, all of the null hypotheses can be rejected and all of the (non-directional) experimental hypotheses can be accepted. Therefore, the full GLM provides the best description of the experimental data.

5.3 REGRESSION GLMs FOR FACTORIAL ANOVA

As mentioned earlier, comparing full and reduced GLMs is a distilled form of linear modeling made possible by the nature of experimental data. In factorial designs, because factor levels are completely crossed, factor main and interaction effects are orthogonal. This means there is no overlap in the information defining the two factor predictors and the interaction predictor, and so any variance in the dependent variable attributed to one factor will be distinct from any variance in the dependent variable attributed to any other factor or interaction between factors. Consequently, it makes no difference which term first enters the factorial ANOVA GLM—irrespective of entry order, exactly the same results will be obtained.

The regression ANOVA GLM for the factorial (2 × 3) experimental design applied in the memory experiment is

$$Y_i = \beta_0 + \beta_1 X_1 + \beta_2 X_2 + \beta_3 X_3 + \beta_4 X_4 + \beta_5 X_5 + \varepsilon_i \tag{5.23}$$

The effect coding required for the regression GLM to implement the factorial (2 × 3) ANOVA is presented in Table 5.4. As only (p – 1) predictors are required to code p experimental conditions, the two levels of Factor A can be coded by the predictor variable X_1. Similarly, the three levels of Factor B can be coded by the predictor variables X_2 and X_3. Therefore, predictor X_1 represents the main effect of Factor A and predictors X_2 and X_3 represent the main effect of Factor B. The interaction between Factors A and B is coded by the variables X_4 and X_5. Variable X_4 is obtained by multiplying the codes of predictors X_1 and X_2, and variable X_5 is obtained by multiplying the codes of predictors X_1 and X_3. It is worth noting that the number of predictors required to code each main effect and the interaction effect ($A - X_1$, $B - X_2$ and X_3, and A × B – X_4 and X_5) equals the *dfs* for each effect ($A = 1, B = 2$, and A × B = 2).

Table 5.5 presents the ANOVA summary table output from statistical software when the effect coded regression ANOVA GLM is applied to the data presented in Table 5.4. The residual SS in Table 5.5 equals the error SS in Table 5.3. The regression SS in

Table 5.4 Effect Coding for a Two-Factor (2 × 3) Experimental Design

		X_1	X_2	X_3	X_3	X_5	Y
	s1	1	1	0	1	0	7
A1B1	⋮						
	s8	1	1	0	1	0	7
	s9	1	0	1	0	1	7
A1B2	⋮						
	s16	1	0	1	0	1	11
	s17	1	–1	–1	–1	–1	8
A1B3	⋮						
	s24	1	–1	–1	–1	–1	12
	s25	–1	1	0	–1	0	16
A2B1	⋮						
	s31	–1	1	0	–1	0	8
	s32	–1	0	1	0	–1	16
A2B2	⋮						
	s40	–1	0	1	0	–1	12
	s41	–1	–1	–1	1	1	24
A2B3	⋮						
	s48	–1	–1	–1	1	1	24

Table 5.5 ANOVA Summary Table Output from SYSTAT Statistical Software Implementing the Regression ANOVA GLM Described by Equation (5.23) Using Effect Coding

Source	SS	*df*	MS	*F*	*p*
Regression	1328.000	5	265.000	29.356	<0.001
Residual	380.000	42	9.048		
Total	1708.000	47			

R: 0.882; R^2: 0.778; adjusted R^2: 0.751.

Table 5.5 is equivalent to the sum of the SS for Factors A, B, and their interaction. However ANOVA provides a separate SS for each factor and factor interaction. As factorial ANOVA with balanced data ensures orthogonal factors with the result that the variance attributed to each of the three predictors (i.e., Factors A, B, and their interaction) is unaffected by their order of entry into the ANOVA GLM, one of the simplest ways to obtain the separate factor and factor interaction SS estimates is to carry out an *incremental analysis* (Cohen *et al.*, 2003).

5.4 ESTIMATING EFFECTS WITH INCREMENTAL ANALYSIS

Several names have been applied to describe incremental analysis. One of the most popular names was *hierarchical analysis*, but this label now is applied most frequently to *multilevel analysis*. So, to avoid confusion, the title *incremental analysis* is applied to the form of analysis described below. (Multilevel analysis for related measures is presented in Chapter 12.)

Incremental analysis is employed frequently to cope with multicolinearity. When multicolinearity exists, some or all of the predictors are related. Due to the relations between the predictors, the predictors will be associated with the same variance in the dependent variable. This means that the order of entry into the GLM will determine which predictors accommodate what dependent variable variance. If both predictors A and B are associated with part C of the dependent variable variance, variance part C will be attributed to whichever predictor, A or B, first enters the GLM. As this variance is no longer available to be attributed to the latter predictor, the order of predictor entry into the GLM determines the dependent variable variance attribution to predictors. In incremental analysis predictors (or sets of predictors) are entered into the GLM, cumulatively, in a principled order set by the researcher's knowledge of the topic. After each predictor has entered the GLM, the new model may be compared with the previous model, with any changes attributable to the predictor just included.

The alternative to an incremental analysis is a simultaneous analysis. However, simultaneous analyses attribute to predictors only the dependent variable variance that is unique to each predictor. Any dependent variable variance associated with more than one predictor is not unique and so, it is not attributed to any predictor.

Instead, this variance is relegated to the residual or error term, with the consequence that analysis power is diminished. (See Pedhazur, 1997, for a defence of simultaneous analysis and a criticism of incremental analysis. Cohen, Cohen, West, and Aiken, 2003, present and discuss incremental analysis.)

Incremental analysis may be implemented using experimental design GLMs or regression ANOVA GLMs. The equivalence of both of these approaches is illustrated below.

5.4.1 Incremental Regression Analysis

5.4.1.1 Step 1

Here, the SS for Factor A is obtained. This is accomplished by applying the experimental design GLM

$$Y_{ijk} = \mu + \alpha_j + \varepsilon_{ijk} \tag{5.24}$$

or the regression ANOVA GLM

$$Y_i = \beta_0 + \beta_1 X_1 + \varepsilon_i \tag{5.25}$$

Table 5.6 presents an ANOVA summary table for the equivalent experimental design and regression ANOVA GLMs, which reveals that the Factor A SS is identical to that for Factor A in Table 5.3. However, note that the error/residual SS the error *dfs* and MSe at this step do *not* equal the full two factor independent measures ANOVA error SS, the error *dfs* and MSe.

5.4.1.2 Step 2

Here, the SS for Factor B is obtained. This is accomplished by applying the experimental design GLM

$$Y_{ijk} = \mu + \alpha_j + \beta_k + \varepsilon_{ijk} \tag{5.14, rptd}$$

or the regression ANOVA GLM

$$Y_i = \beta_0 + \beta_1 X_1 + \beta_2 X_2 + \beta_3 X_3 + \varepsilon_i \tag{5.26}$$

Table 5.6 ANOVA Summary Table of the First Step in the Incremental Analysis of the Two-Factor Experiment

Source	SS	*df*	MS	*F*	*p*
Factor A	432.000	1	432.000	15.574	<0.001
Error/residual	1276.000	46	27.739		
Total	1708.000	47	27.739		

R: 0.503; R^2: 0.253; adjusted R^2: 0.237.

Table 5.7 ANOVA Summary Table of the Second Step in the Incremental Analysis of the Two-Factor Experiment

Source	SS	*df*	Mean Square	*F*	*p*
Factors A and B	1104.000	3	368.000	26.808	<0.001
Error/residual	604.000	44	13.727		
Total	1708.000	47	27.739		

R: 0.804; R^2: 0.646; adjusted R^2: 0.622.

Table 5.7 presents an ANOVA summary table for the equivalent experimental design and regression ANOVA GLMs. The reduction in the error/residual SS from Step 1 to Step 2 equals the SS attributable to the inclusion of Factor B. Therefore, $1276.0 - 604.0 = 672.0$. This SS is identical to the SS for Factor B in Table 5.3.

5.4.1.3 Step 3

Here, the interaction between Factors A and B is obtained. This is accomplished by applying the experimental design GLM

$$Y_{ijk} = \mu + \alpha_j + \beta_k + (\alpha\beta)_{jk} + \varepsilon_{ijk} \quad (5.1, \text{rptd})$$

or the regression ANOVA GLM

$$Y_i = \beta_0 + \beta_1 X_1 + \beta_2 X_2 + \beta_3 X_3 + \beta_4 X_4 + \beta_5 X_5 + \varepsilon_i \quad (5.23, \text{rptd})$$

Of course, these are now equivalent full experimental design and full regression ANOVA GLMs—the final addition being terms for the interaction between Factors A and B. Table 5.8 presents an ANOVA summary table for these ANOVA GLMs.

The reduction in the error/residual SS from step 2 to step 3 equals the SS attributable to the inclusion of the Factor A $\times$ Factor B interaction. Therefore, $604.0 - 380.0 = 224.0$. This SS is identical to the SS for Factor A $\times$ Factor B interaction in Table 5.3.

Table 5.8 ANOVA Summary Table of the Third Step in the Incremental Analysis of the Two-Factor Experiment

Source	SS	*df*	MS	*F*	*p*
Factor A, B, and A $\times$ B	1328.000	5	265.000	29.356	<0.001
Error/residual	380.000	42	9.048		
Total	1708.000	47			

R: 0.882; R^2: 0.778; adjusted R^2: 0.751.

5.5 EFFECT SIZE ESTIMATION

In Section 4.3.1, a distinction was made between complete and partial $\hat{\omega}^2$ effect sizes. Similar distinctions can be made with respect to factorial experimental designs. However, the situation with factorial experimental designs is slightly more complicated due to there being omnibus, partial and specific single *df* comparison effect sizes for two or more main effects and one or more interactions.

5.5.1 SOA for Omnibus Main and Interaction Effects

Complete and then partial SOA effect size estimates for main and interaction effects in factorial experimental designs are discussed in this section.

5.5.1.1 Complete $\hat{\omega}^2$ for Main and Interaction Effects

As before, the complete $\hat{\omega}^2$effect size expresses the variance due to the experimental effect as a proportion of the all the variance observed in the experimental scores. Equation (5.27) defines the complete $\hat{\omega}^2$ effect size for a two-factor independent measures study

$$\omega^2 = \frac{\sigma^2_{\text{effect}}}{\sigma^2_A + \sigma^2_B + \sigma^2_{A\times B} + \sigma^2_{\text{error}}} \tag{5.27}$$

This definition of the complete $\hat{\omega}^2$ effect size reveals that in a two factor independent measures design, the total variation in the study is the sum of the main effect, interaction, and error variance. However, most often complete $\hat{\omega}^2$ effect size calculations are based on sample data, with the simplest formulas making use of calculated F-values. The estimate of the complete $\hat{\omega}^2$ effect size for each of the main and interaction effects is defined by

$$\hat{\omega}^2_{\text{effect}} = \frac{df_{\text{effect}}(F_{\text{effect}} - 1)}{df_A(F_A - 1) + df_B(F_B - 1) + df_{A\times B}(F_{A\times B} - 1) + N} \tag{5.28}$$

Applying equation (5.28) to the hypothetical two-factor independent measures experiment and its ANOVA provides

$$\hat{\omega}^2_A = \frac{1(47.747 - 1)}{1(47.747 - 1) + 2(37.137 - 1) + 2(12.379 - 1) + 48} = \frac{46.747}{189.779} = 0.246$$

$$\hat{\omega}^2_B = \frac{2(37.137 - 1)}{1(47.747 - 1) + 2(37.137 - 1) + 2(12.379 - 1) + 48} = \frac{72.274}{189.779} = 0.381$$

$$\hat{\omega}^2_{A\times B} = \frac{2(12.379 - 1)}{1(47.747 - 1) + 2(37.137 - 1) + 2(12.379 - 1) + 48} = \frac{22.758}{189.779} = 0.120$$

(It is worth noting that the effect sizes obtained above for the hypothetical experiment are far greater than typically observed, cf. Table 4.1.)

5.5.1.2 *Partial $\widehat{\omega}^2$ for Main and Interaction Effects*

A partial ω^2 can be defined for omnibus main and interaction effects. In common with Keppel and Wickens (2004), angled brackets are used to identify such partial effect estimates

$$\omega^2_{\langle \text{effect} \rangle} = \frac{\sigma^2_{\text{effect}}}{\sigma^2_{\text{effect}} + \sigma^2_{\text{error}}} \tag{5.29}$$

Equation (5.29) shows that the partial ω^2 estimate of the omnibus effect size expresses the variance due to the effect as a proportion of only the effect variance plus the error variance for just that effect. When based on experimental data, the estimate of the partial $\widehat{\omega}^2$ effect size for each of the main and interaction effects can be defined by

$$\widehat{\omega}^2_{\langle \text{effect} \rangle} = \frac{df_{\text{effect}}(F_{\text{effect}} - 1)}{df_{\text{effect}}(F_{\text{effect}} - 1) + N} \tag{5.30}$$

Applying equation (5.30) to the hypothetical two-factor independent measures experiment and its ANOVA provides

$$\widehat{\omega}^2_{\langle \text{A} \rangle} = \frac{1(47.747 - 1)}{1(47.747 - 1) + 48} = \frac{46.747}{94.747} = 0.493$$

$$\widehat{\omega}^2_{\langle \text{B} \rangle} = \frac{2(37.137 - 1)}{2(37.137 - 1) + 48} = \frac{72.274}{120.274} = 0.601$$

$$\widehat{\omega}^2_{\langle \text{A} \times B \rangle} = \frac{2(12.379 - 1)}{2(12.379 - 1) + 48} = \frac{22.758}{70.758} = 0.322$$

5.5.2 Partial $\widehat{\omega}^2$ for Specific Comparisons

As well as defining partial ω^2 effect sizes for main and interaction effects, it is also possible to define partial ω^2 effect sizes for specific comparisons. In Section 4.3.1, the partial ω^2 effect size for a specific comparisons was defined as

$$\omega^2_{\langle \psi \rangle} = \frac{\sigma^2_{\psi}}{\sigma^2_{\psi} + \sigma^2_{\text{e}}} \tag{4.15, rptd}$$

and exactly the same equation applies to two-factor independent measures experimental designs. The variance attributable to the specific comparison is determined in exactly the same way in both single- and two-factor independent measures experimental designs. Similarly, a single error term is obtained in both designs and this plus the specific comparison variance provides the denominator for the variance attributable to the specific comparison. When estimated from experimental data, the estimate of the partial ω^2 effect size for specific comparisons can be defined by

$$\widehat{\omega}^2_{\langle\psi\rangle} = \frac{\mathrm{F}_\psi - 1}{(F_\psi - 1) + 2N_j} \tag{4.16, rptd}$$

5.6 FURTHER ANALYSES

Two significant main effects and a significant interaction effect were detected, but, in common with single factor studies, further analyses are necessary to determine exactly which experimental conditions differ. When the GLM assumptions are tenable (see Chapter 10), the use of a single error term in fully independent factorial experimental design GLMs makes their further analyses amongst the most easily implemented. Moreover, as described in Section 3.7.1, the most powerful comparisons are obtained when this omnibus MSe is employed in tests of the differences between the experimental condition means.

5.6.1 Main Effects: Encoding Instructions and Study Time

Interpreting the main effect of the *encoding instructions* factor in terms of mean differences is quite simple. As there are only two levels of this factor, the (marginal) means of these two levels of encoding instruction are the only means that can be unequal. Therefore, no further test needs to be applied and all that remains to be done is to determine the direction of the effect by identifying the encoding instruction levels with the larger and smaller means. Plotting pertinent means on a graph is an extremely useful tool in interpreting data from any experiment and the plot of the two encoding instruction marginal means presented in Figure 5.1 reveals the nature of this main effect. The overall free recall memory performance of subjects is significantly greater after story and imagery instructions at encoding than after memorization instructions at encoding.

When an experimental design includes only two levels of a factor, a planned comparison is suggested. This is certainly true for the current experiment, and so the comparison of subjects' free recall memory performance after story and imagery instructions and after memorization instructions should be assessed without any Type 1 error rate adjustment. However, even if this comparison was not planned, no Type 1 error rate adjustment would be applied to this comparison. The reason is the ANOVA convention of treating each main effect and each interaction effect as separate hypothesis families and this is only hypothesis in this main effect family of

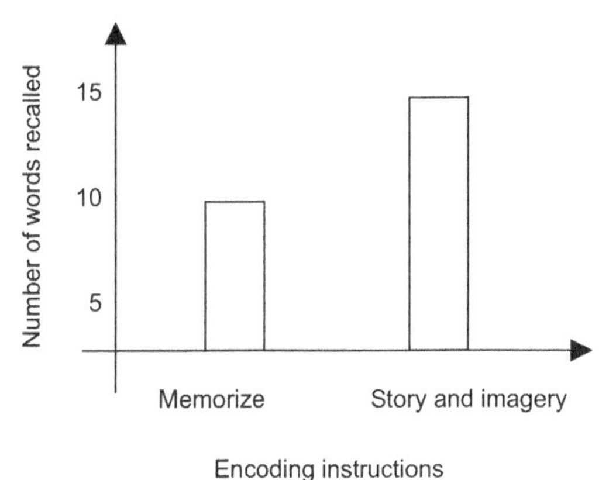

Figure 5.1 Number of words recalled as a function of encoding instructions.

hypotheses (cf. each planned comparison being considered as the sole member of a hypotheses family).

Interpreting the main effect of the *study time* factor in terms of mean differences is slightly more complicated. As this factor has three levels, the pertinent unequal (marginal) means may be any one or more of, b1 versus b2, b2 versus b3, and b1 versus b3, and nonpairwise differences also may contribute to the significant main effect. Therefore, further tests are required to identify exactly which means differ. Figure 5.2 presents the mean number of words recalled as a function of study time and suggests that free recall increases in a linear fashion as study time increases. If this linearity hypothesis is of interest, then a conducting a trend analysis would be appropriate (Howell, 2010; Keppel and Wickens, 2004; Kirk, 1995; Maxwell and Delaney, 2004 provide accounts of this form of analysis).

The next step in the analysis strategy outlined in Section 3.8 is to identify the planned comparison(s). It will be assumed that a comparison of the 30 s versus 180 s experimental conditions was planned and all other comparisons of experimental conditions are unplanned. Therefore, the hypothesis manifest in the comparison of the 30 s versus 180 s experimental condition means is assessed without any Type 1 error

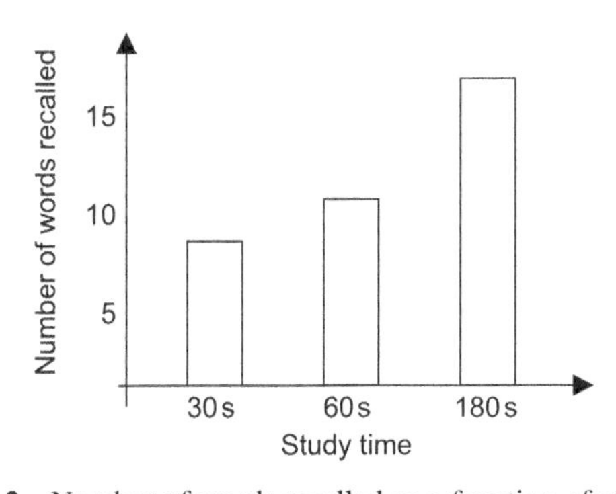

Figure 5.2 Number of words recalled as a function of study time.

Table 5.9 Marginal Means for the Three Study Time Experimental Conditions

Study Time		
$k=1$	$k=2$	$k=3$
b1	b2	b3
30 s	60 s	180 s
8	11	17

rate adjustment, while any other (pairwise or nonpairwise) comparisons are conceived as members of a separate family over which the Type 1 error rate is controlled at $\alpha = 0.05$. For simplicity, it is assumed that the only other comparisons of interest are the two pairwise comparisons between the 30 s versus 60 s and the 60 s versus 180 s experimental conditions.

The planned comparison is examined by applying the appropriate linear contrast and determining the comparison SS. Table 5.9 presents the three marginal study time means. There are $q = 3$ levels of study time, so each population mean is designated by μ_k and each estimate of the population mean is provided by the sample means, $\overline{Y}_k$ where $k = 1$, 2, or 3.

As described in Section 3.4, the linear contrast for the planned comparison expressed in terms of population means is

$$\psi_{pc} = (1)\mu_1 + (0)\mu_2 + (-1)\mu_3$$

The linear contrast for the planned comparison employing the sample means estimators of the population means is

$$\psi_{pc} = (1)8 + (0)11 + (-1)17 = -9$$

Therefore, bearing in mind that each study time marginal mean is based on 16 subjects (i.e., $p\mathrm{N}_{jk} = 2 \times 8$)

$$\mathrm{SS}_{\widehat{\psi}_{pc}} = \frac{p\mathrm{N}_{jk}\widehat{\psi}_{pc}^2}{\sum c_j^2} = \frac{(16)(-9)^2}{(1)^2 + (-1)^2 + (0)^2} = \frac{1296}{2} = 648$$

One df is associated with $\mathrm{SS}_{\widehat{\psi}_{pc}}$, so the mean square for the contrast is

$$\mathrm{MS}_{\widehat{\psi}_{pc}} = \frac{\mathrm{SS}_{\psi_{pc}}}{1} = \frac{648}{1} = 648$$

and so

$$F_{(1,42)} = \frac{\mathrm{MS}_{\psi_{pc}}}{\mathrm{MSe}} = \frac{648}{9.048} = 71.618$$

As the comparison of the 30 s versus 180 s experimental conditions is a planned comparison, no Type 1 error adjustment is made, $F_{(1,42)} = 71.618$, MSe = 9.048, $p < 0.001$. Therefore, the difference between subjects' free recalls after 30 and 180 s is significant.

The procedures applied above to obtain planned comparison F-tests also are applied to obtain the F-values for each of the unplanned comparisons. This provides

30 s marginal mean vs 60 s marginal mean: $F_{(1,42)} = 10.800$, MSe = 9.048, $p = 0.003$.
60 s marginal mean vs 180 s marginal mean: $F_{(1,42)} = 9.600$, MSe = 9.048, $p = 0.004$.

As unplanned comparisons, the two hypotheses are regarded as constituting a family of hypotheses and the Type 1 error rate is controlled over the family. The technique generally recommended for this task is Shaffer's R test (see Section 3.9.2.2). However, consideration of the LRH with respect to the planned and unplanned comparisons can reveal a much reduced risk of Type 1 error.

As explained previously (see Sections 3.6.6.1), with three experimental conditions, there are three unique pairwise comparisons. The significant F-test informs that the overall null hypothesis (i.e., $\mu_1 = \mu_2 = \mu_2$) is false, while the rejection of a pairwise null hypothesis was confirmed by the result of the planned pairwise comparison. Therefore, the appropriate Type 1 error rate control for the next pairwise comparison requires accommodation of only one possibly true pairwise null hypothesis. As this is the classic statistical test protection, no p-value adjustment is necessary and both unplanned comparisons can be accepted as significant at their classic $p = 0.003$ and $p = 0.004$ values.

5.6.2 Interaction Effect: Encoding Instructions × Study Time

An interaction effect indicates that the effect of one factor is not consistent over all of the levels of the other factor(s). This can be seen by the plot of the means of the six experimental conditions presented in Figure 5.3. The two-factor (2 × 3) design

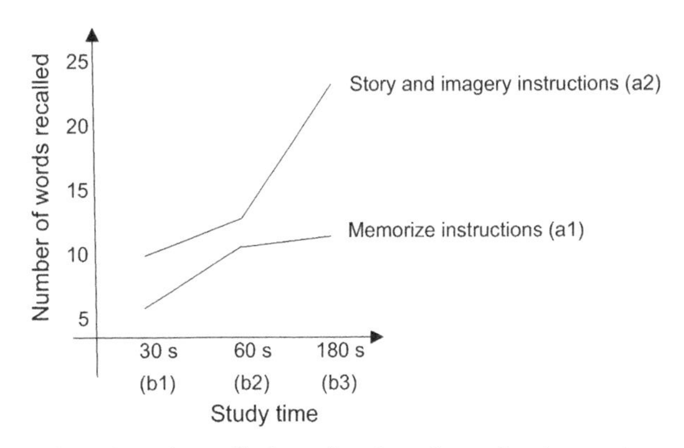

Figure 5.3 Number of words recalled as a function of encoding instructions and study time.

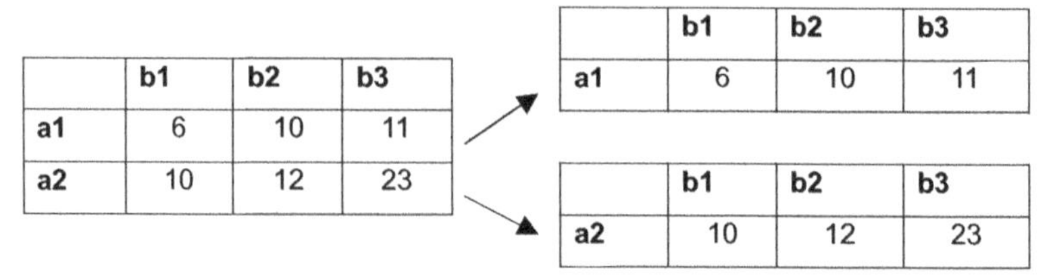

Figure 5.4 Decomposition of the 2 × 3 factor design into two simple effects: the effect of the levels of Factor B (i.e., b1, b2, and b3) at each of the levels of Factor A (i.e., a1 and a2).

applied in the memory experiment can be broken down into the *simple effects* presented in Figures 5.4 and 5.5. The interaction may be described in terms of the simple effects of Factor B (i.e., b1, b2, and b3) at each level of Factor A (i.e., a1 and a2, see Figure 5.4). Alternatively, the interaction may be described in terms of the simple effects of Factor A (i.e., a1 and a2) at each level of Factor B (i.e., b1, b2, and b3, see Figure 5.5). The latter simple effect comparisons are equivalent to three pairwise comparisons. The theoretical or practical issues under investigation determine which simple effects analyses will be applied.

5.6.2.1 Simple Effects: Comparing the Three Levels of Factor B at a1, and at a2

Figure 5.4 presents a schematic illustration of the simple effects of Factor B. The three levels of Factor B are compared at each level of Factor A. The two simple effect analyses can be regarded as two single factor ANOVAs with three levels (i.e., b1, b2, and b3) and in fact, this is exactly what the two simple effect analyses do—they apply two single factor ANOVAs. One ANOVA is applied to the (a1) memorization data and the other ANOVA is applied to the (a2) story and imagery data. Each line (one for the memorization study time conditions and the other for the story and imagery study time conditions) in Figure 5.3 represent the data to be examined by each ANOVA.

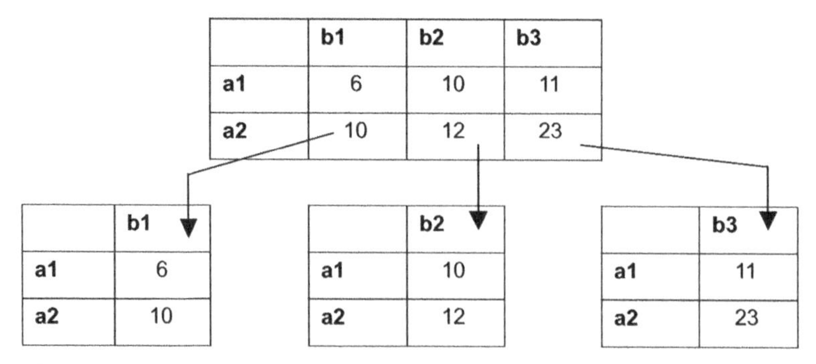

Figure 5.5 Decomposition of the 2 × 3 factor design into three simple effects: the effect of the levels of Factor A (i.e., a1 and a2) at each of the levels of Factor B (i.e., b1, b2, and b3).

The SS calculations for each simple effects analysis are identical to the single factor independent measures ANOVA discussed in Section 2.7.1. Equation (5.31) defines the SS for the Factor B effect at the separate levels of Factor A

$$SS_{\text{B at } j} = \sum_{k=1}^{q} N_{jk}(\overline{Y}_{jk} - \overline{Y}_{j.})^2$$

$$SS_{\text{B at } j=1} = \sum 8(6-9)^2 + 8(10-9)^2 + 8(11-9)^2 = 112 \tag{5.31}$$

$$SS_{\text{B at } j=2} = \sum 8(10-15)^2 + 8(12-15)^2 + 8(23-15)^2 = 784$$

The numerator dfs for each of these simple effects is $(q-1)=(3-1)=2$. These provide the simple effect MS

$$MS_{\text{B at } j} = \frac{SS_{B,j}}{df_{B,j}}$$

$$MS_{\text{B at } j=1} = \frac{SS_{B,1}}{df_{Bk,1}} = \frac{112}{2} = 56.000 \tag{5.32}$$

$$MS_{\text{B at } j=2} = \frac{SS_{B,2}}{df_{B,2}} = \frac{784}{2} = 392.000$$

Finally, to obtain the F-value for the simple effect of B_k at j, the MSe from the full two factor ANOVA is employed. Table 5.10 presents all of this information in the form of an ANOVA summary table.

Given this example involves exactly the same analysis of exactly the same data, it should be no surprise to see the simple effect SS and MS for memorize instructions (a1) is identical to that presented in Tables 2.3, 2.7 and 2.9. However, as the MSe used for the two simple effect analyses is obtained from the two-factor experimental data, it differs from that employed in Tables 2.3, 2.7 and 2.9. The MSe from the two-factor ANOVA employed in Table 5.10 also differs from the two MSe terms based on only the data involved in each separate simple effect analysis. Each simple effect analysis is based on only one-half of the experimental data and so each associated MSe would

Table 5.10 ANOVA Summary Table for the Simple Effect of Factor B at Each Level of Factor A

Source	SS	*df*	MS	*F*	*p*
At a1 (Memorize)					
Study time	112.000	2	56.000	6.189	0.004
Error	380.000	42	9.048		
At a2 (Story and imagery)					
Study time	784.000	2	392.000	43.325	<0.001
Error	380.000	42	9.048		

have only 21 *dfs*, whereas the two-factor omnibus error term is associated with 42 *dfs*. As mentioned, assuming the GLM assumptions are tenable and all else is equal, the two-factor omnibus error term provides the most powerful tests.

The simple effect analyses indicate greater study time effects after story and imagery encoding instructions than after memorization encoding instructions. With balanced data, the greater F-value identifies the greater effect. The graphical presentation of the data in Figure 5.3 suggests, more specifically, that the difference between the 60 and 180 s study times is much greater after encoding using story and imagery than after memorization encoding and it might be expected that this would be the theoretical prediction. Separate pairwise comparisons for the 60 s and 180 s experimental conditions given memorization instructions and given story and imagery instructions should be conducted (just as described for single factor experiments) to establish if this is the case. If it was planned to carry out these specific comparisons to test these hypotheses, no Type 1 error rate adjustment is necessary. However, the Type 1 error rate should be controlled with respect to any unplanned comparisons conducted. The hypotheses tested by the unplanned comparisons should be organized into hypothesis families, or, if separate hypothesis families are not appropriate, all unplanned hypotheses should be included in a single family.

The interaction effect detected indicates the effect of study time with memorization instructions differs from the effect of study time with story and imagery instructions. Two simple effect outcomes are consistent with this interaction. A simple effect of study time may exist at one level of the (memorization or story and imagery) encoding instructions, but not at the other, or alternatively, there may be simple effects of study time with both (memorization or story and imagery) encoding instructions, but one of these simple effects is significantly greater than the other. If the first description is accurate, it is possible for only one of the two simple effect omnibus null hypotheses to be true. If the alternative description is accurate, it is not possible for either of the null hypotheses to be true. These two possible outcomes provide a maximum of one null hypothesis over which Type 1 error control must be exerted. As this is consistent with the classical conception, no adjustment to control the Type 1 error rate is necessary when the two simple effect omnibus F-tests are applied, irrespective of whether the simple effect analyses were planned or unplanned.

The different single factor, simple effect F-values presented in Table 5.10 reveal that neither of the omnibus null hypotheses are true and so attention turns to further analyses of the effect of study time, separately, under memorization and under story and imagery conditions (cf. the unplanned pairwise comparisons between the three study times discussed in Section 5.6.1.). It follows that per memorization and story and imagery encoding condition, only one null hypothesis possibly could be true and so there are a maximum of two possibly true null hypotheses. (If only one of the simple effect omnibus F-tests had been significant, the three pairwise null hypotheses in this instruction condition would be true and with the one possibly true pairwise null hypothesis in the other instruction condition, the maximum number of possibly true pairwise null hypotheses would be four.) If both of these null hypotheses are accommodated within a single hypothesis family, Type 1 error rate control must be exerted over both hypotheses, but, if there are theoretical reasons to consider each hypothesis as the single member of a hypothesis family, then no Type 1 error rate adjustment is necessary.

5.6.2.2 Simple Effects: Comparing the Two Levels of Factor A at b1, at b2, and at b3

Figure 5.5 presents a schematic illustration of the simple effects of Factor A. The simple effects of Factor A are the Factor A effects at each level of Factor B. This means that the two levels of Factor A are compared at each level of Factor B. The three simple effect analyses can be regarded as three *t*-tests or three single factor ANOVAs with two levels (i.e., a1 and a2). One ANOVA is applied to the (b1) 30 s study time data, another ANOVA is applied to the (b2) 60 s study time data and the third ANOVA is applied to the (b3) 180 s study time data. These simple effect analyses compare the two lines (one for the memorization study time conditions and the other for the story and imagery study time conditions) depicted in Figure 5.3 at each of the three study times.

Equation (5.33) defines the SS for the Factor A effect at the separate levels of Factor B

$$\begin{aligned}
SS_{\text{A at }k} &= \sum_{j=1}^{p} N_{j1}(\overline{Y}_{j1} - \overline{Y}_{.1})^2 \\
SS_{\text{A at b1}} &= \sum 8(6-8)^2 + 8(10-8)^2 = 64 \\
SS_{\text{A at b2}} &= \sum 8(10-11)^2 + 8(12-11)^2 = 16 \\
SS_{\text{A at b3}} &= \sum 8(11-17)^2 + 8(23-17)^2 = 576
\end{aligned} \tag{5.33}$$

The numerator *dfs* for each of these simple effects is $(p-1)=(2-1)=1$. These provide the simple effect MS

$$\begin{aligned}
MS_{\text{A at }k} &= \frac{SS_{\text{A},k}}{df_{\text{A},k}} \\
MS_{\text{A at b1}} &= \frac{SS_{\text{A},1}}{df_{\text{A},1}} = \frac{64}{1} = 64.000 \\
MS_{\text{A at b2}} &= \frac{SS_{\text{A},2}}{df_{\text{A},2}} = \frac{16}{1} = 16.000 \\
MS_{\text{A at b3}} &= \frac{SS_{\text{A},3}}{df_{\text{A},3}} = \frac{576}{1} = 576.000
\end{aligned} \tag{5.34}$$

As before, to obtain the *F*-value for the simple effect of Factor A at *k*, the MSe from the full two-factor independent measures ANOVA is employed. All of this information can be presented easily in an ANOVA summary table (Table 5.11).

The simple effect analyses indicate that the greatest difference between encoding conditions (favoring the story and imagery instructions) is observed after 180 s study time. A smaller difference between encoding conditions (still favoring the story and

Table 5.11 ANOVA Summary Table for the Simple Effect of Factor A at Each Level of Factor B

Source	SS	*df*	MS	*F*	*p*
At b1 (30 s)					
Encoding instruction	64.000	1	64.000	7.073	0.011
Error	380.000	42	9.048		
At b2 (60 s)					
Encoding instruction	16.000	1	16.000	1.768	0.191
Error	380.000	42	9.048		
At b3 (180 s)					
Encoding instruction	576.000	1	576.000	63.661	<0.001
Error	380.000	42	9.048		

imagery instructions) is observed after 30 s, whereas no difference between encoding conditions is observed after 60 s.

It is worth mentioning that the SS and MS values for all of the simple effects analyses discussed above can be obtained using statistical software by selecting the appropriate data for each simple effect analysis and applying a single factor ANOVA. Subsequently, the simple effects SS and MS from the single factor ANOVA should be employed with the error term from the two-factor ANOVA to provide the simple effects *F*-tests.

As described for the simple effects of study time, planned and unplanned simple effect comparisons should be distinguished. No Type 1 error rate adjustment is necessary for planned comparisons. However, if it is appropriate, unplanned comparisons should be organized into hypothesis families. If such organization is not appropriate, then a single hypothesis family will contain all of the unplanned comparisons. Type 1 error rate control should be applied over each family of hypotheses, or over the single hypothesis family.

The interaction effect indicates that at least one of the three pairwise comparisons differs from one of the other pairwise comparisons. Therefore, the maximum number of possibly true null hypotheses is two. If both of these null hypotheses are accommodated within a single hypothesis family, Type 1 error rate control must be exerted over both hypotheses. However, if there are theoretical reasons to consider each hypothesis as the single member of a hypothesis family, then no Type 1 error rate adjustment is necessary.

5.7 POWER

5.7.1 Determining the Sample Size Needed to Detect Omnibus Main Effects and Interactions

Three omnibus ANOVA *F*-tests are applied to data from a two-factor independent measures experimental design: the main effects of Factor A and Factor B and the $A \times B$ interaction effect, and it is for these separate *F*-tests that power is set and

Table 5.12 The Numbers of Subjects in the 2 × 3 Experimental Design

	b1	b2	b3	Total
a1	25	25	25	75
a2	25	25	25	75
Total	50	50	50	150

sample size is determined. Essentially, the same procedure described for determining the sample size in single factor independent measures designs is applied separately to each main and interaction effect F-test, employing the partial $\widehat{\omega}^2$ for main and interaction effects discussed in Section 5.4.2. However, this means that the sample size estimated to provide a test operating at a specified power will be the number of subjects required in each of the experimental conditions defined by Factor A, Factor B, or the A × B interaction. For example in Section 4.7.3, the sample size for a single factor independent measures design with three experimental conditions operating with $\alpha = 0.05$ and power = 0.8 to detect a medium effect size was determined to be approximately 150. This situation is presented schematically in terms of the 2 × 3 design in Table 5.12. However, with this design, it follows that if there are 50 subjects in every study time condition, then there will be 25 subjects in each experimental condition and so, for the Factor A effect, the total sample size again will be 150, but $N_j = 75$. Using the same method, the sample size required to detect a medium sized Factor A effect, with $\alpha = 0.05$ and power = 0.8 can be determined

$$\begin{aligned}
\phi &= \sqrt{\frac{\omega^2}{1-\omega^2}}\sqrt{N_j} \\
\phi &= \sqrt{\frac{0.06}{1-0.06}}\sqrt{75} \\
\phi &= 0.25(8.66) \\
\phi &= 2.17
\end{aligned} \qquad (4.22, \text{rptd})$$

Examination of the power function chart (see Appendix C) for numerator *dfs* $(\nu_1) = 1$, $\alpha = 0.05$, denominator *dfs* $(\nu_2) = p \times N_j - p = 2 \times 75 - 2 = 148$, and $\phi = 2.17$ reveals power = 0.86. Therefore, a sample size of 150 would provide adequate power to detect the two main effects. The same assessment can be made for the A × B interaction. Again, a medium-sized effect is to be detected, with $\alpha = 0.05$ and power = 0.8. However, Table 5.12 shows that with 150 subjects in a two-factor independent measures design, the interaction effect $N_j = 25$. Applying equation (4.22) provides

$$\begin{aligned}
\phi &= \sqrt{\frac{0.06}{1-0.06}}\sqrt{25} \\
\phi &= 0.25(5) \\
\phi &= 1.25
\end{aligned}$$

Examination of the power function chart for numerator *dfs* $(\nu_1) = 5$, $\alpha = 0.05$, denominator *dfs* $(\nu_2) = p \times N_j - p = 2 \times 75 - 2 = 148$, and $\phi = 1.25$ reveals power $= 0.65$ (see Appendix C). Therefore, a sample size of 150 would not provide adequate power to detect the interaction effect. In fact, the sample size required to detect a medium-sized interaction effect with $\alpha = 0.05$ and power $= 0.8$, is 216, so $N_j = 36$.

Keppel and Wickens (2004) point out that as the approach to power analysis and sample size determination just described assumes separate single factor analyses, the power estimates obtained will be slightly greater than would be obtained when the power is estimated properly for a factorial design. To compensate for this overestimate, they suggest simply adding one or two subjects to each experimental condition. Of course, if power analysis and sample size determination is implemented using the statistical software mentioned, then power can be estimated precisely for the experimental designs described.

Finally, it is very worthwhile noting that for the Factor A, Factor B, and the interaction effects examined in two-factor independent measures design, α, and effect size remained constant, but to detect the same sized effects at the same level of power, different numbers of subjects were required. In other words, when the α, the effect size and the sample size are fixed, the power to detect the different omnibus effects in factorial experimental designs varies, with greater power associated with those effects with the greater N_j. This means that in all factorial experimental designs with balanced data, those factors with fewer levels will be assessed with greater power and interactions will be assessed with least power.

5.7.2 Determining the Sample Size Needed to Detect Specific Effects

The omnibus null hypothesis is rarely the hypothesis in which there is interest. Usually, the real interest is in the hypotheses manifest in specific pairwise comparisons of particular experimental condition means (see Section 3.2). Therefore, the key piece of information required for the power analysis to determine the required sample size is a partial ω^2 (or an equivalent f) estimate of the specific comparison under consideration (Section 5.5.2). Once the partial ω^2 for the specific comparison is obtained, the procedure for determining the required sample size proceeds as discussed for the omnibus effects (Section 5.7.1).

CHAPTER 6

GLM Approaches to Related Measures Designs

6.1 INTRODUCTION

A defining feature of related measures designs is the accommodation of the relations between the dependent variable measures recorded across the experimental conditions. Although several different types of related measures designs are available (see below), related measures analyses applying least squares-based estimation procedures employ the same technique to accommodate the relations between the dependent variable measures (see Section 6.3 and Keselman, Algina, and Kowalchuk, 2001 for a review of repeated measures designs and analyses). Accommodating the data relations in this way requires the data to be grouped on the basis of the related aspect. In turn, this requires the researcher to know which aspect of the study is responsible for the relation between the dependent variable measures and to be able to group the data appropriately for the related analysis. Unfortunately, however, related data also can arise without the researcher's awareness and even if a researcher becomes aware of these relations, it may not be possible after the study is complete to group the data appropriately for the related measures analysis. As prevention is far better than cure, the emphasis should be on applying appropriate designs to avoid or to accommodate relations between the dependent variable measures (also see Section 10.4.1.3).

In the following sections, the common organizational structure for the different types of related measures designs provided by Kirk's (1968) account of randomized blocks will be described. Subsequently, the focus will be repeated measure related designs. Repeated measures designs are applied much more frequently in psychological research than other randomized block designs, but as all conditions are experienced by subjects participating in the experiment, problematic effects can arise due to the order in which the experimental conditions are experienced. These problems are identified and a number of ways of dealing with them are described, before the GLM approach to repeated measures designs is presented.

ANOVA and ANCOVA: A GLM Approach, Second Edition. By Andrew Rutherford.

Table 6.1 A Single Factor Randomized Block Design with Six Subject Blocks and Three-Factor Levels

Block	Condition 1	Condition 2	Condition 3
1	s1 19 years	s2 19 years	s3 19 years
2	s4 25 years	s5 25 years	s6 25 years
3	s7 30 years	s8 30 years	s9 30 years
4	s10 37 years	s11 37 years	s12 37 years
5	s13 47 years	s14 47 years	s15 47 years
6	s16 55 years	s17 55 years	s18 55 years

6.1.1 Randomized Block Designs

Randomized block designs are employed to deal with subject variables which are likely to affect subjects' performance in an experiment. For example, as people get older, their reaction times tend to get a little slower. Therefore, if the dependent variable in an experiment is to be reaction time measured in milliseconds, the subject's age is very likely to influence this performance measure and is likely to be labeled a nuisance variable. One way of dealing with nuisance variables is to let the random sampling and random allocation of subjects distribute older and younger subjects randomly across the experimental conditions, so the age-related RT effects are distributed randomly (i.e., unsystematically) across experimental conditions. However, another approach is to recruit sets of similarly aged subjects and allocate one subject from each set to each experimental condition. This would provide an experiment where each subject in Condition 1 was matched with a similar subject in each of the other conditions. Such designs are known as randomized block designs. Table 6.1 presents a randomized block design with six blocks of subjects and three experimental conditions. Within any block, subjects are matched on one or more of the nuisance variables thought to affect the dependent variable measure. The label, *randomized block design*, refers to the random allocation of subjects within any block to the experimental conditions.

Of course, there is no reason to limit such designs to experiments with only three conditions. However, as the number of experimental conditions increases, recruiting the required number of exactly matched subjects will become increasingly difficult. In such circumstances, defining matching age (or other variable) ranges may be a more efficient strategy (see Table 6.2).

Table 6.2 A Single Factor Randomized Block Design with Six Subject Blocks and Four-Factor Levels

Block	Condition 1	Condition 2	Condition 3	Condition 4
1	s1 19–24 years	s2 19–24 years	s3 19–24 years	s4 19–24 years
2	s5 25–29 years	s6 25–29 years	s7 25–29 years	s8 25–29 years
3	s9 30–36 years	s10 30–36 years	s11 30–36 years	s12 30–36 years
4	s13 37–46 years	s14 37–46 years	s15 37–46 years	s16 37–46 years
5	s17 47–54 years	s18 47–54 years	s19 47–54 years	s20 47–54 years
6	s21 55–59 years	s22 55–59 years	s23 55–59 years	s24 55–59 years

Table 6.3 A Single Factor Matched Samples Design with Two-Factor Levels

Block	Condition 1	Condition 2
1	s1 19 years	s2 19 years
2	s3 25 years	s4 25 years
3	s5 30 years	s6 30 years
4	s7 37 years	s8 37 years
5	s9 47 years	s10 47 years
6	s11 55 years	s12 55 years

6.1.2 Matched Sample Designs

Typically, a matched sample design is a randomized block design that employs only two experimental conditions. Table 6.3 presents a typical matched samples design. Although blocks are unlikely to be mentioned when such a design is described, it can be seen that this is simply a reduced version of Table 6.1.

6.1.3 Repeated Measures Designs

Repeated measures designs refer to situations where each subject experiences and provides dependent variable measures in two or more experimental conditions. Therefore, repeated measures designs can be conceived as particular instances of randomized block designs where there is only one subject per block. Table 6.4 presents this design, where the same subjects are measured, repeatedly, under each of the three experimental conditions. As Table 6.4 illustrates, all of the subjects experience all of the experimental conditions. Therefore, the experimental variable may be described as being manipulated within the same group of subjects and so, repeated measures designs also are known as within subjects designs.

Like independent measures designs, the purpose of repeated measures designs is to determine the effect of different experimental conditions on a dependent variable. Although repeated measures designs provide information about individual subjects' performance over all of the experimental conditions, describing individual subjects' performance is not an objective. Indeed, a random factor is used to organize subject information, so emphasizing that the experiment aims to generalize to the population from which the subjects are drawn, rather than focus on the individual subjects providing the data.

Table 6.4 A Single Factor Repeated Measures Design with Three Levels—Equivalent to a Randomized Block Design with Each Block Containing Only One Subject

Block	Condition 1	Condition 2	Condition 3
1	s1 19 years	s1 19 years	s1 19 years
2	s2 25 years	s2 25 years	s2 25 years
3	s3 30 years	s3 30 years	s3 30 years
4	s4 37 years	s4 37 years	s4 37 years
5	s5 47 years	s5 47 years	s5 47 years
6	s6 55 years	s6 55 years	s6 55 years

The labels related measures design, randomized block design, repeated measures design and within subjects design might be considered sufficient. However, defining the experimental factor as a fixed effect and the subject factor as a random effect has the consequence that these designs also can labeled as mixed designs.

Repeated measures designs offer a number of advantages over comparable randomized block and independent measures designs. Using the same subjects in all experimental conditions eliminates the sometimes considerable number of subjects and effort required to identify sets of matched subjects for randomized block designs. Comparable independent measures designs also require more subjects than repeated measures designs to provide the same number of data points. For example, the single factor repeated measures design with three conditions employing 6 subjects, illustrated in Table 6.4, provides as much data as 18 subjects in an equivalent independent measures design. Another advantage of repeated measures designs (shared with other randomized block designs) is a reduction in error variance. It would be expected that the total amount of score variation will be less with three subjects each performing under three experimental conditions, than with three subjects performing under one condition, another three subjects performing under the second condition, and yet another three subjects performing under the third experimental condition. (It is often said that subjects act as their own controls in repeated measures designs. However, as many experiments do not use a standard control condition, this statement can be rather obtuse. Perhaps a more accurate and more easily understood description is that subjects provide their own comparisons: each subject's performance can be compared across all levels of the experimental factor.) The reduction in variation is due to the greater similarity of the scores provided by the same subjects (or the matched subjects in other randomized block designs) compared to the scores provided by different subjects. In other words, scores from the same subject (or the matched subjects) are assumed to be correlated. The advantages (and some disadvantages) of related designs accrue from the correlations between these scores. Utilizing these correlations not only requires more complicated statistical analyses, but also more restrictive statistical assumptions than are required for independent measures designs (see Chapter 10).

The presentation order of the experimental conditions also can effect repeated measures designs. These effects are categorized as incidental effects, carryover effects, contrast effects, and context effects (Keppel and Wickens, 2004). Consider the design outlined in Table 6.4. If all subjects experience all experimental conditions, each subject must experience Conditions 1, 2, and 3 in some order and the particular order experienced may affect the subject's performance in each of the experimental conditions. For instance, if all subjects experienced Condition 1, then Condition 2, and then Condition 3, performance may increase from Conditions 1 to 3 due to the practise subjects receive as they perform the same experimental task under different conditions. Alternatively, and particularly if the task is very demanding, subjects' performance might decrease from Conditions 1 to 3 due to fatigue. Following Keppel and Wickens (2004), practice and fatigue effects are labeled as instances of incidental effects. Incidental effects cannot be eliminated from subjects' performance, but by applying appropriate controls with respect to the order in which subjects receive the experimental conditions, systematic influences can be avoided. In other words, steps can be taken to try and ensure subjects' score variance attributable to practice or fatigue is accommodated by the error term and not confounded with the score variance attributed to an experimental factor.

Carryover effects occur when an experimental condition exerts a transient influence over the next (or subsequent) experimental condition(s) experienced. For example, a mood induction procedure may be used to examine the cognitive consequences of mild depression, but testing subjects in the normal mood condition before the effects of the depressed mood induction had worn off (or a normal mood induced) would provide data contaminated by the subjects' still depressed moods. A contrast effect (termed a differential order effect by Maxwell and Delaney, 2004) occurs when two specific experimental conditions interact in a specific fashion. For example, in a memory experiment comparing two encoding strategies, differences between conditions may be diminished because subjects employing a semantic encoding strategy may continue to apply it (intentionally or unintentionally) in a graphemic encoding strategy condition. However, the converse, applying the graphemic encoding strategy in the semantic encoding condition is less likely to occur, resulting in an interaction between the experimental conditions and the presentation orders. A context effect occurs when subjects' experience and behavior in one experimental condition determines their behavior in subsequent experimental conditions. For example, measuring unintentional learning with a surprise memory test cannot be done repeatedly because, despite instructions, the subjects' earlier experience of performing a memory test is likely to change their previously unintentional learning into intentional learning in subsequent experimental conditions.

Conventional repeated measures analyses are inappropriate where carryover effects, contrast effects or context effects can occur. However, as carryover, contrast and context effects arise as a consequence of the order in which subjects experience all of the experimental conditions, applying one of the randomized block designs described in Sections 6.1.1 and 6.1.2 can avoid these problems and still provide many of the benefits of a repeated measures design. When there is only a single set of matched subjects per block (as illustrated in Tables 6.1, 6.2, 6.3 and 6.4), the procedures described in Section 6.3 can be applied directly to the experimental data, with the term representing the subject effect being reinterpreted as denoting the matched subject effect (i.e., the effect of matched subjects on the dependent variable scores obtained across all of the experimental conditions). Procedures for designs where more than one set of matched subjects are included in each randomized block also are available (e.g., Kirk, 1995).

If any carryover, contrast and context effects become apparent over the course of a study employing a repeated measures design, rather than abandoning any experimental data collected, Keppel and Wickens (2004) suggest it may be possible to apply an independent measures analysis to each subject's first experimental condition performance – this experimental condition cannot be affected by a prior experimental condition because there is no prior experimental condition. Further discussion of carryover effects, contrast effects and context effects is provided by Greenwald (1976) and Maxwell and Delaney (2004).

6.2 ORDER EFFECT CONTROLS IN REPEATED MEASURES DESIGNS

6.2.1 Randomization

The basic statistical theory underlying repeated measures designs assumes the order of experimental conditions experienced by subjects is determined randomly. Randomly generating and implementing an order for each subject means any order is equally likely to be generated and experienced by a subject. Therefore, it is extremely unlikely that there will be a sufficient accumulation of orders necessary to produce systematic biases in the data. Although randomization does not exclude the possibility of a systematic bias arising, it does make it extremely unlikely. Consequently, randomization may be regarded as providing an approximation to counterbalancing (see below). When randomization is applied to control order effects, the error term accommodates the score variance attributable to the different orders experienced by the subjects. However, designs employing counterbalancing allow this variance to be identified and removed from the error term, so increasing the power of analysis.

6.2.2 Counterbalancing

6.2.2.1 Crossover Designs

One way to control incidental order effects is to allocate at least one subject to every order of conditions possible in the experiment. In the current example, there are three conditions (labeled A, B, and C for clarity) and so, ($3! = 3 \times 2 \times 1$) six different order permutations are possible—ABC, ACB, BCA, BAC, CAB, and CBA. Allocating a subject to each order of conditions does not eliminate the effect of any particular order, as each individual subject's performance in each experiment continues to be influenced by the order in which they experience the conditions. Nevertheless, in a full crossover counterbalanced design, all orders are experienced by equal numbers of subjects and so, any performance benefit arising as a consequence of a particular order of conditions is counterbalanced by the effect of its counterorder (e.g., ABC and CBA). Moreover, including all orders in the experiment has the consequence that the orders are crossed with the other experimental factors. Provided more than one subject is allocated to each presentation order, it is possible to construct a factorial model of the data that explicitly represents the score variance due to the order effects and, if desired, the interaction between this factor and all the other factors in the experiment. A contrast effect should manifest as an interaction between the presentation order and the experimental factor, but carryover and context effects are more likely to manifest as a reduction in the influence of the experimental factor. As the experimental purpose is to determine if a factor exerts an influence, unequivocally identifying carryover or context effects can be difficult. Nevertheless, despite the potential benefit of enabling a check on contrast effects, examination of the psychological literature reveals that GLMs applied to repeated measures designs typically rely on the quality of the experimental design to prevent carryover, contrast, and context effects and omit those terms representing the presentation orders. Therefore, the GLMs described here for repeated measures designs also will omit these terms. However, if presentation order can be coded as a factor, it would be beneficial to do so, as the most efficient data analyses are obtained when the experimental design is reflected fully in the GLM applied.

6.2.2.2 Latin Square Designs

With only three experimental conditions, it is quite feasible to allocate two, three, or more subjects to each experimental order. Even with 5 subjects per order, a total of only 30 subjects is required. However, as the number of experimental conditions increases, the number of order permutations increases exponentially. For example, six experimental conditions provide ($6! = 6 \times 5 \times 4 \times 3 \times 2 \times 1$) 720 order permutations. It is very unlikely that an experimenter would want to run 720 subjects in a single experiment. (Worse still, just two subjects per presentation order requires 2×720 subjects.) However, there are alternatives to fully counterbalanced crossover designs that may be applied before the number of order permutations requires very large numbers of subjects. Rather than assign subjects to each of all possible orders, it is possible to determine a smaller set of orders in which each experimental condition occurs once in each order position. This arrangement is termed a Latin square design. An example of a Latin square for the four experimental conditions A, B, C, and D, is provided in Table 6.5.

As Latin square designs employ a small set of orders to represent all of the order permutations, the selection of the orders constituting the Latin square is an important consideration. The ideal is a digram-balanced Latin square, as presented in Table 6.5. A digram-balanced Latin square is obtained when each experimental condition both precedes and follows all others. The main disadvantage of digram-balanced Latin squares is that they can be applied only when there are an even number of experimental conditions. Two digram-balanced Latin squares can be employed when there are odd numbers of experimental conditions, or a randomly permuted Latin square may be applied. Although a randomly permuted Latin square does not have the property that each experimental condition both precedes and follows all others, they are a reasonable compromise (see Kirk, 1995; Maxwell and Delaney, 2004). In any event, the sort of Latin square to avoid constructing is a cyclic square. These Latin squares arise when the same sequence of experimental conditions occurs in each order employed (e.g., ABCD, BCDA, CDAB, and DABC). Although each of the conditions

Table 6.5 A Latin Square for Single Factor Repeated Measures Design with Four Levels (i.e., A, B, C, and D)

	Position in Order			
Order	P1	P2	P3	P4
1	A	B	C	D
2	C	A	D	B
3	B	D	A	C
4	D	C	B	A

occupy all positions once in the four orders, the same sequence or partial sequence of A followed by B followed by C followed by D is maintained.

Kirk (1995) and Maxwell and Delaney (2004) present models for Latin square designs that represent score variance due to order effects. However, Latin square designs employ only a particular set of the experimental condition orders (unlike crossover designs) and consequently, only order main effects can be estimated, so order interaction effects cannot be used to check the repeated measures assumption of constant order effects. Although Tukey's (1949, 1955) test for additivity (see Kirk, 1995) can provide some assurance that contrast effects are not present, a simple and parsimonious assessment employing residuals is described in Chapter 10.

Although conceived as single factor designs, all repeated measures designs are analyzed as factorial designs. This is because an additional factor is employed to represent the influence of each individual subject on each dependent variable measure. Moreover, crossover and Latin square counterbalanced designs employ yet another factor to represent the presentation orders. Therefore, a "single factor" repeated measures design actually can involve three factors. However, rather than embark upon the description of such "multifactor" designs, the simplest single factor repeated measures design, where the order of experimental conditions experienced by subjects is randomized, is considered.

6.3 THE GLM APPROACH TO SINGLE FACTOR REPEATED MEASURES DESIGNS

Imagine the experiment described in Chapter 2 had been obtained from a single factor repeated measures design and not a single factor independent measures design. Rather than observing data from 24 different subjects divided equally over three conditions, the performance of the same 8 subjects would be observed under each of the three experimental conditions. Table 6.6 presents the data from Table 2.2 as if it had been obtained from a single factor repeated measures design. As the same data is used under the independent and repeated designs conceptions, direct comparisons may be drawn between these two designs.

The GLM underlying a single factor repeated measures design ANOVA is described by the equation

$$Y_{ij} = \mu + \pi_i + \alpha_j + (\pi\alpha)_{ij} + \varepsilon_{ij} \tag{6.1}$$

where Y_{ij} is the ith subject's dependent variable score in the jth experimental condition, μ is the general mean of the experimental condition population means, π_i represents the random effect of the ith subject, α_j is the effect of the jth experimental condition, $(\pi\alpha)_{ij}$ represents the interaction between the ith subject and the jth experimental condition, and the error term, ε_{ij}, reflects variation due to any uncontrolled source. As usual, equation (6.1) summarizes a system of equations, where each equation describes a single dependent variable score.

Table 6.6 Subjects' Free Recall Scores in Single Factor Repeated Measures Design

Subjects	30 s	60 s	180 s	Marginal Means
s1	7	7	8	7.33
s2	3	11	14	9.33
s3	6	9	10	8.33
s4	6	11	11	9.33
s5	5	10	12	9.00
s6	8	10	10	9.33
s7	6	11	11	9.33
s8	7	11	12	10.00
Marginal means	6	10	11	9.00

In comparison to the experimental design GLM for a single factor independent measures ANOVA, the only differences are the inclusion of the terms π_i and $(\pi\alpha)_{ij}$. The term, π_i, represents the influence of the subjects on the dependent variable scores obtained across all of the experimental conditions. The interaction term, $(\pi\alpha)_{ij}$, represents the varying influence of the subjects on the dependent variable scores per experimental condition. As a major part of the single factor independent measures ANOVA error term is due to differences between subjects, specifically accommodating score variance attributable to different subjects with the π_i term reduces the size of the repeated measures error variance considerably and is one of the reasons for the greater analysis power provided by repeated measures designs. However, the role played by the interaction term $(\pi\alpha)_{ij}$ in repeated measures experimental design GLMs is a little more complicated.

Earlier it was said that repeated measures designs are analyzed as factorial designs. Table 6.7 presents the data from Table 6.6 cast in line with the two-factor conception. Here, there are three levels of the experimental condition factor and eight levels of the subject factor, providing one score per experimental design cell (i.e., one score per subject per experimental condition). However, usually factorial ANOVA designs (see Chapter 3) contain several scores per cell. The mean of the cell scores is taken as the best estimate of the cell score and is used to calculate interaction effects, with the discrepancy between the mean and the actual score providing the estimates of experimental error. If there is only one score per subject per experimental condition, then a mean and its error cannot be calculated per subject per experimental condition and without these

Table 6.7 Data from a Single Factor Repeated Measures Design with Subjects Cast as a Second Factor

	Subjects							
Experimental Conditions	b1	b2	b3	b4	b5	b6	b7	b8
a1	7	3	6	6	5	8	6	7
a2	7	11	9	11	10	10	11	11
a3	8	14	10	11	12	10	11	12

estimates, the experimental error (ε_{ij}) cannot be separated from the interaction effect $(\pi\alpha)_{ij}$. Therefore, a more accurate description of the single factor repeated measures experimental design GLM applied most frequently is

$$Y_{ij} = \mu + \pi_i + \alpha_j + [(\pi\alpha)_{ij} + \varepsilon_{ij}] \tag{6.2}$$

Fortunately, however, the lack of a specific error term does not prevent assessment of the experimental conditions effect (α_j). When a single random factor is included in a model with fixed effects and the fixed effects are to be tested, limiting the interaction of the pertinent fixed factor and the random factor to zero (i.e., setting it to zero) provides an error term appropriate for assessing the fixed factor effects. As the interaction between the subjects and the experimental factors can be set to zero simply by omitting the interaction between these two factors from the GLM, the single factor repeated measures ANOVA experimental design GLM usually is described by the equation

$$Y_{ij} = \mu + \pi_i + \alpha_j + \varepsilon_{ij} \tag{6.3}$$

Equation (6.3) may be used for simplicity, but whenever an interaction between experimental conditions and subjects exists, equation (6.2) describes the data more accurately. Nevertheless, when such an interaction exists and the interaction term is omitted, the expected mean square for the experimental conditions, like the expected mean square for error, includes variation attributable to the interaction between experimental conditions and subjects. Therefore, the F-test of the effect of experimental conditions involves the following expected mean squares

$$F = \frac{E(\mathrm{MS}_{\text{experimental conditions}})}{E(\mathrm{MS}_{\text{error}})} = \frac{\sigma^2_{\text{experimental conditions}} + \sigma^2_{\text{experimental conditions*subjects}} + \sigma^2_{\text{error}}}{\sigma^2_{\text{experimental conditions*subjects}} + \sigma^2_{\text{error}}} \tag{6.4}$$

Therefore, setting the interaction between the fixed effect of experimental conditions and the random effect of subjects to zero, by omitting the interaction term from the single factor repeated measures GLM, provides an appropriate F-test of the fixed effect of experimental conditions, but not of the random effect of subjects (e.g., Howell, 2010; Maxwell and Delaney, 2004).

The lack of an F-test of the random effect of subjects is no great loss. Usually, there is no interest in the subject effect and as a consequence, many statistical software packages do not even report the subject effect in the repeated measures ANOVA output. The real aim in applying repeated measures designs, or any related measures design, is to increase the power of the F-test of the experimental factor (i.e., the test of the experimental manipulation). The subjects factor is included in analyses simply because the variance attributed to the subjects factor is extracted from the F-test error term, resulting in an increase in the power of the F-test of the experimental factor.

Comparing the effect of the experimental conditions against the error plus interaction variation estimate makes intuitive sense. The interaction represents the extent to which the experimental condition effect varies across subjects. The greater this inconsistency in relation to the effect of the experimental conditions, the less likely it is that the experimental condition effect is reliable. However, as might be expected, there is a cost associated with this approach. Only when the experimental conditions covariance matrix is spherical will the F-ratio mean square biases cancel out and provide a valid and accurate F-test. If the experimental conditions covariance matrix is not spherical, both the F-ratio mean square numerator and denominator will be biased and will provide a biased F-test (see Section 10.2.2).

As always, the model component of the GLM equation describes the predicted scores

$$\widehat{Y}_{ij} = \mu + \pi_i + \alpha_j \tag{6.5}$$

As μ is a constant, variation in prediction arises not only from the influence of the experimental conditions (α_j) but also from which subject provides the scores (π_i). Consequently, the repeated measures experimental design GLM can predict a different score for each subject in each experimental condition. However, as there is no interaction effect, the predicted scores for each experimental condition are equal to the mean of all subjects' scores per condition, as given by the marginal means at the bottom of Table 6.6.

The estimate of the single factor repeated measures experimental design GLM parameter, μ, is defined as the general mean of the dependent variable scores

$$\widehat{\mu} = \frac{\sum_{i=1}^{N} \sum_{j=1}^{p} Y_{ij}}{Np} = \overline{Y}_{G} \tag{6.6}$$

Notice the symbol, N, is employed in equation (6.6). Typically, N represents the total number of subjects in the experiment and analysis, and N with an appropriate subscript or n is used to represent the number of subjects in an experimental condition. However, in repeated measures designs all subjects participate in all experimental conditions so N, N with an appropriate subscript, and n are all equivalent. Applying equation (6.6) to the data in Table 6.6 provides

$$\widehat{\mu} = \overline{Y}_G = \frac{7 + 3 + \cdots + 11 + 12}{8(3)} = \frac{216}{24} = 9$$

With balanced designs, μ also may be defined as the mean of the experimental condition means

$$\mu = \frac{\sum_{j=1}^{p} \mu_j}{p} \tag{2.19, rptd}$$

Applied to the data in Table 6.6 provides

$$\widehat{\mu} = \overline{Y}_G = \frac{6 + 10 + 11}{3} = 9$$

Given that $\left(\sum_{i=1}^{N} Y_{ij}/N\right)$ are the experimental condition means ($\overline{Y}_j$), the experimental effect estimates are defined by

$$\widehat{\alpha}_j = \left(\frac{\sum_{i=1}^{n} Y_{ij}}{N}\right) - \overline{Y}_{\mathrm{G}} \tag{6.7}$$

Applying formula (6.8) to the data in Table 6.6 provides

$$\begin{aligned}
\widehat{\alpha}_1 &= 6 - 9 = -3 \\
\widehat{\alpha}_2 &= 10 - 9 = 1 \\
\widehat{\alpha}_3 &= 11 - 9 = 2 \\
\hline
\textstyle\sum_{j=1}^{p} \widehat{\alpha}_j &\quad = 0
\end{aligned}$$

From these estimates of experimental effects, the Experimental Condition SS can be calculated as

$$\text{Experimental Conditions SS} = \sum_{j=1}^{p} N(\mu_j - \mu)^2 \tag{2.28, rptd}$$

$$\begin{aligned}
&= \textstyle\sum 8(-3^2) + 8(1^2) + 8(2^2) \\
&= 72 + 8 + 32 \\
&= 112
\end{aligned}$$

Therefore, the estimate of the effect of experimental conditions in the single factor repeated measures design is identical to that obtained in the single factor independent measures design. Given that $\left(\sum_{j=1}^{p} Y_{1,j}/p\right)$ is the mean of the scores provided by each subject, the subject effect estimates are defined by

$$\widehat{\pi}_i = \left(\frac{\sum_{j=1}^{p} Y_{1,j}}{p}\right) - \overline{Y}_{\mathrm{G}} \tag{6.8}$$

Applying formula (6.8) to the data in Table 6.6 provides

$$\begin{aligned}
\widehat{\pi}_1 &= 7.333 - 9 = -1.667 \\
\widehat{\pi}_2 &= 9.333 - 9 = 0.333 \\
\widehat{\pi}_3 &= 8.333 - 9 = -0.667 \\
\widehat{\pi}_4 &= 9.333 - 9 = 0.333 \\
\widehat{\pi}_5 &= 9.000 - 9 = 0.000 \\
\widehat{\pi}_6 &= 9.333 - 9 = 0.333 \\
\widehat{\pi}_7 &= 9.333 - 9 = 0.333 \\
\widehat{\pi}_8 &= 10.000 - 9 = 1.000 \\
\hline
\textstyle\sum_{i=1}^{N} \widehat{\pi}_i &\quad = 0.000
\end{aligned}$$

As with the experimental effect estimates, it is possible to calculate the subject SS

$$\begin{aligned}\text{Subjects SS} = \sum_{i=1}^{N} p(\mu_j - \mu)^2 &= 3(-1.667^2) + 3(0.333^2) + 3(-0.667^2)\\ &\quad + 3(0.333^2) + 3(0^2) + 3(0.333^2)\\ &\quad + 3(0.333^2) + 3(1.000)\\ &= 8.337 + 0.333 + 1.332 + 0.333\\ &\quad + 0 + 0.333 + 0.333 + 3.000\\ &= 14.000\end{aligned}$$

Using each of the parameter estimates in equation (6.5) provides the predicted scores presented in Table 6.8.

Finally, the error estimate is provided by the discrepancy between each observed score (see Table 6.4) and each predicted score (see Table 6.6)

$$\varepsilon_{ij} = Y_{ij} - \widehat{Y}_{ij} \tag{6.9}$$

Table 6.9 presents the calculation of the errors and the sum of the squared errors. This is the estimate of the error sum of squares (SS_{error}) for the single factor repeated measures GLM described by equation (6.3). Whereas the SS for the experimental conditions equaled that obtained with a single factor independent measures design, the single factor repeated measures design error SS is smaller, with the difference between the two SS errors being that SS attributable to subjects (i.e., $52 - 38 = 14$).

Having calculated the SS for both experimental conditions and error, the next step is to determine the degrees of freedom. The logic determining the experimental conditions *dfs* is identical to that for independent measures designs. Therefore

Table 6.8 Scores Predicted by the Single Factor Repeated Measures Experimental Design GLM

Subjects	30 s	60 s	180 s	Means
s1	4.333	8.333	9.333	7.333
s2	6.333	10.333	11.333	9.333
s3	5.333	9.333	10.333	8.333
s4	6.333	10.333	11.333	9.333
s5	6.000	10.000	11.000	9.000
s6	6.333	10.333	11.333	9.333
s7	6.333	10.333	11.333	9.333
s8	7.000	11.000	12.000	10.000
Means	6.00	10.00	11.00	9.00

Table 6.9 Calculation of the Errors Per Experimental Condition Per Subject and the Sum of the Squared Errors

Subjects	30 s	60 s	180 s
s1	7 − 4.333 = 2.667	7 − 8.333 = −1.333	8 − 9.333 = −1.333
s2	3 − 6.333 = −3.333	11 − 10.333 = 0.667	14 − 11.333 = 2.667
s3	6 − 5.333 = 0.667	9 − 9.333 = −0.333	10 − 10.333 = −0.333
s4	6 − 6.333 = −0.333	11 − 10.333 = 0.667	11 − 11.333 = −0.333
s5	5 − 6.000 = −1.000	10 − 10.000 = 0.000	12 − 11.000 = 1.000
s6	8 − 6.333 = 1.667	10 − 10.333 = −0.333	10 − 11.333 = −1.333
s7	6 − 6.333 = −0.333	11 − 10.333 = 0.667	11 − 11.333 = −0.333
s8	7 − 7.000 = 0.000	11 − 11.000 = 0.000	12 − 12.000 = 0.000
$\sum_{i=1}^{N} \varepsilon_{ij}^2$	= 22.667	= 3.333	= 12.000
$\sum_{i=1}^{N} \sum_{i=1}^{p} \varepsilon_{ij}^2 =$		38.000	

$$df_{\text{experimental conditions}} = p - 1 = 3 - 1 = 2 \tag{6.10}$$

As for error *df*s, a separate mean is employed in each experimental condition, so a *df* is lost from the *N* scores of each condition. However, a separate mean also is employed to describe every set of *p* scores a subject provides, so for every set of *p* scores a *df* is lost. Therefore

$$df_{\text{error}} = (N - 1)(p - 1) = (8 - 1)(3 - 1) = 14 \tag{6.11}$$

All of this information can be placed in an ANOVA summary table, as in Table 6.10. However, the subject effect reported in Table 6.10 may not be presented, as it is generally of little interest and due to the lack of an appropriate MSe, an *F*-value cannot be calculated and its significance determined. The tabled critical *F*-values presented in Appendix B may be used to determine significance if hand calculation is employed or the statistical software employed does not output the required *p*-values.

Table 6.10 Single Factor Repeated Measures ANOVA Summary Table

Source	SS	*df*	MS	*F*	*p*
Subjects	14.000	7	2.000		
Experimental conditions	112.000	2	56.000	20.634	<0.001
Error	38.000	14	2.714		
Total	164.000	23			

6.4 ESTIMATING EFFECTS BY COMPARING FULL AND REDUCED REPEATED MEASURES DESIGN GLMs

The full single factor repeated measures experimental design GLM was described by equation (6.3). The reduced GLM is similar, but excludes experimental conditions. Therefore, the GLMs are

$$\text{Reduced GLM:} \quad Y_{ij} = \mu + \pi_i + \varepsilon_{ij} \tag{6.12}$$

$$\text{Full GLM:} \quad Y_{ij} = \mu + \pi_i + \alpha_j + \varepsilon_{ij} \tag{6.3, rptd}$$

The reduced GLM manifests the null hypothesis

$$\alpha_j = 0 \tag{2.33, rptd}$$

which states that the experimental condition effects equal 0, that is, the experimental conditions exert no effect. The full GLM manifests the nondirectional experimental hypothesis

$$\alpha_j \neq 0 \quad \text{for some } j \tag{2.30, rptd}$$

This states that the effect is not zero for some experimental conditions. In other words, the experimental conditions exert an effect. A convenient formula for the reduced GLM error SS is

$$SSE_{\text{RGLM}} = \sum_{i=1}^{N}\sum_{j=1}^{p}(Y_{ij} - \overline{Y}_j)^2 \tag{6.13}$$

Applying equation (6.13) to the data in Table 6.5 provides the calculations presented in Table 6.11.

A convenient formula for the full GLM SS_{error} is

$$SSE_{\text{FGLM}} = \sum_{i=1}^{N}\sum_{j=1}^{p}(Y_{ij} - \overline{Y}_j - \overline{Y}_i + \overline{Y}_G)^2 \tag{6.14}$$

Applying equation (6.14) to the data in Table 6.4 provides the calculations presented in Table 6.12.

An F-test of the error component sum of squares, attributed to the inclusion of the experimental condition effects, is given by

$$F = \frac{(\text{SSE}_{\text{RGLM}} - \text{SSE}_{\text{FGLM}})/(df_{\text{RGLM}} - df_{\text{FGLM}})}{\text{SSE}_{\text{FGLM}}/df_{\text{FGLM}}} \tag{2.42, rptd}$$

Table 6.11 Calculation of the SS Error for the Reduced GLM

Subjects	a1 30 s	a2 60 s	a3 180 s
s1	7 − 7.333 = −0.333	7 − 7.333 = −0.333	8 − 7.333 = 0.667
s2	3 − 9.333 = −6.333	11 − 9.333 = 1.667	14 − 9.333 = 4.667
s3	6 − 8.333 = −2.333	9 − 8.333 = 0.667	10 − 8.333 = 1.667
s4	6 − 9.333 = −3.333	11 − 9.333 = 1.667	11 − 9.333 = 1.667
s5	5 − 9.000 = −4.000	10 − 9.000 = 1.000	12 − 9.000 = 3.000
s6	8 − 9.333 = −1.333	10 − 9.333 = 0.667	10 − 9.333 = 0.667
s7	6 − 9.333 = −3.333	11 − 9.333 = 1.667	11 − 9.333 = 1.667
s8	7 − 10.000 = −3.000	11 − 10.000 = 1.000	12 − 10.000 = 2.000
		$\sum_{i=1}^{N}\sum_{j=1}^{p}\varepsilon_{ij}^2 = 150.000$	

Therefore

$$F = \frac{(150 - 38)/(16 - 14)}{38/14} = \frac{56}{2.714}$$

$$F(2, 14) = 20.634$$

The full and reduced models under consideration are

$$\text{Reduced GLM:} \quad Y_{ij} = \mu + \pi_i + \varepsilon_{ij} \qquad (6.12, \text{rptd})$$

$$\text{Full GLM:} \quad Y_{ij} = \mu + \pi_i + \alpha_j + \varepsilon_{ij} \qquad (6.3, \text{rptd})$$

As the only difference between these two models is the term representing the experimental conditions, so the difference between these two models provides an

Table 6.12 Calculation of the SS Error for the Full GLM

Subjects	30 s	60 s	180 s
s1	7 − 4.333 = 2.667	7 − 8.333 = −1.333	8 − 9.333 = −1.333
s2	3 − 6.333 = −3.333	11 − 10.333 = 0.667	14 − 11.333 = 2.667
s3	6 − 5.333 = 0.667	9 − 9.333 = 0.333	10 − 10.333 = 0.333
s4	6 − 6.333 = −0.333	11 − 10.333 = 0.667	11 − 11.333 = −0.333
s5	5 − 6.000 = −1.000	10 − 10.000 = 0.000	12 − 11.000 = 1.000
s6	8 − 6.333 = 1.667	10 − 10.333 = −0.333	10 − 11.333 = −1.333
s7	6 − 6.333 = −0.333	11 − 10.333 = 0.667	11 − 11.333 = −0.333
s8	7 − 7.000 = 0.000	11 − 11.000 = 0.000	12 − 12.000 = 0.000
		$\sum_{i=1}^{N}\sum_{j=1}^{p}\varepsilon_{ij}^2 = 38.00$	

estimate of the experimental effect. However, to obtain an estimate of the subject effect requires an even more reduced model that does not include the subject term

$$\text{Even more reduced GLM:} \quad Y_{ij} = \mu + \varepsilon_{ij} \qquad (2.32, \text{rptd})$$

In fact, this model may be recognized as the single factor independent measures ANOVA reduced GLM. The SS error for the even more reduced GLM is provided in Section 2.7.2 (also see Section 6.6) and is equal to 164. This is the SS error when all experimental data are described only by the general mean of all scores. Therefore, an estimate of the subject effect can be obtained from the difference between the (even more) reduced GLM SS error and the reduced GLM SS error

$$\text{SSE}_{(\text{EM})\text{RGLM}} - SSE_{\text{RGLM}} = 164 - 150 = 14$$

It is convenient to construct an ANOVA summary table, as presented as Table 6.13, as the various calculations are completed.

It is useful to compare the effect estimates provided by the single factor repeated measures ANOVA GLM presented in Table 6.13, with those provided by the single factor independent measures ANOVA GLM presented in Tables 2.3 and 2.7. This comparison demonstrates that irrespective of whether a repeated or an independent measures ANOVA GLM is applied, the SS estimates of the experimental effect are identical. The main difference between the repeated and independent measures ANOVA GLMs is the inclusion of a term to represent the subject effect in the repeated measures ANOVA GLM. Variance that can be accommodated only by the error term in the independent measures ANOVA GLM is attributed to the subject term and so the repeated measures ANOVA GLM error term is reduced. This is demonstrated by the difference between the full single factor independent measures ANOVA GLM error term SS (52) and the full single factor repeated measures ANOVA GLM error term SS (38), which is equal to the subject effect SS (14).

Table 6.13 Single Factor Repeated Measures ANOVA Summary Table

Source	SS	*df*	MS	*F*	*p*
Subjects	14.000	7	2.000		
Error reduction due to experimental conditions	112.000	2	56.000	20.634	<0.001
FGLM error	38.000	14	2.714		
Total	164.000	23			

6.5 REGRESSION GLMs FOR SINGLE FACTOR REPEATED MEASURES DESIGNS

The experimental design GLM equation (6.3) may be compared with the equivalent regression equation

$$Y_i = \beta_0 + \beta_1 X_{i,1} + \beta_2 X_{i,2} + \beta_3 X_{i,3} + \beta_4 X_{i,4} + \beta_5 X_{i,5} + \beta_6 X_{i,6} + \beta_7 X_{i,7} + \beta_8 X_{i,8} + \beta_9 X_{i,9} + \varepsilon_i \tag{6.15}$$

where Y_i represents the *i*th dependent variable score (not the *i*th subject), β_0 is a constant, β_1 is the regression coefficient for the predictor variable X_1, and β_2 is the regression coefficient for the predictor variable X_2. However, in repeated measures design, the subjects providing the repeated measures also are represented. The N levels of the subject factor are represented by $(N-1)$ variables. Therefore, the eight levels (i.e., subjects) are represented by the first seven variables (X_1–X_7). Similarly, the p levels of the experimental factor are represented by $(p-1)$ variables. Therefore, the three experimental conditions are represented by the last two variables (X_8 and X_9). Again, the random variable ε_i represents error.

Table 6.14 presents effect coding for the single factor repeated measures regression GLM. This coding scheme shows scores associated with subjects 1–7 are identified by the presence of a 1 in the variable column representing the subject, while subject 8's scores are identified by a -1 across all (X_1–X_7) subject variables. As in GLM equation (6.3), terms representing the interaction between experimental conditions and subjects are omitted.

Consistent with the incremental strategy (see Section 5.4) and estimating effects by comparing full and reduced GLMs, the first regression carried out here is that for the full single factor repeated measures experimental design GLM, when all subject and experimental condition predictor variables are included (i.e., variables X_1–X_9). Table 6.15 presents the predictor variable regression coefficients and standard deviations, the standardized regression coefficients, and significance tests (t- and p-values) of the regression coefficient. From the repeated measures ANOVA perspective, the analysis results in this table are of little interest, although it is worth noting that the regression coefficient estimates are equivalent to the subject effect estimates calculated earlier. Table 6.16 presents the ANOVA summary table for the regression GLM describing the complete single factor repeated measures ANOVA. As the residual SS is that obtained when both subject and experimental conditions are included in the regression, this is the error term obtained when the single factor repeated measures ANOVA GLM is applied.

The next aim is to determine by how much the residual SS increases when the predictor variables representing the experimental conditions are omitted. To do this, a regression GLM corresponding with the reduced single factor repeated measures experimental design GLM described by equation (6.14) is constructed. This regression GLM employs only those variables representing the subjects (variables X_1–X_7) as predictors. As subject and experimental condition variables are orthogonal, the

Table 6.14 Effect Coding for the Single Factor Repeated Measures Regression GLM

Subjects	Y	X_1	X_2	X_3	X_4	X_5	X_6	X_7	X_8	X_9
s1	7	1	0	0	0	0	0	0	1	0
s2	3	0	1	0	0	0	0	0	1	0
s3	6	0	0	1	0	0	0	0	1	0
s4	6	0	0	0	1	0	0	0	1	0
s5	5	0	0	0	0	1	0	0	1	0
s6	8	0	0	0	0	0	1	0	1	0
s7	6	0	0	0	0	0	0	1	1	0
s8	7	−1	−1	−1	−1	−1	−1	−1	1	0
s1	7	1	0	0	0	0	0	0	0	1
s2	11	0	1	0	0	0	0	0	0	1
s3	9	0	0	1	0	0	0	0	0	1
s4	11	0	0	0	1	0	0	0	0	1
s5	10	0	0	0	0	1	0	0	0	1
s6	10	0	0	0	0	0	1	0	0	1
s7	11	0	0	0	0	0	0	1	0	1
s8	11	−1	−1	−1	−1	−1	−1	−1	0	1
s1	8	1	0	0	0	0	0	0	−1	0
s2	14	0	1	0	0	0	0	0	−1	0
s3	10	0	0	1	0	0	0	0	−1	0
s4	11	0	0	0	1	0	0	0	−1	0
s5	12	0	0	0	0	1	0	0	−1	0
s6	10	0	0	0	0	0	1	0	−1	0
s7	11	0	0	0	0	0	0	1	−1	0
s8	12	−1	−1	−1	−1	−1	−1	−1	−1	0

Table 6.15 SYSTAT Output Pertinent to Multiple Regression Equation for Effect Coding

Variable	Coefficient	Standard Error	Standard Coefficient	t	p (Two-Tailed)
Constant	9	0.336	<0.001	26.762	<0.001
X_1	−1.667	1.654	−0.319	−1.008	0.329
X_2	0.333	1.654	0.064	0.202	0.843
X_3	−0.667	1.654	−0.128	−0.403	0.692
X_4	0.333	1.654	0.064	0.202	0.843
X_5	<0.001	1.654	<0.001	<0.001	1.000
X_6	0.333	1.654	0.064	0.202	0.843
X_7	0.333	1.654	0.064	0.202	0.843
X_8	−3.000	0.476	−0.937	−6.308	<0.001
X_9	1.000	0.476	0.312	2.103	0.054

Table 6.16 ANOVA Summary Table for Subject and Experimental Condition Effect Regression

Source	SS	*df*	MS	*F*	*p*
Regression	126.000	9	14.000	5.158	0.003
Residual	38.000	14	2.714		

R: 0.877; R^2: 0.768; adjusted R^2: 0.619.

predictor variable regression coefficients, their standard deviations, the standardized regression coefficients, and the significance tests (t- and p-values) of the regression coefficients provided by this analysis are identical to those presented in Table 6.15. Therefore, of most interest is the ANOVA summary presented in Table 6.17. This presents the residual SS for the reduced single factor repeated measures experimental design GLM. (As the residual SS contains both SS for experimental conditions and SS error, the F-test is irrelevant.) The experimental condition effect can be estimated by subtracting the full GLM residual SS from the reduced GLM residual SS

SS error for reduced regression GLM	=	150.000
SS error for full regression GLM	=	−38.000
SS error reduction due to experimental conditions	=	112.000

Another regression would be applied if information on the subject effect was required. The GLM for this regression would exclude the terms for both subjects (variables X_1–X_7) and experimental conditions (variables X_8 and X_9). With all of these variables excluded, the only prediction of a subject's score is provided by the general mean $\overline{Y}_G$. This model may be recognized as the single factor independent measures ANOVA reduced GLM, the SS error of which is equal to 164 (see Section 2.8.2). An alternative method of calculating the SS error for this GLM is to subtract $\overline{Y}_G$ from each score and then square and sum the residuals. Table 6.18 presents these calculations. The subject effect can be estimated by subtracting the reduced GLM residual SS from the even more reduced GLM (equivalent to the single factor independent measures ANOVA reduced GLM)

SS error for even more reduced regression GLM	=	164.000
SS error for reduced regression GLM	=	−150.000
SS error reduction due to subjects	=	14.000

Table 6.17 ANOVA Summary Table for Subject Effect Regression

Source	SS	*df*	MS	*F*	*p*
Regression	14.000	7	2.000	0.213	0.977
Residual	150.000	16	9.375		

R: 0.292; R^2: 0.085; adjusted R^2: 0.000.

Table 6.18 Single Factor Independent Measures ANOVA Reduced GLM SS Error Calculation

Subjects	Y	$\overline{Y}_G$	$Y - \overline{Y}_G$	$(Y - \overline{Y}_G)^2$
s1	7	9	−2	4
s2	3	9	−6	36
s3	6	9	−3	9
s4	6	9	−3	9
s5	5	9	−4	16
s6	8	9	−1	1
s7	6	9	−3	9
s8	7	9	−2	4
s1	7	9	−2	4
s2	11	9	2	4
s3	9	9	0	0
s4	11	9	2	4
s5	10	9	1	1
s6	10	9	1	1
s7	11	9	2	4
s8	11	9	2	4
s1	8	9	−1	1
s2	14	9	5	25
s3	10	9	1	1
s4	11	9	2	4
s5	12	9	3	9
s6	10	9	1	1
s7	11	9	2	4
s8	12	9	3	9
$\sum$				164

Putting this information and the corresponding *df*s in Table 6.19 essentially recasts Table 6.16, but separates the regression SS and *df*s into Experimental Condition SS and *df*s, and subject SS and *df*s.

Repeated measures ANOVA also may be implemented by a regression GLM that uses a single criterion scaled variable, rather than $(N - 1)$ variables, to accommodate the subject effect (e.g., Pedhazur, 1982). One advantage of this approach is the reduction in predictors required, especially with larger numbers of participants. This

Table 6.19 ANOVA Summary Table for Single Factor Repeated Measures ANOVA

Source	SS	*df*	MS	*F*	*p*
Subjects	14.000	7	2.000		
Experimental conditions	112.000	2	56.000	20.634	<0.001
Error	38.000	14	2.714		

was particularly useful when statistical software was limited in the number of predictor variables that could be accommodated in a regression analysis. However, as the capability of most statistical software now far exceeds the demands likely to be made by most repeated measures designs, this is no longer a serious concern.

6.6 EFFECT SIZE ESTIMATION

Effect size estimation with repeated measures designs is a more complicated business than for independent designs due to the consequences of accommodating subject variance. As described below, this results in different types of repeated measures SOA effect size estimates developed to serve different purposes.

6.6.1 A Complete $\widehat{\omega}^2$ SOA for the Omnibus Effect Comparable Across Repeated and Independent Measures Designs

As described in Section 6.4, repeated measures experimental design GLMs accommodate score variance attributable to the subjects. As subject differences is a major contributor to the single factor independent measures ANOVA error term, accommodating score variance attributable to subjects (with the π_i term) usually reduces the repeated measures ANOVA GLM error variance considerably. As said in Section 4.5, one purpose of estimating effect size is to enable comparisons across studies free of the influence of sample size, but the ability of repeated measures ANOVA GLMs to reduce error estimates by accommodating subject variance with the π_i term is a major source of discrepancy between independent and repeated measures effect size estimates. However, if the π_i term (and its interactions with other GLM terms) is omitted from the GLM, then an independent ANOVA GLM is applied to the repeated measures data and this will provide effect size estimates comparable across repeated and independent designs.

In Section 4.3, the complete omnibus ω^2 for a single factor independent measures ANOVA GLM was defined in full and reduced model comparison terms and in traditional ANOVA summary table terms. As two reduced models (the reduced model and the even more reduced model) and one full model are involved in estimating repeated measures effect sizes, it is simpler to describe the estimation of repeated measures effect sizes using the traditional ANOVA summary table terms definition of ω^2. This defines the omnibus ω^2 as

$$\widehat{\omega}^2 = \frac{\text{Experimental conditions SS} - (p-1)\text{MSe}}{\text{Total SS} + \text{MSe}} \tag{4.11, rptd}$$

The omnibus ω^2 may also be defined in terms of the number of experimental conditions, the F-statistic and the number of subjects per condition (in a balanced design)

$$\widehat{\omega}^2 = \frac{(p-1)(F-1)}{(p-1)(F-1) + pN} \tag{4.12, rptd}$$

It was seen in Section 6.5 that the same experimental effect SS of 112 is obtained with independent or repeated measures ANOVA GLMs. Similarly, irrespective of whether independent or repeated measures ANOVA GLMs are applied, there are $p = 3$ experimental conditions, so $p - 1 = 2$. However, as equation (4.11) is for independent measures ANOVA GLMs, the MSe mentioned is the independent measures ANOVA GLM MSe. Tables 2.3 and 2.7 show this to be 2.476, while Tables 2.3, 2.7, 6.10, 6.13, 6.17, and 6.19 show that the total SS is 164. Inserting these values into equation (4.11) provides

$$\widehat{\omega}^2 = \frac{112 - (2)2.476}{164 + 2.476} = 0.64$$

Therefore, 64% of the total population variation is explained by the experimental effect. The purpose of the effect size estimation above is to allow comparison across independent and repeated measures experimental designs free of the error term reduction due to accommodating score variance attributable to the different subjects by omitting this feature from the effect size estimation. Consequently, an independent measures ω^2 estimate was obtained from the repeated measures data. The same hypothetical data was analyzed under independent and repeated measures conceptions and so the equality of the ω^2 estimate above and the ω^2 estimate obtained in Section 4.3 demonstrates the efficacy of the approach.

6.6.2 A Partial $\widehat{\omega}^2$ SOA for the Omnibus Effect Appropriate for Repeated Measures Designs

Although the ability to compare effect size estimates across repeated and independent designs has its place, there is also a need for effect size estimates that utilize the full ability of repeated designs to reduce error variance. These estimates not only reveal the actual effect sizes observed in repeated measures designs, but also they are required for power analysis determination of the appropriate sample sizes for repeated measures designs.

A complete $\widehat{\omega}^2$ SOA for the omnibus effect is defined as

$$\omega^2_{\text{expt. effect}} = \frac{\sigma^2_{\text{expt. effect}}}{\sigma^2_{\text{total}}} \tag{6.16}$$

Equation (6.16) makes it clear that the complete $\widehat{\omega}^2$ for the omnibus effect is a proportion of the total experimental variation. However, as the total experimental variation includes subject variance and one of the reasons a repeated design will have been employed is to remove this variance, it makes little sense to employ a complete $\widehat{\omega}^2$ effect size estimate for the omnibus effect. A partial $\widehat{\omega}^2$ effect size for the omnibus effect is much more appropriate.

The partial $\widehat{\omega}^2$ SOA for the omnibus effect ignores subject variance and employs the reduced error term. The partial ω^2 estimate of the omnibus effect is defined as

$$\omega^2_{(\text{expt. effect})} = \frac{\sigma^2_{\text{expt. effect}}}{\sigma^2_{\text{expt. effect}} + \sigma^2_{\text{error}}} \tag{6.17}$$

and its estimate may be obtained from

$$\widehat{\omega}^2_{(\text{expt. effect})} = \frac{(p-1)(F_{\text{expt. effect}} - 1)}{(p-1)(F_{\text{expt. effect}} - 1) + pN} \tag{6.18}$$

Applying equation (6.18) to the data in Tables 6.10, 6.13, 6.17 and 6.19 provides

$$\widehat{\omega}^2_{(\text{expt. effect})} = \frac{(2)(20.634-1)}{(2)(20.634-1) + 3(8)} = \frac{39.268}{63.268} = 0.62$$

Therefore, the experimental effect accounts for 62% of the experimental effect and the error population variation. (Keep in mind that the amount of variance of which this SOA estimate of the experimental effect is a proportion is considerably smaller than that employed by the complete omnibus effect size estimate in Section 6.6.1.)

6.6.3 A Partial ω^2 SOA for Specific Comparisons Appropriate for Repeated Measures Designs

As with independent designs, there is likely to be greatest interest in the SOA for particular comparisons between experimental conditions in repeated measures designs. For example, just as it was in the hypothetical independent design, the SOA between the 30 and 180 s conditions in the hypothetical repeated measures study time experiment is of interest.

The partial ω^2 described in Section 4.3.1 expresses the specific comparison variance only as a proportion of the specific comparison variance plus error variance. For single factor-repeated measures designs, this partial ω^2 for specific comparisons can be defined as

$$\widehat{\omega}^2_{(\psi)} = \frac{F_\psi - 1}{F_\psi - 1 + 2N} \tag{6.19}$$

Applying equation (6.19) to the data in Table 6.22 provides

$$\widehat{\omega}^2_{(\psi)} = \frac{21.212 - 1}{21.212 - 1 + 2(8)} = \frac{20.212}{36.212} = 0.56$$

Therefore, 56% of the variance in the 30 and 180 s populations is explained by the comparison between these two experimental conditions.

6.7 FURTHER ANALYSES

As described for the independent experimental design ANOVAs, a significant omnibus F-test leads to rejection of the omnibus null hypothesis

$$\mu = \mu_j \tag{2.34, rptd}$$

Table 6.20 Marginal Means for the Three Study Time Experimental Conditions

Study Time	30 s: Condition 1	60 s: Condition 2	180 s: Condition 3
$\overline{X}$	6	10	11

and acceptance of the experimental or alternate hypothesis

$$\mu \neq \mu_j \quad \text{for some } j \qquad (2.31, \text{rptd})$$

As with the independent single and multifactor experimental designs, the next step in the analysis strategy outlined in Section 3.9 is to identify the planned comparison(s). As before it is assumed the comparison of the 30 s versus 180 s experimental conditions is planned and all other experimental condition comparisons are unplanned. Therefore, the hypothesis manifest in the comparison of the 30 s versus 180 s experimental condition means is assessed without any Type 1 error rate adjustment, while any other (pairwise or nonpairwise) comparisons are conceived as members of a separate family (or families if appropriate) over which the Type 1 error rate is controlled at $\alpha = 0.05$. For simplicity, it again is assumed only the two pairwise comparisons - the 30 s versus 60 s and 60 s versus 180 s experimental conditions - are of any interest as unplanned comparisons. Table 6.20 presents the three study time means. There are $p = 3$ study time levels, each population mean is designated by μ_j and each estimate of the population mean is provided by the sample means, $\overline{Y}_j$ where $j = 1, 2$, or 3.

Calculation of the planned or unplanned comparison sums of squares ($SS_{\psi_{pc}}$) for the single factor repeated measures design is identical to that for the single factor independent measures design. The linear contrast for the planned comparison expressed in terms of population means (see Section 3.4) is

$$\psi_{pc} = (-1)\mu_1 + (0)\mu_2 + (1)\mu_3$$

Replacing the population means with the sample mean estimators provides

$$\psi_{pc} = (-1)6 + (0)10 + (1)11 = 5$$

However, the formula for the repeated measures design $SS_{\psi_{pc}}$ employs the number of subjects participating in each condition, which is also the total number of subjects participating in the whole experiment. Therefore

$$SS_{\widehat{\psi}_{pc}} = \frac{N\widehat{\psi}_{pc}^2}{\sum c_j^2} = \frac{(8)(5)^2}{(-1)^2 + (0)^2 + (1)^2} = \frac{200}{2} = 100$$

One df is associated with $SS_{\widehat{\psi}_{pc}}$, so the mean square for the contrast is

$$MS_{\widehat{\psi}_{pc}} = \frac{SS_{\psi_{pc}}}{1} = \frac{100}{1} = 100$$

The next requirement for the planned comparison F-test is an appropriate error term. For independent measures designs, the omnibus ANOVA error term was the relatively simple choice for a number of reasons (see Section 3.7.1). However, as mentioned in Section 6.2, only when the error covariance matrix is spherical will a valid and accurate F-test be obtained. Spherical experimental condition covariance matrices are obtained when the variance of the differences between the subjects' experimental condition scores are homogeneous (see Section 10.2.2). As psychological data often violates the sphericity assumption and research has shown that even small sphericity violations can exert a large influence on the outcome of pairwise and nonpairwise comparisons (e.g., Boik, 1981), a frequent recommendation is to employ an error term based on only that data contributing to the pairwise (or nonpairwise) comparison. (This also explains the absence of complete ω^2 SOAs for specific comparisons with repeated designs.) This error term can be calculated in a variety of ways. For example, the procedures described in Sections 6.4–6.6 can be applied to only two experimental conditions and would provide both experimental effect and error term estimates.

However, there is a quicker way to calculate the error term for the comparison between the two experimental conditions that has much in common with the way in which a repeated measures t-test is calculated. A repeated measures t-test converts the two experimental condition scores obtained from each subject into a single difference score simply by subtracting one from the other in a manner consistent with the linear contrast. Table 6.21 presents each subjects' scores from the 30 and 180 s experimental conditions, the difference score for each subject, the means of the experimental condition scores, the mean of the difference score, the sum of the difference scores, and the sum of the squared difference scores.

Table 6.21 Differences Between Subjects' Scores Across 30 and 180 s Experimental Conditions

Subjects	180 s	30 s	180 s − 30 s ($\widehat{\psi}_i$)
s1	8	7	1
s2	14	3	11
s3	10	6	4
s4	11	6	5
s5	12	5	7
s6	10	8	2
s7	11	6	5
s8	12	7	5
$\sum \psi_i$			40
$\overline{\psi}_i$	11	6	5
$\sum \widehat{\psi}_i^2$			266

The error SS for the reduced GLM when only the two experimental conditions being compared are considered is provided by

$$\text{SSE}_{\text{RGLM}} = \frac{\sum \widehat{\psi}_i^2}{\sum c_j^2} = \frac{266}{-1^2 + 1^2} = 133$$

The SS for the planned comparison between the 30 and 180 s experimental conditions, $SS_{\psi_{pc}}$, calculated above was 100. As

$$\text{SS}_{\psi_{pc}} = \text{SSE}_{\text{RGLM}} - \text{SSE}_{\text{FGLM}}$$

it follows that

$$\text{SSE}_{\text{FGLM}} = \text{SSE}_{\text{RGLM}} - SS_{\psi_{pc}}$$

Therefore

$$\text{SSE}_{\text{FGLM}} = \frac{\sum \widehat{\psi}_i^2}{\sum c_j^2} - SS_{\widehat{\psi}_{pc}} = \frac{266}{-1^2 + 1^2} - 100 = 33$$

It is convenient to cast the SS results in an ANOVA summary table to complete the calculations with the planned pairwise comparison $dfs = (p - 1)$ and the error term $dfs = (N - 1)(p - 1)$, as done in Table 6.22. As the comparison of the 30 and 180 s experimental condition means was planned, no Type 1 error adjustment is necessary. Therefore, the difference between the subjects' free recalls after 30 and 180 s is declared significant.

Identical calculations can be carried out to assess the two unplanned comparisons involving the 30 and 60 s, and the 60 and 180 s experimental condition means. The linear contrast for the two unplanned comparison expressed in terms of population means are

$$\psi_{30 \text{ vs. } 60}(-1)\mu_1 + (1)\mu_2 + (0)\mu_3$$

$$\psi_{60 \text{ vs. } 180}(0)\mu_1 + (-1)\mu_2 + (1)\mu_3$$

Replacing the population means with the sample mean estimators provides

$$\psi_{30 \text{ vs. } 60} = (-1)6 + (1)10 + (0)11 = 4$$

$$\psi_{60 \text{ vs. } 180} = (0)6 + (-1)10 + (1)11 = 1$$

Table 6.22 ANOVA Summary Table for the Planned Pairwise Comparison Between the 30 and 180 s Experimental Condition Means

Source	SS	*df*	MS	*F*	*p*
30 s vs. 180 s PC	100.000	1	100.000	21.212	0.002
Error	33.000	7	4.714		

The SS for these two unplanned comparisons are

$$\mathrm{SS}_{\widehat{\psi}_{30\ \mathrm{vs.}\ 60}} = \frac{N\widehat{\psi}^{2}_{30\ \mathrm{vs.}\ 60}}{\sum c_j^2} = \frac{(8)(4)^2}{(-1)^2 + (0)^2 + (1)^2} = \frac{128}{2} = 64$$

$$\mathrm{SS}_{\widehat{\psi}_{60\ \mathrm{vs.}\ 180}} = \frac{N\widehat{\psi}^{2}_{60\ \mathrm{vs.}\ 180}}{\sum c_j^2} = \frac{(8)(1)^2}{(-1)^2 + (0)^2 + (1)^2} = \frac{8}{2} = 4$$

As one *df* is associated with each unplanned comparison SS, the contrast mean squares are

$$\mathrm{MS}_{\widehat{\psi}_{30\ \mathrm{vs.}\ 60}} = \frac{\mathrm{SS}_{\psi_{30\ \mathrm{vs.}\ 60}}}{1} = \frac{64}{1} = 64$$

$$\mathrm{MS}_{\widehat{\psi}_{60\ \mathrm{vs.}\ 180}} = \frac{\mathrm{SS}_{\psi_{60\ \mathrm{vs.}\ 180}}}{1} = \frac{4}{1} = 4$$

The values in Table 6.23 allow calculation of the reduced GLM SSE and together with the SS for each comparison, (see Table 6.24). The calculations for the full GLM SSE for each of the two comparisons are presented below

$$\mathrm{SSE}_{\mathrm{RGLM}(30\ \mathrm{vs.}\ 60)} = \frac{\sum \widehat{\psi}^{2}_{i,30\ \mathrm{vs.}\ 60}}{\sum c_j^2} = \frac{168}{-1^2 + 1^2} = 84$$

$$\mathrm{SSE}_{\mathrm{RGLM}(60\ \mathrm{vs.}\ 180)} = \frac{\sum \widehat{\psi}^{2}_{i,60\ \mathrm{vs.}\ 180}}{\sum c_j^2} = \frac{16}{-1^2 + 1^2} = 8$$

$$\mathrm{SSE}_{\mathrm{FGLM}(30\ \mathrm{vs.}\ 60)} = \frac{\sum \widehat{\psi}^{2}_{i,30\ \mathrm{vs.}\ 60}}{\sum c_j^2} - \mathrm{SS}_{\widehat{\psi}_{\mathrm{pc}}} = \frac{168}{-1^2 + 1^2} - 64 = 20$$

$$\mathrm{SSE}_{\mathrm{FGLM}(60\ \mathrm{vs.}\ 180)} = \frac{\sum \widehat{\psi}^{2}_{i,60\ \mathrm{vs.}\ 180}}{\sum c_j^2} - \mathrm{SS}_{\widehat{\psi}_{\mathrm{pc}}} = \frac{16}{-1^2 + 1^2} - 4 = 4$$

Of course, it is much easier to implement all of these comparison calculations using statistical software. This can be done simply by applying a standard repeated measures ANOVA for each comparison. For each comparison, only that data obtained under the pertinent two experimental conditions is selected and analyzed.

As the last two comparisons are unplanned comparisons and there is no theoretical reason to consider these hypotheses separately, the two hypotheses constitute a single hypothesis family over which Type 1 error rate control is exerted. However, consideration of the LRH is always worthwhile and as the situation here mirrors that described in Section 5.6.1, there can be only one possibly true null hypothesis, and as the classic statistical test protection is for one possibly true null hypothesis, and as this is the classic statistical test protection, no *p*-value adjustment is necessary to

Table 6.23 Differences Between Subjects' Scores Across the 30 and 60 s, and the 60 and 180 s Experimental Conditions

Subjects	60 s	30 s	60 s − 30 s ($\widehat{\psi}_i$)	180 s	60 s	180 s − 60 s ($\widehat{\psi}_i$)
s1	7	7	0	8	7	1
s2	11	3	8	14	11	3
s3	9	6	3	10	9	1
s4	11	6	5	11	11	0
s5	10	5	5	12	10	2
s6	10	8	2	10	10	0
s7	11	6	5	11	11	0
s8	11	7	4	12	11	1
$\sum \psi_i$			32			8
$\overline{\psi}_i$			4			1
$\sum \widehat{\psi}_i^2$			168			16

maintain the appropriate control of Type 1 error rate. Therefore, the unplanned comparisons above require no p-value adjustment and both can be accepted as significant at their classic $p = 0.002$ and $p = 0.033$ values.

It have been noticed that breaking the experiment up into separate pairs of conditions for analysis results in a considerable loss of *df*s from the separate comparison error terms. Only 7 *df*s are associated with each of the separate comparison error terms, while 14 *df*s are associated with the error term in the omnibus repeated measures ANOVA. This reduction in error *df*s is likely to diminish the power of the separate comparisons.

Section 10.2.2 considers the spherical covariance matrix assumption made by univariate repeated measures ANOVA. Most statistical software packages provide two statistics (Geisser and Greenhouse, 1958; Huynh and Feldt, 1976) that attempt to estimate Box's (1954) parameter, ε, which varies between 0 and 1 and indexes the extent of the violation of the sphericity assumption, with lower values indicating greater sphericity violation. Of the two, Huynh and Feldt's $\tilde{\varepsilon}$ is superior estimate of ε. Therefore, if Huynh and Feldt's $\tilde{\varepsilon}$ is very close to 1, using the omnibus MSe with the greater associated *df*s may be worth considering if an increase in comparison power is required.

Table 6.24 ANOVA Summary Table for the Unplanned Pairwise Comparisons Between the 30 Versus 60 s and the 60 Versus 180 s Experimental Condition Means

Source	SS	*df*	MS	*F*	*p*
30 s vs. 60 s PC	64.000	1	64.000	22.400	0.002
Error	20.000	7	2.857		
60 s vs. 180 s PC	4.000	1	4.000	7.000	0.033
Error	4.000	7	0.571		

6.8 POWER

6.8.1 Determining the Sample Size Needed to Detect the Omnibus Effect

The method for determining the required sample size to achieve a specific level of power in a repeated measures design is a simple generalization of the method described for independent measures designs. The four pieces of information required are

- The significance level (or Type 1 error rate)
- The power required
- The numerator *dfs*
- The effect size

The usual $\alpha = 0.05$ is employed and the convention in psychology is to set power at 0.8. The numerator *dfs* are set by the number of experimental conditions. For the hypothetical single factor repeated measures design experiment presented here, the numerator *dfs* $(\nu_1) = (p - 1) = (3 - 1) = 2$. It is also important to note that when ω^2 effect sizes observed in previous research or pilot studies, and power analyses are conducted to determine the sample size needed to detect an omnibus effect, the partial ω^2 SOA for the omnibus effect size estimate (Section 6.6.2) or the partial ω^2 SOA for specific comparisons (Section 6.6.3) should be used. Here, it will be assumed that a medium effect size, $\omega^2 = 0.06$, is to be detected.

When power charts are used (see Appendix C), the next step is to make an educated guess as to how many subjects might be needed and equation (4.22) is applied

$$\phi = \sqrt{\frac{\omega^2}{1 - \omega^2}}\sqrt{N} \qquad (4.22, \text{rptd})$$

The educated guess is to use a sample size of 50 subjects per experimental condition. Applying equation (4.22) with $\omega^2 = 0.06$ and $N = 50$ provides

$$\begin{aligned} \phi &= \sqrt{\frac{0.06}{1 - 0.06}}\sqrt{50} \\ &= 0.25(7.07) \\ &= 1.77 \end{aligned}$$

It should be noted that when repeated measures designs are considered, ν_2, the denominator *df*, is defined differently to when independent measures designs are considered. Specifically, with 50 subjects participating in each of the experimental conditions, the repeated measures denominator *dfs* $(\nu_2) = (N - 1)(p - 1) = (50 - 1)(3 - 1) = 98$. Examination of the power charts reveals that with $N = 50$, $\phi = 1.77$, and denominator *dfs* $(\nu_2) = 98$ (i.e., 100 in charts), power ~ 0.8. As well as being a good (or lucky) educated guess, the example also illustrates how many subjects can be required to obtain higher power levels even when repeated measures designs are employed. Applying G*Power to this data, to obtain a more accurate estimate, indicates that with $N = 50$ (and the other values defined as above), power $= 0.76$. A power of 0.8 is achieved when $N = 54$.

It is also possible to apply a power analysis to determine sample size that employs the correlation between subjects' scores across the experimental conditions (see Keppel and Wickens, 2004; Kirk, 1995; Maxwell and Delaney, 2004). This correlation may be known or estimated from previous research or pilot studies. However, in most situations, researchers are unlikely to know this correlation and in any novel situations, any previously observed correlations are unlikely to apply.

6.8.2 Determining the Sample Size Needed to Detect Specific Effects

As mentioned before, the omnibus null hypothesis is rarely the hypothesis in which there is real interest. Usually, the real interest is with regard to the hypotheses manifest in specific pairwise comparisons between the specific experimental condition means of means.

The key piece of information required for the power analysis to determine the required sample size is a partial ω^2 (or an equivalent f) estimate of the specific comparison under consideration. The pertinent partial ω^2 (or an equivalent f) estimate is the partial SOA for specific comparisons described in Section 6.6.3. This is the pertinent effect size estimate because the study to be designed will employ repeated measures and so the detection of this effect will be assisted by the error reduction resulting from the accommodation of the variance attributable to subjects. After the partial ω^2 for the specific comparison is obtained, the procedure for determining the required sample size continues as described above for the omnibus effect in Section 6.8.1.

CHAPTER 7

The GLM Approach to Factorial Repeated Measures Designs

7.1 FACTORIAL RELATED AND REPEATED MEASURES DESIGNS

Factorial repeated measures designs are amongst the most popular research designs employed in psychological research. This popularity has much to do with the combined advantages of factorial and repeated measures randomized block designs (see Sections 5.1, 5.2 and 6.1.3). Other forms of randomized blocking (see Sections 6.1.1 and 6.1.2) offer benefits similar to repeated measures designs without any risk of the problems that can arise from experimental condition presentation orders. However, these designs also require much more effort and many more subjects from which to identify the matching subjects, particularly in factorial studies where each block of subjects must be matched across all the levels of all the factors. Indeed, the number of subjects and effort required to obtain similar matching blocks for factorial studies probably is a major reason why so few factorial nonrepeated measures randomized block designs are applied. Therefore, the focus of this and the next chapter is factorial repeated measures designs. Nevertheless, as noted in Section 6.1.3, the procedures described here for repeated measures randomized block designs are appropriate for all randomized block designs employing only one matched set of subjects per randomized block, while Kirk (1995) presents procedures for designs employing more than one matched set of subjects per randomized block.

Repeated measures factors can be combined with independent factors, or with other repeated measures factors. The combination of an independent measures factor and a repeated measures factor is addressed in Chapter 8. The present chapter address the combination of two repeated measures factors. When only repeated measures factors are employed, the design can be labelled a fully repeated measures factorial design. In fully repeated measures factorial designs, every subject experiences every experimental condition.

ANOVA and ANCOVA: A GLM Approach, Second Edition. By Andrew Rutherford.

Table 7.1 A Fully Related Two-Factor (2 × 3) Design

Factor A	a1			a2		
Factor B	b1	b2	b3	b1	b2	b3
	s1	s1	s1	s1	s1	s1
	s2	s2	s2	s2	s2	s2
	⋮	⋮	⋮	⋮	⋮	⋮
	s8	s8	s8	s8	s8	s8

7.2 FULLY REPEATED MEASURES FACTORIAL DESIGNS

A two-factor fully repeated measures design is presented schematically in Table 7.1. In common with single factor repeated measures designs, appropriate controls are required to address the potential confounding posed by order effects. In fully repeated measures factorial designs, there are likely to be more experimental conditions under which the same subjects provide scores, so there will be a greater number of order permutations and so a more extensive implementation of order controls is required.

As with all factorial designs, there are a greater number of main and interaction effects in factorial repeated measures designs compared with single factor repeated measures designs. With fully related factorial designs, there is also an increase in the number of "error" terms. In fact, there is a separate "error" term for each fixed experimental factor and interaction between fixed experimental factors.

The GLM for a fully related two-factor ANOVA is described by the equation

$$Y_{ijk} = \mu + \pi_i + \alpha_j + \beta_k + (\pi\alpha)_{ij} + (\pi\beta)_{ik} + (\alpha\beta)_{jk} + \varepsilon_{ijk} \tag{7.1}$$

where Y_{ijk} is the dependent variable score for the ith subject at the jth level of Factor A and the kth level of Factor B, μ is the general mean of the experimental condition population means, π_i is a parameter representing the random effect of the ith subject, α_j is the effect of the jth level of Factor A, β_k is the effect of the kth level of Factor B, $(\pi\alpha)_{ij}$ is the effect of the interaction between the ith subject and the jth level of Factor A, $(\pi\beta)_{ik}$ is the effect of the interaction between the ith subject and the kth level of Factor B, $(\alpha\beta)_{jk}$ is the interaction effect of the jth level of Factor A and the kth level of Factor B, and, as always, ε_{ijk} represents the random error associated with the ith subject in the jth level of Factor A and the kth level of Factor B.

As with single factor repeated measures designs, due to there being only one score per subject per experimental condition, the error term and the interaction between the two experimental factors and subjects cannot be separated and so, ε_{ijk} is written more accurately as $[(\pi\alpha\beta)_{ijk} + \varepsilon_{ijk}]$ and often is refered to simply as $(\pi\alpha\beta)_{ijk}$, or (S × A × B). However, the variation associated with the error term ε_{ijk} is used only to assess the effect of the interaction between the two fixed experimental factors. As described with respect to the single factor repeated measures design, when a single random factor is included in a model with fixed effects and the fixed effects are to be tested, limiting the interaction of the pertinent fixed factor(s) and the random factor to zero provides an appropriate error term. In fully repeated measures factorial designs, the term representing the interaction of the fixed Factors A and B is to be tested, while the only random factor represents the influence of the subjects. Therefore, setting the $[(\pi\alpha\beta)_{ijk}]$ interaction to zero simply by omitting this term from the GLM provides an appropriate error term to assess the interaction between the two fixed experimental Factors A and B.

The strategy just described seems to leave the main effects of Factors A and B without an error term denominator for the F-test. However, in the fully repeated measures factorial design, the variation associated with the interaction between Factor A and subjects $[(\pi\alpha)_{ij}]$ is used to assess the effect of the Factor A manipulation, while the variation associated with the interaction between Factor B and subjects $[(\pi\beta)_{ik}]$ is used to assess the effect of the Factor B manipulation (For F-test numerator and denominator expected mean squares, see Howell, 2010). As in the single factor repeated measures design, using these variation estimates to assess the main effects of Factor A and B makes intuitive sense. In both instances, the interactions represent the extent to which the factor effect is inconsistent across different subjects and the greater this inconsistency in relation to the factor effect, the less likely is the factor effect to be reliable. Nevertheless, due to the way in which the interactions between Factor A and Subjects, and Factor B and Subjects are calculated (the fully factorial GLM description of factor and subject factor interactions are based on single mean scores – the means of each subject's repeated measures under the pertinent factor levels), these interaction terms also accommodate error components. Consequently, all of the fully repeated measures factorial ANOVA F-tests are accurate and valid only when the variance of differences across the factor levels is homogeneous, i.e., when the sphericity assumption is tenable (see Section 10.2.2).

Consider the experimental data presented in Chapter 5 cast as a fully repeated measures factorial design, as presented in Table 7.2. In this design, all subjects are presented with and attempt to recall different sets of words under all 6 experimental conditions, with each subject receiving a randomized presentation order of the 6 experimental conditions. However, randomizing presentation orders has the consequence that there will be many instances where a story and imagery instruction condition is followed by a memorization condition. A particular concern is that subjects may continue to employ a story and imagery encoding strategy in subsequent memorization conditions (because story and imagery instructions are more specific and seem more effective than memorization instructions), so diminishing the distinction between these factor levels. As it is less likely that subjects asked to memorize and then asked to construct stories from the words and imagine the story events would employ the memorize only strategy rather than the story and imagery strategy, a contrast effect is likely. Therefore, a fully repeated measures factorial design would be inappropriate in these circumstances. Nevertheless, for the sake of illustrating the analysis of a fully repeated measures factorial design, assume the data in Table 7.2 had been obtained from just 8 subjects participating in all conditions and no contrast effect occurred.

Table 7.2 Experimental Data from a Fully Repeated Measures Two-Factor (2 × 3) Design

Encoding Instructions	a1 Memorize			a2 Story and Image			
	b1	b2	b3	b1	b2	b3	Subject
Study Time	30 s	60 s	180 s	30 s	60 s	180 s	Means
s1	7	7	8	16	16	24	13.000
s2	3	11	14	7	10	29	12.333
s3	6	9	10	11	13	10	9.833
s4	6	11	11	9	10	22	11.500
s5	5	10	12	10	10	25	12.000
s6	8	10	10	11	14	28	13.500
s7	6	11	11	8	11	22	11.500
s8	7	11	12	8	12	24	12.333
$\overline{X}$	6	10	11	10	12	23	12.000

As was described for the single factor repeated measures design, the manner of calculating experimental condition effects remains the same as in independent measures designs, emphasizing that repeated measures designs have consequence only for the error estimates. As the estimates of μ, and the α_j, β_k, and $(\alpha\beta)_{jk}$effects are defined just as for the independent measures factorial design, their definitions are not repeated here.

The mean of the scores provided by each subject is

$$\widehat{\mu}_i = \left(\frac{\sum_{j=1}^{p} \sum_{k=1}^{q} Y_{i,j}}{pq} \right) \tag{7.2}$$

and so the subject effects are

$$\widehat{\pi}_i = \mu_i - \mu \tag{7.3}$$

Applying formula (7.3) to the data in Table 7.2 provides

$$\begin{aligned}
\widehat{\pi}_1 &= 13.000 - 12 = 1.000 \\
\widehat{\pi}_2 &= 12.333 - 12 = 0.333 \\
\widehat{\pi}_3 &= 9.833 - 12 = -2.167 \\
\widehat{\pi}_4 &= 11.500 - 12 = -0.500 \\
\widehat{\pi}_5 &= 12.000 - 12 = 0.000 \\
\widehat{\pi}_6 &= 13.500 - 12 = 1.500 \\
\widehat{\pi}_7 &= 11.500 - 12 = -0.500 \\
\widehat{\pi}_8 &= 12.333 - 12 = 0.333 \\
\hline
\sum_{i=1}^{N} \widehat{\pi}_i &= 0.000
\end{aligned}$$

The subject SS is given by

$$\begin{aligned} SS_{\text{subjects}} &= pq\sum_{i=1}^{N}(\mu_i - \mu)^2 \\ &= 6[(1.000)^2 + (0.333)^2 + (-2.167)^2 + (-0.500)^2 \\ &\quad + (0)^2 + (1.500)^2 + (-0.500)^2 + (0.333)^2] \\ SS_{\text{subjects}} &= 52.008 \end{aligned}$$

The subject × Factor A interaction effects are defined by

$$(\pi\alpha)_{ij} = \mu_{ij} - (\mu + \pi_i + \alpha_j) \tag{7.4}$$

which reveals each interaction effect to be the extent to which each subject mean within each level of Factor A diverges from the additive pattern of subject and Factor A main effects. Applying formula (7.4) to the data in Table 7.2 provides

$(\pi\alpha)_{1,1} = 7.333 - (12.000 + 1.000 - 3) = -2.667$	$(\pi\alpha)_{1,2} = 18.667 - (12.000 + 1.000 + 3) = 2.667$
$(\pi\alpha)_{2,1} = 9.333 - (12.000 + 0.333 - 3) = 0.000$	$(\pi\alpha)_{2,2} = 15.333 - (12.000 + 0.333 + 3) = 0.000$
$(\pi\alpha)_{3,1} = 8.333 - (12.000 - 2.167 - 3) = 1.500$	$(\pi\alpha)_{3,2} = 11.333 - (12.000 - 2.167 + 3) = -1.500$
$(\pi\alpha)_{4,1} = 9.333 - (12.000 - 0.500 - 3) = 0.833$	$(\pi\alpha)_{4,2} = 13.667 - (12.000 - 0.500 + 3) = -0.833$
$(\pi\alpha)_{5,1} = 9.000 - (12.000 + 0.000 - 3) = 0.000$	$(\pi\alpha)_{5,2} = 15.000 - (12.000 + 0.000 + 3) = 0.000$
$(\pi\alpha)_{6,1} = 9.333 - (12.000 + 1.500 - 3) = -1.167$	$(\pi\alpha)_{6,2} = 17.667 - (12.000 + 1.500 + 3) = 1.167$
$(\pi\alpha)_{7,1} = 9.333 - (12.000 - 0.500 - 3) = 0.833$	$(\pi\alpha)_{7,2} = 13.667 - (12.000 - 0.500 + 3) = -0.833$
$(\pi\alpha)_{8,1} = 10.000 - (12.000 + 0.333 - 3) = 0.667$	$(\pi\alpha)_{8,2} = 14.667 - (12.000 + 0.333 + 3) = -0.666$

$$\sum_{i=1}^{N}\sum_{j=1}^{P}(\pi\alpha)_{ij} = 0.000$$

The subject × Factor A SS is given by

$$SS_{\text{subjects}\times\text{Factor A}} = q\sum_{i=1}^{N}[\mu_{ij} - (\mu + \pi_i + \alpha_j)^2] \tag{7.5}$$

or alternatively

$$SS_{\text{subjects}\times\text{Factor A}} = q\sum_{i=1}^{N}(\pi\alpha)_{ij}^2 \tag{7.6}$$

Therefore

$$\begin{aligned} SS_{\text{subjects}\times\text{Factor A}} &= 3[(-2.667^2) + (0^2) + (1.500^2) + (0.833^2) + (0^2) \\ &\quad + (-1.167^2) + (0.833^2) + (0.667^2) + (2.667) \\ &\quad + (0^2) + (-1.500^2) + (-0.833^2) + (0^2) \\ &\quad + (1.167^2) + (-0.833^2) + (-0.666^2)] \\ &= 75.333 \end{aligned}$$

Similarly, the subject × Factor B interaction effects are defined by

$$(\pi\beta)_{ik} = \mu_{ik} - (\mu + \pi_i + \beta_k) \tag{7.7}$$

which reveals each interaction effect to be the extent to which each subject mean within each level of Factor B diverges from the additive pattern of subject and Factor B main effects. Applying formula (7.5) to the data in Table 7.2 provides

$$
\begin{aligned}
(\pi\beta)_{1,1} &= 11.500 - (12.000 + 1.000 - 4) = 2.500\\
(\pi\beta)_{2,1} &= 5.000 - (12.000 + 0.333 - 4) = -3.333\\
(\pi\beta)_{3,1} &= 8.500 - (12.000 - 2.167 - 4) = 2.667\\
(\pi\beta)_{4,1} &= 7.500 - (12.000 - 0.500 - 4) = 0.000\\
(\pi\beta)_{5,1} &= 7.500 - (12.000 + 0.000 - 4) = -0.500\\
(\pi\beta)_{6,1} &= 8.500 - (12.000 + 1.500 - 4) = -1.000\\
(\pi\beta)_{7,1} &= 7.000 - (12.000 - 0.500 - 4) = -0.500\\
(\pi\beta)_{8,1} &= 7.500 - (12.000 + 0.333 - 4) = -0.833
\end{aligned}
$$

$$
\begin{aligned}
(\pi\beta)_{1,2} &= 11.500 - (12.000 + 1.000 - 1) = -0.500\\
(\pi\beta)_{2,2} &= 10.500 - (12.000 + 0.333 - 1) = -0.833\\
(\pi\beta)_{3,2} &= 11.000 - (12.000 - 2.167 - 1) = 2.167\\
(\pi\beta)_{4,2} &= 10.500 - (12.000 - 0.500 - 1) = 0.000\\
(\pi\beta)_{5,2} &= 10.000 - (12.000 + 0.000 - 1) = -1.000\\
(\pi\beta)_{6,2} &= 12.000 - (12.000 + 1.500 - 1) = -0.500\\
(\pi\beta)_{7,2} &= 11.000 - (12.000 - 0.500 - 1) = 0.500\\
(\pi\beta)_{8,2} &= 11.500 - (12.000 + 0.333 - 1) = 0.167
\end{aligned}
$$

$$
\begin{aligned}
(\pi\beta)_{1,3} &= 16.000 - (12.000 + 1.000 + 5) = -2.000\\
(\pi\beta)_{2,3} &= 21.500 - (12.000 + 0.333 + 5) = 4.167\\
(\pi\beta)_{3,3} &= 10.000 - (12.000 - 2.167 + 5) = -4.833\\
(\pi\beta)_{4,3} &= 16.500 - (12.000 - 0.500 + 5) = 0.000\\
(\pi\beta)_{5,3} &= 18.500 - (12.000 + 0.000 + 5) = 1.500\\
(\pi\beta)_{6,3} &= 19.000 - (12.000 + 1.500 + 5) = 0.500\\
(\pi\beta)_{7,3} &= 16.500 - (12.000 - 0.500 + 5) = 0.000\\
(\pi\beta)_{8,3} &= 18.000 - (12.000 + 0.333 + 5) = 0.667
\end{aligned}
$$

$$\sum_{i=1}^{N}\sum_{k=1}^{q}(\pi\beta)_{ik} = 0.000$$

The subject × Factor B SS is given by

$$\mathrm{SS}_{\text{subjects}\times\text{Factor B}} = p\sum_{i=1}^{N}[\mu_{ik} - (\mu + \pi_i + \beta_k)^2] \tag{7.8}$$

or alternatively

$$\mathrm{SS}_{\text{subjects}\times\text{Factor B}} = p\sum_{i=1}^{N}(\pi\beta)_{ik}^2 \tag{7.9}$$

$$
\begin{aligned}
\mathrm{SS}_{\text{subjects}\times\text{Factor B}} = 2[&(2.500^2) + (-3.333^2) + (2.667^2) + (0^2)\\
&+ (-0.500^2) + (-1.000^2) + (-0.500^2) + (-0.833^2)\\
&+ (-0.500^2) + (-0.833^2) + (2.167^2) + (0^2)\\
&+ (-1.000^2) + (-0.500^2) + (0.500^2) + (0.167^2)\\
&+ (-2.000^2) + (4.167^2) + (-4.833^2) + (0^2)\\
&+ (1.500^2) + (0.500^2) + (0^2) + (0.667^2)]
\end{aligned}
$$

$$\mathrm{SS}_{\text{subjects}\times\text{Factor B}} = 161.000$$

Table 7.3 Predicted Scores for the Fully Repeated Measures Two-Factor (2 × 3) Experiment

	a1			a2		
	b1	b2	b3	b1	b2	b3
s1	6.833	7.833	7.333	16.167	15.167	24.667
s2	3.000	9.500	15.500	7.000	11.500	27.500
s3	8.000	11.500	5.500	9.000	10.500	14.500
s4	6.333	10.333	11.333	8.667	10.667	21.667
s5	5.500	9.000	12.500	9.500	11.000	24.500
s6	6.333	9.833	11.833	12.667	14.167	26.167
s7	5.833	10.833	11.333	8.167	11.167	21.667
s8	6.167	11.167	12.667	8.833	11.833	23.333

Based on the model component of the fully repeated measures two-factor experimental design GLM equation, predicted scores are given by

$$\widehat{Y}_{ijk} = \mu + \pi_i + \alpha_j + \beta_k + (\pi\alpha)_{ij} + (\pi\beta)_{ik} + (\alpha\beta)_{jk} \tag{7.10}$$

Using the parameter estimates in this formula provides the predicted scores per subject per experimental conditions. The final parameters for the fully repeated measures two-factor experimental design GLM, the error terms, which represent the discrepancy between the actual scores observed (Table 7.2) and the scores predicted by the two-factor GLM (Table 7.3), are defined as

$$\varepsilon_{ijk} = Y_{ijk} - \widehat{Y}_{ijk} \tag{7.11}$$

and are presented by subject and experimental condition in Table 7.4.

Table 7.4 Error Terms for the Fully Repeated Measures Two-Factor (2 × 3) Experiment

	a1			a2		
	b1	b2	b3	b1	b2	b3
s1	0.167	−0.833	0.667	−0.167	0.833	−0.667
s2	−0.000	1.500	−1.500	0.000	−1.500	1.500
s3	−2.000	−2.500	4.500	2.000	2.500	−4.500
s4	−0.333	0.667	−0.333	0.333	−0.667	0.333
s5	−0.500	1.000	−0.500	0.500	−1.000	0.500
s6	1.667	0.167	−1.833	−1.667	−0.167	1.833
s7	0.167	0.167	−0.333	−0.167	−0.167	0.333
s8	0.833	−0.167	−0.667	−0.833	0.167	0.667
$\sum_{i=1}^{N} \varepsilon_{ijk}^2$	7.889	10.722	27.223	7.889	10.723	27.222
$\sum_{i=1}^{N} \sum_{j=1}^{p} \sum_{k=1}^{q} \varepsilon_{ijk}^2$			91.668			

Degrees of freedom are required next. For the subject effect

$$df\mathrm{s}_{\mathrm{subject}} = N - 1$$

This reflects how the subject effect is calculated from the deviation of N means from μ, which is the mean of these N means. Therefore, as described before, only $(N-1)$ of the component means are free to vary. As the subject factor has N levels, the subject × Factor A interaction effect *dfs* are given by

$$df_{\mathrm{subject} \times \mathrm{Factor\ A}} = (N-1)(p-1)$$

and the subject × Factor B interaction effect *dfs* are given by

$$df_{\mathrm{subject} \times \mathrm{Factor\ B}} = (N-1)(q-1)$$

For the error *dfs*, as a separate mean is employed in each experimental condition, a *df* is lost from the N scores of each condition. Moreover, a separate mean is employed to describe every set of p scores a subject provides, so for every set of p scores a *df* is lost, and similarly, a separate mean is employed to describe every set of q scores a subject provides, so for every set of q scores a *df* is lost. Therefore

$$df_{\mathrm{error}} = (N-1)(p-1)(q-1)$$

The error/interaction SS estimates, along with the SS for the experimental Factors A and B (calculation of which is identical to that described for the independent Factors A and B in Section 5.2.1) and the *dfs* are presented in an ANOVA Summary Table (Table 7.5). If hand calculation is employed or the statistical software employed does not output the required *p*-values, then the tabled critical *F*-values presented in Appendix B may be used to determine significance.

Table 7.5 Fully Repeated Measures Two-Factor ANOVA Summary Table

Source	Sum of Squares	*df*	MS	*F*	*p*
Subject	52.000	7	7.429		
Encoding instructions (A)	432.000	1	432.000	40.141	<0.001
Subjects × Encode instructions (S × A)	75.333	7	10.762		
Study time (B)	672.000	2	336.000	29.217	<0.001
Subjects × Study time (S × B)	161.000	14	11.500		
Encode instructions × Study time (A × B)	224.000	2	112.000	17.105	<0.001
Subjects × Encode instructions × Study time (S × A × B)	91.668	14	6.548		

7.3 ESTIMATING EFFECTS BY COMPARING FULL AND REDUCED EXPERIMENTAL DESIGN GLMs

The independent factors ANOVA was estimated by comparing full and reduced experimental design GLMs, so the hypotheses concerning the main effect of the repeated measures Factor A, the main effect of the repeated measures Factor B and the effect of the interaction between the repeated measures Factors A and B may be assessed by constructing three reduced GLMs, which manifest data descriptions under the respective null hypotheses, and comparing their error components with the full model. Again this approach is simplified by virtue of all the subject, experimental factors, and their interactions being orthogonal. As the effect estimates are completely distinct, omitting, or including any particular effect has no consequence for the estimates of the other effects.

The main effect of Factor A is assessed by constructing the reduced experimental design GLM

$$Y_{ijk} = \mu + \pi_i + \beta_k + (\pi\alpha)_{ij} + (\pi\beta)_{ik} + (\alpha\beta)_{jk} + \varepsilon_{ijk} \tag{7.12}$$

This model manifests the null hypothesis that the p levels of Factor A do not influence the data. More formally this is expressed as

$$\alpha_j = 0 \tag{7.13}$$

The main effect of Factor B is assessed by constructing the reduced experimental design GLM

$$Y_{ijk} = \mu + \pi_i + \alpha_j + (\pi\alpha)_{ij} + (\pi\beta)_{ik} + (\alpha\beta)_{jk} + \varepsilon_{ijk} \tag{7.14}$$

This model manifests the null hypothesis that the q levels of Factor B do not influence the data. More formally, this is expressed as

$$\beta_k = 0 \tag{7.15}$$

Finally, the reduced GLM for assessing the effect of the interaction between Factors A and B is

$$Y_{ijk} = \mu + \pi_i + \alpha_j + \beta_k + (\pi\alpha)_{ij} + (\pi\beta)_{ik} + \varepsilon_{ijk} \tag{7.16}$$

This reduced GLM manifests the data description under the null hypothesis that the interaction between the levels of Factors A and B do not influence the data and is expressed more formally as

$$(\alpha\beta)_{jk} = 0 \tag{7.17}$$

Nevertheless, when fully repeated measures two-factor ANOVAs are carried out by hand, the strategy of comparing different experimental design GLM residuals is very laborious, as there are so many reduced experimental design GLMs. In addition to the full experimental design GLM error term, reduced experimental design GLM error

Table 7.6 Formulas for the (Balanced) Fully Repeated Measures Two-Factor ANOVA Effects

Effect	Formula
Subject	$pq\sum_{i=1}^{N}(\overline{Y}_i-\overline{Y}_G)^2$
A	$qN\sum_{j=1}^{p}(\overline{Y}_i-\overline{Y}_G)^2$
S × A	$q\sum_{i=1}^{N}\sum_{j=1}^{p}(\overline{Y}_{ij}-\overline{Y}_i-\overline{Y}_j+\overline{Y}_G)^2$
B	$pN\sum_{k=1}^{q}(\overline{Y}_k-\overline{Y}_G)^2$
S × B	$p\sum_{i=1}^{N}\sum_{k=1}^{Q}(\overline{Y}_{ik}-\overline{Y}_i-\overline{Y}_j+\overline{Y}_G)^2$
A × B	$N\sum_{j=1}^{P}\sum_{k=1}^{q}(\overline{Y}_{ik}-\overline{Y}_j-\overline{Y}_k+\overline{Y}_G)^2$
A × B × S (or error)	$\sum_{i=1}^{N}\sum_{j=1}^{p}\sum_{k=1}^{q}(Y_{ijk}-\overline{Y}_{ij}-\overline{Y}_{ik}-\overline{Y}_{jk}+\overline{Y}+\overline{Y}_j+\overline{Y}_k-\overline{Y}_G)^2$

terms have to be calculated for each of the effects (A, B, and AB), and then to obtain the error terms for the main effect of Factor A (S × A), the main effect of Factor B (S × B) the interaction effect (S × A × B), further reduced experimental design GLMs must be constructed. Therefore, when hand calculations are employed, instead of calculating the error SS associated with each of these reduced experimental design GLMs and comparing them with the full experimental design GLM, it is more efficient to calculate directly the SS for each of the effects and errors. Formulas for calculating all of the fully repeated measures two-factor ANOVA effects directly, which are more convenient than those used to define and illustrate the SS calculation, are provided in Table 7.6. However, as described below, the strategy of comparing different experimental design GLM residuals to estimate fully repeated measures two-factor ANOVA effects is a simple way to implement related ANOVAs using regression GLMs.

7.4 REGRESSION GLMs FOR THE FULLY REPEATED MEASURES FACTORIAL ANOVA

The fully repeated measures two-factor (2 × 3) experimental design GLM equation (7.1) may be compared with the equivalent regression equation

$$\begin{aligned}Y_i = \beta_0 &+ \beta_1X_{i,1} + \beta_2X_{i,2} + \beta_3X_{i,3} + \beta_4X_{i,4} + \beta_5X_{i,5} + \beta_6X_{i,6} + \beta_7X_{i,7}\\ &+ \beta_8X_{i,8} + \beta_9X_{i,9} + \beta_{10}X_{i,10} + \beta_{11}X_{i,11} + \beta_{12}X_{i,12} + \beta_{13}X_{i,13} + \beta_{14}X_{i,14}\\ &+ \beta_{15}X_{i,15} + \beta_{16}X_{i,16} + \beta_{17}X_{i,17} + \beta_{18}X_{i,18} + \beta_{19}X_{i,19} + \beta_{20}X_{i,20} + \beta_{21}X_{i,21}\\ &+ \beta_{22}X_{i,22} + \beta_{23}X_{i,23} + \beta_{24}X_{i,24} + \beta_{25}X_{i,25} + \beta_{26}X_{i,26} + \beta_{27}X_{i,27} + \beta_{28}X_{i,28}\\ &+ \beta_{29}X_{i,29} + \beta_{30}X_{i,30} + \beta_{31}X_{i,31} + \beta_{32}X_{i,32} + \beta_{33}X_{i,33} + \varepsilon_i \qquad (7.18)\end{aligned}$$

where Y_i represents the ith dependent variable score (not the ith subject), β_0 is a constant, β_1 is the regression coefficient for the predictor variable X_1, and β_2 is the regression coefficient for the predictor variable X_2, and so on. As with the single factor repeated measures regression GLM, there are 7 variables that represent scores from individual subjects (from X_1 to X_7), 3 variables that represent experimental factors (from X_8 to X_{10}), and 21 variables that represent interactions between the subjects and the experimental factors (from X_{11} to X_{31}), and 2 variables that represent the interaction between the experimental factors (from X_{32}, X_{33}). Clearly, equation (7.18) is unwieldy and the earlier mention of the proliferation of predictor variables required for repeated measures designs can be appreciated. Nevertheless, once the effect coding scheme has been established in a computer data file, it is relatively simple to carry out the fully repeated measures factorial ANOVA. Effect coding applied to the data in Table 7.2 is presented in Table 7.7.

Applying a regression GLM to implement a fully repeated measures factors ANOVA may be done in a manner consistent with incremental analyses and estimating effects by comparing full and reduced GLMs. As all of the variables representing effects are orthogonal in a balanced design, the order in which SSs are estimated is of no consequence. The first regression carried out is that for the full fully repeated measures factorial experimental design GLM—when all subject and experimental condition predictor variables are included (variables X_1 to X_{33}). Although information about each of the predictor variables will be provided by linear regression software, as most of the experimental design effects are represented by two or more regression predictor variables, generally, information about the individual predictor coefficients, and so on, is of little interest. Of much more interest is the ANOVA summary presented in Table 7.8, which provides the full GLM residual SS. This may be compared with the fully repeated measures factorial experimental design GLM error term in Table 7.4.

Having obtained the full GLM residual SS, the next stages involve the implementation of the various reduced GLMs to obtain their estimates of residual SS. The reduced GLM for the effect of the subject factor is obtained by carrying out the regression analysis again, but omitting the predictors representing the (variables X_1 to X_7). The summary of this ANOVA, presented in Table 7.9, provides the subjects reduced GLM residual SS.

Therefore

			dfs
Subjects factor reduced GLM residual SS	=	143.667	21
Full GLM residual SS	=	91.667	14
SS attributable to subjects factor	=	52.000	7

The reduced GLM for the effect of Factor A is applied by omitting only the predictor representing the Factor A experimental conditions—variable X_8. The summary of this ANOVA, presented in Table 7.10, provides the Factor A reduced GLM residual SS.

Table 7.7 Effect Coding for a Fully Repeated Measures Two-Factor (2 × 3) Experimental Design

		S	A	B	S × A	S × B	A × B																											
s	Y	X_1	X_2	X_3	X_4	X_5	X_6	X_7	X_8	X_9	X_{10}	X_{11}	X_{12}	X_{13}	X_{14}	X_{15}	X_{16}	X_{17}	X_{18}	X_{19}	X_{20}	X_{21}	X_{22}	X_{23}	X_{24}	X_{25}	X_{26}	X_{27}	X_{28}	X_{29}	X_{30}	X_{31}	X_{32}	X_{33}
s1	7	1	0	0	0	0	0	0	1	1	0	1	0	0	0	0	0	0	1	0	0	0	0	0	0	0	0	0	0	0	0	0	1	0
s2	3	0	1	0	0	0	0	0	1	1	0	0	1	0	0	0	0	0	0	1	0	0	0	0	0	0	0	0	0	0	0	0	1	0
s3	6	0	0	1	0	0	0	0	1	1	0	0	0	1	0	0	0	0	0	0	1	0	0	0	0	0	0	0	0	0	0	0	1	0
s4	6	0	0	0	1	0	0	0	1	1	0	0	0	0	1	0	0	0	0	0	0	1	0	0	0	0	0	0	0	0	0	0	1	0
s5	5	0	0	0	0	1	0	0	1	1	0	0	0	0	0	1	0	0	0	0	0	0	1	0	0	0	0	0	0	0	0	0	1	0
s6	8	0	0	0	0	0	1	0	1	1	0	0	0	0	0	0	1	0	0	0	0	0	0	1	0	0	0	0	0	0	0	0	1	0
s7	6	0	0	0	0	0	0	1	1	1	0	0	0	0	0	0	0	1	0	0	0	0	0	0	1	0	0	0	0	0	0	0	1	0
s8	7	−1	−1	−1	−1	−1	−1	−1	1	1	0	−1	−1	−1	−1	−1	−1	−1	−1	−1	−1	−1	−1	−1	−1	0	0	0	0	0	0	0	1	0
s1	7	1	0	0	0	0	0	0	1	0	1	1	0	0	0	0	0	0	0	0	0	0	0	0	0	1	0	0	0	0	0	0	0	1
s2	11	0	1	0	0	0	0	0	1	0	1	0	1	0	0	0	0	0	0	0	0	0	0	0	0	0	1	0	0	0	0	0	0	1
s3	9	0	0	1	0	0	0	0	1	0	1	0	0	1	0	0	0	0	0	0	0	0	0	0	0	0	0	1	0	0	0	0	0	1
s4	11	0	0	0	1	0	0	0	1	0	1	0	0	0	1	0	0	0	0	0	0	0	0	0	0	0	0	0	1	0	0	0	0	1
s5	10	0	0	0	0	1	0	0	1	0	1	0	0	0	0	1	0	0	0	0	0	0	0	0	0	0	0	0	0	1	0	0	0	1
s6	10	0	0	0	0	0	1	0	1	0	1	0	0	0	0	0	1	0	0	0	0	0	0	0	0	0	0	0	0	0	1	0	0	1
s7	11	0	0	0	0	0	0	1	1	0	1	0	0	0	0	0	0	1	0	0	0	0	0	0	0	0	0	0	0	0	0	1	0	1
s8	11	−1	−1	−1	−1	−1	−1	−1	1	0	1	−1	−1	−1	−1	−1	−1	−1	0	0	0	0	0	0	0	−1	−1	−1	−1	−1	−1	−1	0	1
s1	8	1	0	0	0	0	0	0	1	−1	−1	1	0	0	0	0	0	0	−1	0	0	0	0	0	0	−1	0	0	0	0	0	0	−1	−1
s2	14	0	1	0	0	0	0	0	1	−1	−1	0	1	0	0	0	0	0	0	−1	0	0	0	0	0	0	−1	0	0	0	0	0	−1	−1
s3	10	0	0	1	0	0	0	0	1	−1	−1	0	0	1	0	0	0	0	0	0	−1	0	0	0	0	0	0	−1	0	0	0	0	−1	−1
s4	11	0	0	0	1	0	0	0	1	−1	−1	0	0	0	1	0	0	0	0	0	0	−1	0	0	0	0	0	0	−1	0	0	0	−1	−1
s5	12	0	0	0	0	1	0	0	1	−1	−1	0	0	0	0	1	0	0	0	0	0	0	−1	0	0	0	0	0	0	−1	0	0	−1	−1
s6	10	0	0	0	0	0	1	0	1	−1	−1	0	0	0	0	0	1	0	0	0	0	0	0	−1	0	0	0	0	0	0	−1	0	−1	−1
s7	11	0	0	0	0	0	0	1	1	−1	−1	0	0	0	0	0	0	1	0	0	0	0	0	0	−1	0	0	0	0	0	0	−1	−1	−1
s8	12	−1	−1	−1	−1	−1	−1	−1	1	−1	−1	−1	−1	−1	−1	−1	−1	−1	1	1	1	1	1	1	1	1	1	1	1	1	1	1	−1	−1

s1	16	1	0	0	0	0	0	0	-1	1	0	-1	0	0	0	0	0	0	1	0	0	0	0	0	0	0	0	0	0	0	0	0	-1	0
s2	7	0	1	0	0	0	0	0	-1	1	0	0	-1	0	0	0	0	0	0	1	0	0	0	0	0	0	0	0	0	0	0	0	-1	0
s3	11	0	0	1	0	0	0	0	-1	1	0	0	0	-1	0	0	0	0	0	0	1	0	0	0	0	0	0	0	0	0	0	0	-1	0
s4	9	0	0	0	1	0	0	0	-1	1	0	0	0	0	-1	0	0	0	0	0	0	1	0	0	0	0	0	0	0	0	0	0	-1	0
s5	10	0	0	0	0	1	0	0	-1	1	0	0	0	0	0	-1	0	0	0	0	0	0	1	0	0	0	0	0	0	0	0	0	-1	0
s6	11	0	0	0	0	0	1	0	-1	1	0	0	0	0	0	0	-1	0	0	0	0	0	0	1	0	0	0	0	0	0	0	0	-1	0
s7	8	0	0	0	0	0	0	1	-1	1	0	0	0	0	0	0	0	-1	0	0	0	0	0	0	1	0	0	0	0	0	0	0	-1	0
s8	8	-1	-1	-1	-1	-1	-1	-1	-1	1	0	1	1	1	1	1	1	1	-1	-1	-1	-1	-1	-1	-1	0	0	0	0	0	0	0	-1	0
s1	16	1	0	0	0	0	0	0	-1	0	1	-1	0	0	0	0	0	0	0	0	0	0	0	0	0	1	0	0	0	0	0	0	0	-1
s2	10	0	1	0	0	0	0	0	-1	0	1	0	-1	0	0	0	0	0	0	0	0	0	0	0	0	0	1	0	0	0	0	0	0	-1
s3	13	0	0	1	0	0	0	0	-1	0	1	0	0	-1	0	0	0	0	0	0	0	0	0	0	0	0	0	1	0	0	0	0	0	-1
s4	10	0	0	0	1	0	0	0	-1	0	1	0	0	0	-1	0	0	0	0	0	0	0	0	0	0	0	0	0	1	0	0	0	0	-1
s5	10	0	0	0	0	1	0	0	-1	0	1	0	0	0	0	-1	0	0	0	0	0	0	0	0	0	0	0	0	0	1	0	0	0	-1
s6	14	0	0	0	0	0	1	0	-1	0	1	0	0	0	0	0	-1	0	0	0	0	0	0	0	0	0	0	0	0	0	1	0	0	-1
s7	11	0	0	0	0	0	0	1	-1	0	1	0	0	0	0	0	0	-1	0	0	0	0	0	0	0	0	0	0	0	0	0	1	0	-1
s8	12	-1	-1	-1	-1	-1	-1	-1	-1	0	1	1	1	1	1	1	1	1	0	0	0	0	0	0	0	-1	-1	-1	-1	-1	-1	-1	0	-1
s1	24	1	0	0	0	0	0	0	-1	-1	-1	-1	0	0	0	0	0	0	-1	0	0	0	0	0	0	-1	0	0	0	0	0	0	1	1
s2	29	0	1	0	0	0	0	0	-1	-1	-1	0	-1	0	0	0	0	0	0	-1	0	0	0	0	0	0	-1	0	0	0	0	0	1	1
s3	10	0	0	1	0	0	0	0	-1	-1	-1	0	0	-1	0	0	0	0	0	0	-1	0	0	0	0	0	0	-1	0	0	0	0	1	1
s4	22	0	0	0	1	0	0	0	-1	-1	-1	0	0	0	-1	0	0	0	0	0	0	-1	0	0	0	0	0	0	-1	0	0	0	1	1
s5	25	0	0	0	0	1	0	0	-1	-1	-1	0	0	0	0	-1	0	0	0	0	0	0	-1	0	0	0	0	0	0	-1	0	0	1	1
s6	28	0	0	0	0	0	1	0	-1	-1	-1	0	0	0	0	0	-1	0	0	0	0	0	0	-1	0	0	0	0	0	0	-1	0	1	1
s7	22	0	0	0	0	0	0	1	-1	-1	-1	0	0	0	0	0	0	-1	0	0	0	0	0	0	-1	0	0	0	0	0	0	-1	1	1
s8	24	-1	-1	-1	-1	-1	-1	v1	-1	-1	-1	1	1	1	1	1	1	1	1	1	1	1	1	1	1	1	1	1	1	1	1	1	1	1

Table 7.8 ANOVA Summary Table for the Fully Repeated Measures Factorial Experimental Design GLM (Subjects and Experimental Condition Effects Regression)

Source	SS	*df*	MS	*F*	*p*
Regression	1616.333	33	48.980	7.481	<0.001
Residual	91.667	14	6.548		

R: 0.973; R^2: 0.946; adjusted R^2: 0.820.

Table 7.9 ANOVA Summary Table for the Reduced GLM that Omits the Subjects Factor

Source	SS	*df*	MS	*F*	*p*
Regression	1564.333	26	60.167	8.795	
Residual	143.667	21	6.841		

R: 0.957; R^2: 0.916; adjusted R^2: 0.812.

Table 7.10 ANOVA Summary Table for the Reduced GLM that Omits the Factor A Experimental Conditions

Source	SS	*df*	Mean Square	*F*	*p*
Regression	1184.333	32	37.010	1.060	0.470
Residual	523.667	15	34.911		

R: 0.833; R^2: 0.693; adjusted R^2: 0.039.

Therefore

			dfs
Factor A reduced GLM residual SS	=	523.667	15
Full GLM residual SS	=	91.667	14
SS attributable to Factor A	=	432.000	1

The subjects × Factor A reduced GLM is applied by omitting only the predictors representing the subjects × Factor A interaction—variables from X_{11} to X_{17}. The summary of this ANOVA, presented in Table 7.11, provides the subject × Factor A reduced GLM residual SS. Therefore

			dfs
Subject × Factor A interaction reduced GLM residual SS	=	167.000	21
Full GLM residual SS	=	91.667	14
SS attributable to subject × Factor A interaction	=	75.333	7

Next, the Factor B reduced GLM is applied by omitting only the predictors representing the Factor B experimental conditions—variables X_9 and X_{10}. The

Table 7.11 ANOVA Summary Table for the Reduced GLM that Omits the Subject × Factor A Interaction

Source	SS	*df*	Mean Square	*F*	*p*
Regression	1541.000	26	59.269	7.453	<0.001
Residual	167.000	21	7.952		

R: 0.950; R^2: 0.902; adjusted R^2: 0.781.

summary of this ANOVA, presented in Table 7.12, provides the Factor B reduced GLM residual SS. Therefore

			dfs
Factor B reduced GLM residual SS	=	763.667	16
Full GLM residual SS	=	91.667	14
SS attributable to Factor B	=	672.000	2

The subjects × Factor B reduced GLM is applied by omitting only the predictors representing the subjects × Factor B interaction—variables from X_{18} to X_{31}. The summary of this ANOVA, presented in Table 7.13, provides the subject × Factor B reduced GLM residual SS. Therefore

			dfs
Subject × Factor B interaction reduced GLM residual SS	=	252.667	28
Full GLM residual SS	=	91.667	14
SS attributable to subject × Factor B interaction	=	161.000	14

The Factor A × Factor B interaction reduced GLM is applied by omitting only the predictors representing the Factor A × Factor B interaction—variables from X_{32}

Table 7.12 ANOVA Summary Table for the Reduced GLM that Omits the Factor B Experimental Conditions

Source	SS	*df*	MS	*F*	*p*
Regression	944.333	31	30.462	0.638	0.862
Residual	763.667	16	47.729		

R: 0.744; R^2: 0. 553; adjusted R^2: 0.000.

Table 7.13 ANOVA Summary Table for the Reduced GLM that Omits the Subject × Factor B Interaction

Source	SS	*df*	MS	*F*	*p*
Regression	1455.333	19	76.596	8.488	<0.001
Residual	252.667	28	9.024		

R: 0.925; R^2: 0.852; adjusted R^2: 0.752.

Table 7.14 ANOVA Summary Table for the Reduced GLM that Omits the Factor A × Factor B Interaction

Source	SS	*df*	MS	*F*	*p*
Regression	1392.333	31	44.914	2.277	0.042
Residual	315.667	16	19.729		

R: 0.903; R^2: 0.815; adjusted R^2: 0.457.

and X_{33}. The summary of this ANOVA presented in Table 7.14 provides the Factor A × Factor B reduced GLM residual SS. Therefore

			dfs
Factors A × B interaction reduced GLM residual SS	=	315.667	16
Full GLM residual SS	=	91.667	14
SS attributable to Factor A × Factor B interaction	=	224.000	2

Using the SS and *dfs* calculated for each effect by comparing full and reduced GLMs, the ANOVA summary Table 7.5 can be reconstructed.

7.5 EFFECT SIZE ESTIMATION

The more complicated nature of effect size estimation with repeated measures measures designs becomes particularly apparent with fully factorial repeated measures designs. Effect size estimates that may be compared across independent and repeated measures designs also are required for fully factorial repeated measures designs and it is that to which attention is turned first.

7.5.1 A Complete $\widehat{\omega}^2$ SOA for Main and Interaction Omnibus Effects Comparable Across Repeated Measures and Independent Designs

In Section 5.5.1, the omnibus ω^2 for a factorial independent measures ANOVA GLM was defined by equation (5.29) in terms of the number of experimental conditions, the F-statistic, and the number of subjects per condition (in balanced designs). Applying this equation to the data in Table 7.5 provides the complete $\widehat{\omega}^2$ SOA for the main and interaction omnibus effects.

$$\widehat{\omega}^2_{\mathrm{A}} = \frac{1(40.141-1)}{1(40.141-1)+2(29.217-1)+2(17.105-1)+48} = \frac{39.141}{175.785} = 0.22$$

$$\widehat{\omega}^2_{\mathrm{B}} = \frac{2(29.217-1)}{1(40.141-1)+2(29.217-1)+2(17.105-1)+48} = \frac{56.434}{175.785} = 0.32$$

$$\widehat{\omega}^2_{\mathrm{A\times B}} = \frac{2(17.105-1)}{1(40.141-1)+2(29.217-1)+2(17.105-1)+48} = \frac{32.210}{175.785} = 0.18$$

7.5.2 A Partial $\widehat{\omega}^2$ SOA for the Main and Interaction Omnibus Effects Appropriate for Repeated Measures Designs

As Keppel and Wickens (2004) describe, there is no difficulty defining a partial $\widehat{\omega}^2$ SOA for main and interaction omnibus effects in terms of population parameters. For example, partial $\widehat{\omega}^2$ SOA for main and interaction omnibus effects can be defined as

$$\begin{aligned} \omega^2_{\langle \mathrm{A} \rangle} &= \frac{\sigma^2_{\text{expt. effect}}}{\sigma^2_{\mathrm{A}} + \sigma^2_{\mathrm{A \times S}} + \sigma^2_{\text{error}}} \\ \omega^2_{\langle \mathrm{B} \rangle} &= \frac{\sigma^2_{\mathrm{B}}}{\sigma^2_{\mathrm{B}} + \sigma^2_{\mathrm{B \times S}} + \sigma^2_{\text{error}}} \\ \omega^2_{\langle \mathrm{A \times B} \rangle} &= \frac{\sigma^2_{\mathrm{A \times B}}}{\sigma^2_{\mathrm{A \times B}} + \sigma^2_{\mathrm{A \times B \times S}} + \sigma^2_{\text{error}}} \end{aligned} \tag{7.19}$$

However, complications and difficulty arise when an attempt is made to estimate these population parameters. Essentially, the problem is the GLM used for fully repeated measures designs does not provide a specific estimate of σ^2_{error} because it incorporates experimental factor and subjects interactions (i.e., $\mathrm{A \times S}$, $\mathrm{B \times S}$, and $\mathrm{A \times B \times S}$). The ANOVA circumvents this problem by using the subject interaction terms as the F-test denominators. Nevertheless, as equations (7.19) make clear, this solution cannot work for the partial $\widehat{\omega}^2$ estimates because both interactions and error term estimates are required to calculate the partial $\widehat{\omega}^2$ estimates. Dodd and Schultz (1973) point out that a range can be specified for each main effect partial $\widehat{\omega}^2$ estimate, but usually these ranges are a bit too large to be of great practical value. For example

$$\widehat{\omega}^2_{\langle A \rangle} \text{ ranges between } \frac{df_A(F_{\mathrm{A}} - 1)}{df(F_{\mathrm{A}} - 1) + pqN} \text{ and } \frac{df_A(F_{\mathrm{A}} - 1)}{df(F_{\mathrm{A}} - 1) + pN}$$

Applied to the data in Table 7.5, this provides a partial $\widehat{\omega}^2$ between 0.45 and 0.71.

$$\widehat{\omega}^2_{\langle B \rangle} \text{ ranges between } \frac{df_{\mathrm{B}}(F_{\mathrm{B}} - 1)}{df(F_{\mathrm{B}} - 1) + pqN} \text{ and } \frac{df_{\mathrm{B}}(F_{\mathrm{B}} - 1)}{df(F_{\mathrm{B}} - 1) + qN}$$

Applied to the data in Table 7.5, this provides a partial $\widehat{\omega}^2$ between 0.54 and 0.71. However, as the nature of the ranges above suggest, a specific value can be obtained for the interaction partial $\widehat{\omega}^2$ estimate

$$\widehat{\omega}^2_{\langle \mathrm{A \times B} \rangle} = \frac{df_{\mathrm{A \times B}}(F_{\mathrm{A \times B}} - 1)}{df(F_{\mathrm{A \times B}} - 1) + pqN}$$

Applied to the data in Table 7.5 provides a partial $\widehat{\omega}^2 = 0.40$.

7.5.3 A Partial $\widehat{\omega}^2$ SOA for Specific Comparisons Appropriate for Repeated Measures Designs

As with independent designs, there will be interest in the SOA for particular comparisons between experimental conditions in repeated measures designs. For example, the SOA between the memorization 30 and 180 s conditions in the hypothetical fully repeated measures two-factor study time experiment again is of interest.

The partial ω^2 for a specific comparison expresses the comparison variance as a proportion of only the specific comparison variance plus error variance. For fully repeated measures two-factor designs, the partial ω^2 for specific comparisons can be defined as

$$\widehat{\omega}_{\psi}^2 = \frac{F_{\psi} - 1}{F_{\psi} - 1 + 2N} \tag{7.20}$$

Given the F_{Ψ} for the memorization 30 and 180 s conditions = 21.212, applying equation (7.20) provides

$$\widehat{\omega}_{\psi}^2 = \frac{21.212 - 1}{21.212 - 1 + 2(8)} = \frac{20.212}{36.212} = 0.56$$

As might be expected given the nature of the partial ω^2 for a specific comparison and its application to the same experimental data, exactly the same result is obtained here as was obtained with the design and the analysis presented in Section 6.6.3. Fifty-six percent of the variance in the 30 and 180 s populations is explained by the comparison between these two experimental conditions.

7.6 FURTHER ANALYSES

7.6.1 Main Effects: Encoding Instructions and Study Time

As described in Section 5.6.1, interpreting the main effect of the encoding instructions factor in terms of mean differences is simple due to there being only two levels or conditions of this factor and so only two means that can be unequal. No further test need be applied and all that remains to be done is to determine the direction of the effect by identifying the encoding instruction levels with the larger and smaller means, ideally by plotting these means on a graph (see Figure 5.1). All that was said about planned and unplanned comparisons with regard to the independent factorial design in Section 5.6.1 also applies here.

As described for the independent factorial design, interpreting the main effect of the *study time* factor in terms of mean differences is slightly more complicated and is carried out in a different fashion with repeated measures designs. First, as this factor has three levels, the pertinent unequal (marginal) means may be any one or more of, b1 versus b2; b2 versus b3; and b1 versus b3, and non pairwise differences also may contribute to the significant main effect, so further tests are required to identify exactly which means differ. Plotting pertinent means on a graph is an extremely useful tool in interpreting the experimental data (see Figure 5.2).

In terms of analysis strategy, the approach to the repeated measures design does not differ from the approach described for the two-factor independent design. However, although the same hypotheses are tested and the same approach to Type 1 error rate control for planned and unplanned comparisons are applied in independent and related designs, different statistical analyses are motivated by concerns about the consequences of an untenable sphericity assumption. As described in Section 6.8, the approach adopted for most further analyses with related designs is to analyze the specific comparisons of interest separately.

Table 5.10 presents the marginal means for the three study times averaged over the memorization and story and imagery conditions. Further analysis of the main effect of the 30, 60, and 180 s Factor B study times proceeds as if each pairwise comparison constituted a separate experiment. Therefore, a single factor repeated measures ANOVA is applied to all of the data pertinent to the comparison between the 30 and 60 s conditions, a separate single factor repeated measures ANOVA is applied to all of the data pertinent to the comparison between the 30 and 180s conditions, and a separate single factor repeated measures ANOVA is applied to all of the data pertinent to the comparison between the 60 and 180 s conditions.

Planned or unplanned comparison sum of squares ($SS_{\psi_{pc}}$) calculation in the fully repeated measures design is identical to that described for other designs. Again, it is assumed that the planned comparison involves the 30 and 180 s study time conditions. The linear contrast for this planned comparison expressed in terms of population means (see Section 3.4) is

$$\psi = (-1)\mu_1 + (0)\mu_2 + (1)\mu_3$$

Replacing the population means with the sample mean estimators from Table 5.10 provides

$$\psi = (-1)8 + (0)11 + (1)17 = 9$$

However, the formulas for the repeated measures design $SS_{\psi_{pc}}$ employ the number of subjects participating in each condition, which is also the total number of subjects participating in the whole experiment. Therefore,

$$SS\widehat{\psi} = \frac{N\widehat{\psi}^2_{pc}}{\sum c_j^2} = \frac{(8)(9)^2}{(-1)^2 + (0)^2 + (1)^2} = \frac{648}{2} = 324.000$$

One df is associated with $SS_{\widehat{\psi}}$, so the mean square for the contrast is

$$MS_{\widehat{\psi}} = \frac{SS_{\psi_{pc}}}{1} = \frac{324}{1} = 324.000$$

As before, the next requirement for the planned comparison F-test is the error term pertinent to the separate and specific comparison of the 30 and 60 s study time conditions. To calculate this error term, the procedure applied to obtain the specific

Table 7.14 (a) Data and Means of Memorize (M) and Story and Imagery (S&I) 30 and 180 s Experimental Conditions.

	M-30 s	S&I-30 s	30 s $\overline{X}$	M-180 s	S&I-180 s	180 s $\overline{X}$
s1	7	16	11.5	8	24	16.0
s2	3	7	5.0	14	29	21.5
s3	6	11	8.5	10	10	10.0
s4	6	9	7.5	11	22	16.5
s5	5	10	7.5	12	25	18.5
s6	8	11	9.5	10	28	19.0
s7	6	8	7.0	11	22	16.5
s8	7	8	7.5	12	24	18.0

(b) Means of Memorize (M) and Story and Imagery (S&I) 30 and 180 s Experimental Conditions and their Difference Scores

	180 s $\overline{X}$	30 s $\overline{X}$	180 s $\overline{X}$ – 30 s $\overline{X}$
s1	16.0	11.5	4.5
s2	21.5	5.0	16.5
s3	10.0	8.5	1.5
s4	16.5	7.5	9.0
s5	18.5	7.5	11.0
s6	19.0	9.5	9.5
s7	16.5	7.0	9.5
s8	18.0	7.5	10.5
$\sum \overline{\psi}_i$	136	64	72
$\overline{\psi}_i$	17	8	9
$\sum \widehat{\psi}_i^2$			787.5

comparison error term in the single factor repeated measures design is applied, after a slight modification. The modification is to take an average of each subjects' scores over the factor levels not involved in the comparison (i.e., the memorization and story and imagery instruction conditions). This is illustrated in Table 7.14a. The lower part of the table (Table 7.14b) presents these mean scores and applies exactly those procedures applied with respect to the single factor repeated measures design. Essentially, these procedures provide a shortcut method of obtaining estimates from a single factor repeated measures ANOVA GLM applied to the means of the 30 s condition means and the 180 s condition means.

The SS for the planned comparison between the 30 and 180s experimental conditions, $SS_{\psi_{pc}}$, calculated above was 324. Therefore

$$\text{SSE}_{\text{FGLM}} = \frac{\sum \widehat{\psi}_i^2}{\sum c_j^2} - SS\widehat{\psi} = \frac{787.5}{-1^2 + 1^2} - 324 = 69.750$$

Table 7.15 ANOVA Summary Table for the Planned Pairwise Main Effect Comparison Between the 30 and 180 s Experimental Condition Means

Source	SS	*df*	MS	*F*	*p*
30 s versus 180 s	324.000	1	324.000	32.517	<0.001
Error	69.750	7	9.964		

The planned pairwise comparison *dfs* $= (p - 1)$ and the error term *dfs* $= (N_j - 1)(p - 1)$. As before, it is convenient to lay all of this information out in an ANOVA summary table and complete the *F*-test calculation, as presented in Table 7.15. The 30 versus 180 s experimental condition means comparison was planned, so no Type 1 error adjustment is necessary and the difference between the subjects' free recalls after 30 s and 180 s is declared significant.

Identical calculations can be carried out to assess the two unplanned study time main effect comparisons involving the 30 and 60 s, and the 60 and 180 s experimental condition means. All of the issues discussed with respect to the planned and the unplanned comparisons conducted with respect to the single factor repeated measures design apply to the this fully repeated measures two-factor analysis. It is worth noting that breaking the two-factor experiment up into separate pairs of conditions for analysis results in the same loss of *dfs* from the separate comparison error terms as in the single factor case. Only 7 *dfs* are associated with each of the separate comparison error terms, whereas 14 *dfs* are associated with the S × B error term in the omnibus repeated measures ANOVA. This reduction in error *dfs* is likely to diminish the power of these separate comparisons.

7.6.2 Interaction Effect: Encoding Instructions × Study Time

The aims of data analysis in factorial repeated measures designs are the same as those when factorial independent designs are analyzed. The factorial experimental data breaks down in a similar fashion to that described for the independent design (see Section 5.6.2) and the choice of simple effects analysis again depends on the research issues and experimental hypotheses to be assessed, with the result that usually only one set of simple effects analyses are applied.

7.6.2.1 Simple Effects: Comparison of Differences Between the Three Levels of Factor B (Study Time) at Each Level of Factor A (Encoding Instructions)

The three levels of Factor B, study time, are examined at each level of Factor A, encoding instructions. Therefore, each simple effect analysis is actually a single factor repeated measures ANOVA applied to the three study time conditions (30, 60, and 180 s). One such simple effect single factor repeated measures ANOVA is applied to the scores obtained from three study time conditions under memorization instructions and another simple effect single factor repeated measures ANOVA is applied to the scores obtained from three study time conditions under story and imagery instructions (see Figure 5.4). It is worth emphasizing that the relevant SS calculations are for a single factor repeated measures ANOVA and not for a specific comparison that employs a linear contrast.

As each simple effect analysis is a single factor repeated measures ANOVA, the SS calculations for the study time effect will be identical to those described for the single factor repeated measures ANOVA in Section 6.4.

$$\text{Experimental conditions SS} = \sum_{j=1}^{p} N_j(\mu_j - \mu)^2 \qquad (2.28, \text{rptd})$$

Therefore, the SS for the effect of study time under memorization (i.e., when $j = 1$) is

$$SS_{B\ \mathrm{a}j=1} = \sum 8(6-9)^2 + 8(10-9)^2 + 8(11-9)^2 = 112$$

and the SS for the effect of study time under story and imagery (i.e., when $j = 2$) is

$$SS_{B\ \mathrm{a}j=2} = \sum 8(10-15)^2 + 8(12-15)^2 + 8(23-15)^2 = 784$$

Unlike the fully independent measures two-factor design, the omnibus error term is not employed due to concerns about the tenability of the sphericity assumption and so, it is necessary to calculate a separate error term based on only the data involved in each of the single factor ANOVAs. Again a quick way to obtain the SS error per ANOVA is to apply the procedures outlined in Table 7.14b. The calculation of the ANOVA error SS for the study times with memorization is presented in Table 7.16 and below.

$$\frac{\sum_{k=1}^{q} \sum \widehat{\psi}_i^2}{q} = \frac{168 + 266 + 16}{3} = 150.000$$

Table 7.16 Calculation of Error SS for Single Factor ANOVA for Study Time with Memorization

Memorization	30 s	60 s	180 s	60 – 30	180 – 30	180 – 60
s1	7	7	8	0.000	1.000	1.000
s2	3	11	14	8.000	11.000	3.000
s3	6	9	10	3.000	4.000	1.000
s4	6	11	11	5.000	5.000	0.000
s5	5	10	12	5.000	7.000	2.000
s6	8	10	10	2.000	2.000	0.000
s7	6	11	11	5.000	5.000	0.000
s8	7	11	12	4.000	5.000	1.000
$\sum \psi_i$				32.000	40.000	8.000
$\overline{\psi}_i$				4.000	5.000	1.000
$\sum \widehat{\psi}_i^2$				168.000	266.000	16.000

Table 7.17 Calculation of Error SS for Single Factor ANOVA for Study Time with Story and Imagery

Story and Imagery	30 s	60 s	180 s	60 s– 30 s	180 s– 30	180 – 60
s1	16	16	24	0.000	8.000	8.000
s2	7	10	29	3.000	22.000	19.000
s3	11	13	10	2.000	−1.000	−3.000
s4	9	10	22	1.000	13.000	12.000
s5	10	10	25	0.000	15.000	15.000
s6	11	14	28	3.000	17.000	14.000
s7	8	11	22	3.000	14.000	11.000
s8	8	12	24	4.000	16.000	12.000
$\sum \psi_i$				16.000	104.000	88.000
$\overline{\psi}_i$				2.000	26.000	11.000
$\sum \widehat{\psi}_i^2$				48.000	1684.000	1264.000

Therefore, the error SS for the single factor repeated measures ANOVA GLM applied to the data obtained from the three study time conditions with memorization encoding instructions is

$$\text{SSE}_{\text{FGLM}} = 150 - 112 = 38.000$$

Corroboration of the error SS is provided by the equivalent error SS reported in Tables 6.9 and 6.12, when the study time data with memorization instructions was conceived and analyzed as a single factor repeated measures ANOVA.

For the story and imagery instructions ANOVA error SS, see Table 7.17.

$$\frac{\sum_{k=1}^{q} \sum \widehat{\psi}_i^2}{q} = \frac{48 + 1684 + 1264}{3} = 998.667$$

Therefore, the error SS for the single factor repeated measures ANOVA GLM applied to the data obtained from the three study time conditions with story and imagery instructions is

$$\text{SSE}_{FGLM} = 998.667 - 784 = 214.667$$

Table 7.18 presents all of this information in the form of an ANOVA summary table. If further analysis of the simple effect single factor repeated measures ANOVA is decided upon, then the procedures described in Section 6.8 for the further analysis of single repeated measures ANOVAs are appropriate.

7.6.2.2 Simple Effects: Comparison of Differences Between the Two Levels of Factor A (Encoding Instructions) at Each Level of Factor B (Study Time)

The simple effect analyses of the effect of encoding instruction at each of the three study times (30 s, 60 s, and 180 s) involve pairwise comparisons between the two

Table 7.18 ANOVA Summary Table for the Simple Effect of Factor B at Each Level of Factor A

Source	SS	*df*	MS	*F*	*p*
At a1: Memorize					
Study time	112.000	2	56.000	20.634	<0.001
Error	38.000	14	2.714		
At a2: Story and imagery					
Study time	784.000	2	392.000	25.566	<0.001
Error	214.667	14	15.333		

levels of Factor A (memorization and story and imagery). Each of these comparisons can be conceived as a related *t*-test, a single factor repeated measures ANOVA, or as a linear contrast between two experimental condition means. For consistency, the latter conception is applied. The simple linear contrast for the comparisons is

$$\psi_{\text{A at bk}} = (-1)\mu_1 + (1)\mu_2$$

where μ_1 represents the population mean of subjects' scores with memorization and μ_2 represents the population mean of subjects' scores with story and imagery. Substituting the population parameters with the sample mean estimators for the 30 s experimental conditions from Table 7.2 provides

$$\psi_{\text{A at b1}} = (-1)6 + (1)10 = 4$$

However, the formulas for the repeated measures design $\text{SS}_{\Psi_{\text{A at b}k}}$ employs the number of subjects participating in each condition, and the total number of subjects participating in the experiment. Therefore

$$\text{SS}_{\text{A at b1}} = \frac{N\widehat{\psi}^2_{\text{A at b1}}}{\sum c_j^2} = \frac{(8)(4)^2}{(-1)^2 + (1)^2} = \frac{128}{2} = 64.000$$

$$\text{SS}_{\text{A at b2}} = \frac{N\widehat{\psi}^2_{\text{A at b2}}}{\sum c_j^2} = \frac{(8)(12-10)^2}{(-1)^2 + (1)^2} = \frac{32}{2} = 16.000$$

$$\text{SS}_{\text{A at b3}} = \frac{N\widehat{\psi}^2_{\text{A at b2}}}{\sum c_j^2} = \frac{(8)(23-11)^2}{(-1)^2 + (1)^2} = \frac{1152}{2} = 576.000$$

One *df* is associated with each $\text{SS}_{\Psi_{\text{Aat b}k}}$, so the mean square for the SS_{Aatb1} contrast is

$$\text{MS} = \frac{\text{SS}_{\psi_{\text{A at b1}}}}{1} = \frac{324}{1} = 324.000$$

The next requirement is specific error terms for each of the three pairwise comparisons. Again, a quick way to obtain these error terms is to apply the procedures described previously. The first step in calculating the error terms is to calculate the sum of the squared differences between each subjects' story and imagery and memorization scores with 30 s study time. Table 7.19 presents the data required for these calculations.

The error SS for each of the comparisons between the two levels of the encoding instructions factor: story and imagery and memorization are obtained by subtraction

$$\mathrm{SSE}_{\mathrm{FGLM}} = \frac{\sum \widehat{\psi}_i^2}{\sum c_j^2} - \mathrm{SS}_{\mathrm{A\,at\,bk}}$$

Table 7.19 Differences Between Subjects' Scores Across the Story and Imagery Encoding Instruction Experimental Conditions at 30 s, 60 s, and 180 s Study Time

	30 s Study Time		
	S&I	M	S&I − M ($\widehat{\psi}_i$)
s1	16	7	9
s2	7	3	4
s3	11	6	5
s4	9	6	3
s5	10	5	5
s6	11	8	3
s7	8	6	2
s8	8	7	1
$\sum \psi_i$			32
$\overline{\psi}_i$			4
$\sum \widehat{\psi}_i^2$			170

	60 s Study Time		
	S&I	M	S&I − M ($\widehat{\psi}_i$)
s1	16	7	9
s2	10	11	−1
s3	13	9	4
s4	10	11	−1
s5	10	10	0
s6	14	10	4
s7	11	11	0
s8	12	11	1
$\sum \psi_i$			16
$\overline{\psi}_i$			2
$\sum \widehat{\psi}_i^2$			116

Table 7.19 *(Continued)*

	180 s Study Time		
	S&I	M	S&I − M ($\widehat{\psi}_i$)
s1	24	8	16
s2	29	14	15
s3	10	10	0
s4	22	11	11
s5	25	12	13
s6	28	10	18
s7	22	11	11
s8	24	12	12
$\sum \psi_i$			96
$\overline{\psi}_i$			12
$\sum \widehat{\psi}_i^2$			1360

Therefore, the full GLM error $SS_{\text{A at b1}} = \frac{170}{2} - 64 = 21$, the full GLM error $SS_{\text{A at b2}} = \frac{116}{2} - 16 = 42$, and the full GLM error $SS_{A \text{ at b3}} = \frac{1360}{2} - 576 = 104$.

Table 7.20 presents all of this information in the form of an ANOVA summary table. The significance of planned comparisons is assessed without Type 1 error rate adjustment, whereas unplanned comparisons are assessed with an appropriate Type 1 error rate adjustment. With regard to the unplanned comparison Type 1 error rate adjustment, it should be kept in mind that the prior interaction indicates that at least one of the pairwise comparisons described in Table 7.20 is significant. Therefore, only one of the three pairwise comparisons possibly could be nonsignificant and so, Type 1 error rate needs to controlled over only one null hypothesis (see Table 3.6).

Table 7.20 ANOVA Summary Table for the Simple Effect of Factor B at Each Level of Factor A

Source	SS	*df*	MS	*F*	*p*
At b1 (30 s)					
Encoding instructions	64.000	1	64.000	21.333	0.002
Error	21.000	7	3.000		
At b2 (60 s)					
Encoding instructions	16.000	1	16.000	2.667	0.146
Error	42.000	7	6.000		
At b3 (180 s)					
Encoding instructions	576.000	1	576.000	38.769	<0.001
Error	104.000	7	14.857		

7.7 POWER

To conduct power analysis to determine sample size for two-factor fully repeated measures designs using tables, each omnibus main effect and interaction effect, as well as specific comparison effects are assessed separately. Therefore, the procedures described for the single factor repeated measures design in Section 6.9 also apply to factorial fully repeated measures designs. However, more accurate results will be obtained if one of the statistical software packages available for this purpose is employed. For power analysis of fully repeated measures two-factor measures designs and all other such designs, the free statistical software package G*Power 3 (Faul et al., 2007) is recommended highly.

CHAPTER 8

GLM Approaches to Factorial Mixed Measures Designs

8.1 MIXED MEASURES AND SPLIT-PLOT DESIGNS

Any design that combines at least one fixed factor and at least one random factor can be labelled a mixed design (see Section 1.7). Generally, when the mixed measures design label is applied there is usually theoretical or practical interest in the effects of the fixed factors and the random factors. However, when mixed measures designs (and other randomized block designs) are applied in psychology, usually there is only one random factor – the blocking or subject factor – and very rarely is there any interest in its effect. Typically, theoretical and practical interest is confined to the fixed factor effects and the sole purpose of the random factor is to accommodate the relations between the two or more dependent variable measures provided by the same or matched subjects over study conditions.

The type of mixed design employed most frequently in psychology often is labelled a split-plot design. In common with the majority of the experimental designs and analyses employed in psychology and reported in this text, split-plot designs (the first mixed designs) and their analyses were developed originally by Fisher (1925) for use in agricultural experiments – separate plots of land were split into sections which received different treatment levels of one factor, while whole plots received different treatment levels of another factor.

Figure 8.1 presents a two factor (2×3) split-plot design appropriate for a psychology experiment. Factor A is the independent measures factor – different subjects experience the two levels of this factor – and Factor B is the repeated measures factor – the same subjects experience the different levels of this factor. In psychology experiments, the subjects take the place of the plots of land. The particular nature of these "split-plot" designs should not be forgotten, but from this point forward, the more generally applicable mixed measures design label will be applied.

ANOVA and ANCOVA: A GLM Approach, Second Edition. By Andrew Rutherford.

Factor A	a1			a2		
Factor B	b1	b2	b3	b1	b2	b3
	s1	s1	s1	s9	s9	s9
	s2	s2	s2	s10	s10	s10
	⋮	⋮	⋮	⋮	⋮	⋮
	s8	s8	s8	s16	s16	s16

Figure 8.1 A two-factor (2 × 3) mixed measures/split-plot design.

8.2 FACTORIAL MIXED MEASURES DESIGNS

Consider the fully repeated measures factorial experiment described in Chapter 7. It was suggested in Section 7.2 that the fully repeated measures factorial design could produce a contrast effect. In such circumstances, the factorial mixed measures design presented in Figure 8.1 would be a good alternative. A mixed measures design would allow different groups of subjects to receive memorize and story and imagery instructions. This eliminates the possibility of subjects continuing to apply a story and imagery encoding strategy in memorize conditions, but retains the advantages of measuring the same subjects' performance under all of the study times. Table 8.1 presents the hypothetical experimental data as if they had been obtained from the two-factor mixed measures design just outlined.

The GLM for the two-factor mixed measures design ANOVA is described by the equation

$$Y_{ijk} = \mu + \alpha_j + \beta_k + \pi_{i(j)} + (\alpha\beta)_{jk} + (\pi\beta)_{ik(j)} + \varepsilon_{ijk} \tag{8.1}$$

where Y_{ijk} is the dependent variable score for the ith subject at the jth level of Factor A and the kth level of Factor B, μ is the general mean of the experimental

Table 8.1 Experimental Data from a Two-Factor (2 × 3) Mixed Measures Design

Encoding Instructions	a1 Memorize					a2 Story and Imagery			
	b1	b2	b3	Subject		b1	b2	b3	Subject
Study Time	30 s	60 s	180 s	$\overline{X}$		30 s	60 s	180 s	$\overline{X}$
s1	7	7	8	7.333	s9	16	16	24	18.667
s2	3	11	14	9.333	s10	7	10	29	15.333
s3	6	9	10	8.333	s11	11	13	10	11.333
s4	6	11	11	9.333	s12	9	10	22	13.667
s5	5	10	12	9.000	s13	10	10	25	15.000
s6	8	10	10	9.333	s14	11	14	28	17.667
s7	6	11	11	9.333	s15	8	11	22	13.667
s8	7	11	12	10.000	s16	8	12	24	14.667
$\overline{X}$	6	10	11	9	$\overline{X}$	10	12	23	15

condition population means, α_j is the effect of the jth level of Factor A, β_k is the effect of the kth level of Factor B, $\pi_{i(j)}$ is a parameter representing the random effect of the ith subject within the jth level of Factor A, $(\alpha\beta)_{jk}$ is the interaction effect of the jth level of Factor A and the kth level of Factor B, $(\beta\pi)_{ki(j)}$ is the interaction effect of the kth level of Factor B and the ith subject within the jth level of Factor A, and, as always, ε_{ijk} represents the random error associated with the ith subject in the jth level of Factor A and the kth level of Factor B. Although the subjects factor is crossed with the levels of Factor B, all subjects receive all levels of Factor B—the use of brackets around the subscript j indicates that these effects involve the scores of subjects' nested within the p levels of Factor A (i.e., separate groups of subjects are employed in each of the p levels of Factor A).

As described in Section 6.3, related designs conceive of each subject as a level of a random factor—subject. In the two-factor mixed design, different (groups of) subjects experience each level of the encoding instructions independent factor. Therefore, different subjects (i.e., different levels of the subject factor) are nested within the encoding instructions factor, rather than being crossed with this independent factor. As a result, the interaction between encoding instructions and subjects $(\pi\alpha)_{ij}$ cannot be estimated and so there is no term in the GLM representing this interaction.

Experimental effects in mixed factorial designs are assessed using fewer error terms than their equivalent fully related measures factorial designs, but they employ more error terms than equivalent independent measures factorial designs. Mixed designs may be conceived as separate related designs nested within each of the levels of the independent factors. In the example presented in Table 8.1, the two independent factor levels, a1 and a2, each comprize a single factor related measures design. As each subject provides only one score in each level of the related factor, it is not possible to separate error from the subject and the related factor interaction. Consequently, a more accurate version of the ε_{ijk} term would be $[\varepsilon_{ijk} + (\pi\beta)_{ik(j)}]$. However, in common with fully repeated measures factorial designs, $[\varepsilon_{ijk} + (\pi\beta)_{ik(j)}]$ frequently is referred to as $(\pi\beta)_{ijk}$, or (S×B).

Consider the current two-factor mixed design analyzed as two separate single factor related designs, as presented in Table 8.2. (Apart from the subjects SS, these two separate single factor related measures ANOVAs were calculated as the two simple effect analyses applied to examine the two-factor fully related measures interaction, see Table 7.18.) Comparing these two single factor related measures ANOVAs (Table 8.2) with the two-factor mixed measures ANOVA (Table 8.5) reveals that in balanced designs, both of the mixed measures ANOVA SS error terms are simply the sums of the separate single factor related measures ANOVA SS for subject effects and SS error terms: The sum of the SS subject effects provides the error term for the independent factor ($14 + 113.333 = 127.333$), while the sum of the separate error terms provides the error term for the related factor ($38 + 214.667 = 252.667$). (The small discrepancies are due to rounding error.) However, this is not the case in unbalanced designs, as the error terms are pooled weighted averages (weighted by $N_j - 1$).

Table 8.2 Two-Factor (2 × 3) Mixed Measures Experimental Data Cast as Two Single Factor Related Measures Designs

Encoding Instructions		Study Time		
		b1 30 s	b2 60 s	b3 180 s
	s1	7	7	8
	s2	3	11	14
	s3	6	9	10
Memorize (a1)	s4	6	11	11
	s5	5	10	12
	s6	8	10	10
	s7	6	11	11
	s8	7	11	12

Source	SS	*df*	MS	*F*	*p*
Memorize (a1)					
Subject	14	7			
Study time	112.000	2	56.000	20.634	<0.001
Error	38.000	14	2.714		

Encoding Instructions		Study Time		
		b1 30 s	b2 60 s	b3 180 s
	s9	16	16	24
	s10	7	10	29
	s11	11	13	10
Story and imagery (a2)	s12	9	10	22
	s13	10	10	25
	s14	11	14	28
	s15	8	11	22
	s16	8	12	24

Source	SS	*df*	MS	*F*	*p*
Story and imagery (a2)					
Subject	113.333	7			
Study time	784.000	2	392.000	25.566	<0.001
Error	214.667	14	15.333		

One single factor repeated measures ANOVA is applied to the memorization instructions data and a separate similar ANOVA is applied to the story and imagery instructions data.

The estimates of μ, and the α_j, β_k, and $(\alpha\beta)_{jk}$ effects are defined just as for the independent measures factorial design, so it is unnecessary to repeat their definitions and calculations here. This leaves only the two error terms to be calculated. The simplest of these is the error term used to assess the independent experimental Factor A. This is obtained by taking the average of the scores provided by each subject. Essentially, this eliminates the related experimental Factor B and, with each subject providing a single (average) score and different subjects in each level of the independent factor, produces an independent measures ANOVA.

The mean of the scores provided by each subject is

$$\mu_{i(j)} = \left(\frac{\sum_{k=1}^{q} Y_{i(j)}}{q} \right) \tag{8.2}$$

Therefore, the effect due to the different subjects nested within the levels of the independent factor is

$$\hat{\pi}_{i(j)} = \mu_{i(j)} - \mu_j \tag{8.3}$$

Applying equation (8.2) to the data in Table 8.1 provides

$\hat{\pi}_1 = 7.333 - 9 = -1.667$	$\hat{\pi}_1 = 18.667 - 15 = 3.667$
$\hat{\pi}_2 = 9.333 - 9 = 0.333$	$\hat{\pi}_2 = 15.333 - 15 = 0.333$
$\hat{\pi}_3 = 8.333 - 9 = -0.667$	$\hat{\pi}_3 = 11.333 - 15 = -3.667$
$\hat{\pi}_4 = 9.333 - 9 = 0.333$	$\hat{\pi}_4 = 13.667 - 15 = -1.333$
$\hat{\pi}_5 = 9.000 - 9 = 0.000$	$\hat{\pi}_5 = 15.000 - 15 = 0.000$
$\hat{\pi}_6 = 9.333 - 9 = 0.333$	$\hat{\pi}_6 = 17.667 - 15 = 2.667$
$\hat{\pi}_7 = 9.333 - 9 = 0.333$	$\hat{\pi}_7 = 13.667 - 15 = -1.333$
$\hat{\pi}_8 = 10.000 - 9 = 1.000$	$\hat{\pi}_8 = 14.667 - 15 = -0.333$
$\sum_{i=1}^{N} \pi_{i(j)} = 0.000$	$\sum_{i=1}^{N} \pi_{i(j)} = 0.000$
$\sum_{i=1}^{N} \pi_{i(j)}^2 = 4.667$	$\sum_{i=1}^{N} \pi_{i(j)}^2 = 37.782$
$\sum_{j=1}^{p} \sum_{i=1}^{N} \pi_{i(j)}^2 = 42.449$	

The error SS due to the different subjects nested within the levels of the independent factor is

$$\text{Subject error SS} = q \sum_{j=1}^{p} \sum_{i=1}^{N} \pi_{i(j)}^2 = 3(42.449) = 127.347 \tag{8.4}$$

The last error term required is that based on the subject × Factor B interaction. Based on the model component of the fully related two-factor experimental design GLM equation, predicted scores are given by

$$\hat{Y}_{ijk} = \mu + \pi_{i(j)} + \alpha_j + \beta_k + (\alpha\beta)_{jk} \tag{8.5}$$

Using the parameter estimates determined earlier in this formula provides the predicted scores per subject per experimental condition. These values are presented in Table 8.3.

The error terms for the two-factor mixed measures experimental design GLM, which represent the discrepancy between the actual scores observed (Table 8.1) and the scores predicted by the two-factor GLM (Table 8.3), are defined as

$$\hat{\varepsilon}_{ijk} = Y_{ijk} - \hat{Y}_{ijk} \tag{8.6}$$

Table 8.4 presents the error terms by subject and experimental condition.

Table 8.3 Predicted Scores for the Mixed Two-Factor (2 × 3) Experiment

	a1				a2		
	b1	b2	b3		b1	b2	b3
s1	4.333	8.333	9.333	s9	13.667	15.667	26.667
s2	6.333	10.333	11.333	s10	10.333	12.333	23.333
s3	5.333	9.333	10.333	s11	6.333	8.333	19.333
s4	6.333	10.333	11.333	s12	8.667	10.667	21.667
s5	6.000	10.000	11.000	s13	10.000	12.000	23.000
s6	6.333	10.333	11.333	s14	12.667	14.667	25.667
s7	6.333	10.333	11.333	s15	8.667	10.667	21.667
s8	7.000	11.000	12.000	s16	9.667	11.667	22.667

Table 8.4 Error Terms for the Mixed Two-Factor (2 × 3) Experiment

	b1	b2	b3		b1	b2	b3
s1	2.667	−1.333	−1.333	s9	2.333	0.333	−2.667
s2	−3.333	0.667	2.667	s10	−3.333	−2.333	5.667
s3	0.667	−0.333	−0.333	s11	4.667	4.667	−9.333
s4	−0.333	0.667	−0.333	s12	0.333	−0.667	0.333
s5	−1.000	0.000	1.000	s13	0.000	−2.000	2.000
s6	1.667	−0.333	−1.333	s14	−1.667	−0.667	2.333
s7	−0.333	0.667	−0.333	s15	−0.667	0.333	0.333
s8	0.000	0.000	0.000	s16	−1.667	0.333	1.333
$\sum_{i=1}^{N} \varepsilon_{ijk}^2$	22.667	3.333	11.999		44.446	32.446	137.663
$\sum_{j=1}^{p} \sum_{k=1}^{q} \sum_{i=1}^{N} \varepsilon_{ijk}^2$				252.554			

Degrees of freedom for the two error terms employed in two-factor mixed measures experimental design GLM are also required. For the error SS due to the different subjects nested within the levels of the independent factor, S(A),

$$df = (N - p) = (16 - 2) = 14$$

and for the ε_{ijk} error term

$$df = (N - p)(q - 1) = (16 - 2)(3 - 1) = 14(2) = 28$$

Table 8.5 presents the previously obtained SS and *dfs* for the experimental effects along with the current SS and *dfs* in the form of an ANOVA summary table. The tabled critical F-values presented in Appendix B may be used to determine significance if hand calculation is employed or the statistical software used does not output the required p-values.

Table 8.5 Two-Factor Mixed Measures ANOVA Summary Table

Source	Sum of Squares	*df*	MS	*F*	*p*
Encoding instructions [A]	432.000	1	432.000	47.493	<0.001
Subject error [S(A)]	127.347	14	9.096		
Study time [B]	672.000	2	336.000	37.251	<0.001
Encoding instructions × Study time [A × B]	224.000	2	112.000	12.417	<0.001
Error [S × B]	252.554	28	9.020		

8.3 ESTIMATING EFFECTS BY COMPARING FULL AND REDUCED EXPERIMENTAL DESIGN GLMs

The full experimental design GLM for the mixed two-factor ANOVA was described by equation (8.1). As with all the previous factorial ANOVAs calculated by comparing full and reduced experimental design GLMs, the hypotheses concerning the main effect of Factor A, the main effect of Factor B and the effect of the interaction between Factors A and B are assessed by constructing three reduced GLMs, which manifest data descriptions under the respective null hypotheses, and comparing their error components with the full model. Again this approach is simplified by virtue of all the subject and experimental factors, and their interactions being orthogonal. As all of the effect estimates are completely distinct, omitting or including any particular effect has no consequence for any of the other effect estimates.

The main effect of Factor A is assessed by constructing the reduced experimental design GLM

$$Y_{ijk} = \mu + \pi_{i(j)} + \beta_k + (\alpha\beta)_{jk} + \varepsilon_{ijk} \tag{8.7}$$

This model manifests the data description under the null hypothesis

$$\alpha_j = 0 \tag{8.8}$$

The main effect of Factor B is assessed by constructing the reduced experimental design GLM

$$Y_{ijk} = \mu + \pi_{i(j)} + \alpha_j + (\alpha\beta)_{jk} + \varepsilon_{ijk} \tag{8.9}$$

This model manifests the data description under the null hypothesis

$$\beta_k = 0 \tag{8.10}$$

Finally, the reduced GLM for assessing the effect of the interaction between Factors A and B is

$$Y_{ijk} = \mu + \pi_{i(j)} + \alpha_j + \beta_k + \varepsilon_{ijk} \tag{8.11}$$

Table 8.6 Formulas for the (Balanced) Fully Related Two-Factor ANOVA Effects

Effect	Formula
A	$qN\sum_{j=1}^{p}(\overline{Y}_j - \overline{Y}_G)^2$
S(A)	$q\sum_{i=1}^{N}\sum_{j=1}^{p}(\overline{Y}_{ij} - \overline{Y}_j)^2$
B	$pN\sum_{k=1}^{q}(\overline{Y} - \overline{Y}_G)^2$
A × B	$N\sum_{j=1}^{p}\sum_{k=1}^{q}(\overline{Y}_{jk} - \overline{Y}_j - \overline{Y}_k + \overline{Y}_G)^2$
S × B (error)	$\sum_{i=1}^{N}\sum_{j=1}^{p}\sum_{k=1}^{q}(Y_{ijk} - \overline{Y}_{ij} - \overline{Y}_{jk} + \overline{Y}_j)^2$

This reduced GLM manifests the data description under the null hypothesis

$$(\alpha\beta)_{jk} = 0 \tag{8.12}$$

Although fewer error terms need to be calculated for mixed ANOVAs, when hand calculations are employed, instead of calculating the error SS associated with each of these reduced experimental design GLMs and comparing them with the full experimental design GLM, it may be more efficient to calculate directly the SS for each of the effects and errors. Formulas for calculating all of the mixed two-factor ANOVA effects directly, which are more convenient than those used to define and illustrate the SS calculation, are provided in Table 8.6. However, as will be described, the strategy of comparing different experimental design GLM residuals to estimate mixed two-factor ANOVA effects is a simple way to implement related ANOVAs using regression GLMs.

8.4 REGRESSION GLM FOR THE TWO-FACTOR MIXED MEASURES ANOVA

The mixed factor (2 × 3) experimental design GLM equation (8.1) may be compared with the equivalent regression equation

$$\begin{aligned} Y_i = \beta_0 &+ \beta_1 X_{i,1} + \beta_2 X_{i,2} + \beta_3 X_{i,3} + \beta_4 X_{i,4} + \beta_5 X_{i,5} + \beta_6 X_{i,6} + \beta_7 X_{i,7} \\ &+ \beta_8 X_{i,8} + \beta_9 X_{i,9} + \beta_{10} X_{i,10} + \beta_{11} X_{i,11} + \beta_{12} X_{i,12} + \beta_{13} X_{i,13} + \beta_{14} X_{i,14} \\ &+ \beta_{15} X_{i,15} + \beta_{16} X_{i,16} + \beta_{17} X_{i,17} + \beta_{18} X_{i,18} + \beta_{19} X_{i,19} + \varepsilon_i \end{aligned} \tag{8.13}$$

where Y_i represents the ith dependent variable score, β_0 is a constant, β_1 is the regression coefficient for the predictor variable X_1, β_2 is the regression coefficient for the predictor variable X_2, and so on. As with the fully related factorial design, there are X variables that represent experimental factors and their interactions (from X_1 to X_5), while 14 variables (from X_6 to X_{19}) identify the subjects providing the scores. The first seven subject variables (from X_6 to X_{12}) identify those subjects providing scores in the condition labeled as a1, memorize, while the second set of seven subject variables identify those subjects providing scores in the condition labeled as a2, story and imagery. Equation (8.13) is not particularly wieldy, but once the effect coding scheme has been established in a computer data file, it is relatively simple to carry out the mixed factorial ANOVA. Table 8.7 presents effect coding applied to the data in Table 8.1.

Applying a regression GLM to implement a mixed factors ANOVA may also be done in a manner consistent with incremental analysis and estimating effects by comparing full and reduced GLMs. In balanced designs, the predictor variables representing experimental factors and those identifying subjects' scores again are orthogonal and so the order in which the SS are calculated is of no consequence.

The first regression carried out is that for the full mixed factorial experimental design GLM, when all experimental condition and subject predictor variables are included (i.e., variables from X_1 to X_{19}). Of interest is the ANOVA summary presented in Table 8.8, which provides the full GLM residual SS. This may be compared with the mixed factorial experimental design GLM error term in Table 8.18.

Having obtained the full GLM residual SS, the next stages involve implementing the various reduced GLMs to obtain their estimates of residual SS. First, the reduced GLM for the effect of the subjects nested within Factor A is obtained by carrying out the regression analysis again, but omitting the predictors identifying the subjects providing the scores (from X_6 to X_{19}). The summary of this ANOVA, presented in Table 8.9, provides the subjects nested within Factor A reduced GLM residual SS. Therefore

			*df*s
Subjects (Factor A) reduced GLM residual SS	=	380.000	42
Full GLM residual SS	=	252.667	28
SS attributable to subjects nested within Factor A	=	127.333	14

The reduced GLM for the effect of Factor A is applied by omitting only the predictor variable representing the Factor A experimental conditions (X_1). The summary of this ANOVA presented in Table 8.10 provides the Factor A reduced GLM residual SS. Therefore

			*df*s
Factor A reduced GLM residual SS	=	684.667	29
Full GLM residual SS	=	252.667	28
SS attributable to Factor A	=	432.000	1

The Factor B reduced GLM is applied by omitting only the predictor variables representing the Factor B experimental conditions (X_2 and X_3). The summary of this

Table 8.7 Effect Coding for the Mixed (2 × 3) Factorial ANOVA

Subject	Y	A	B		A × B		A = 1 subjects							A = −1 subjects						
	Y	X_1	X_2	X_3	X_4	X_5	X_6	X_7	X_8	X_9	X_{10}	X_{11}	X_{12}	X_{13}	X_{14}	X_{15}	X_{16}	X_{17}	X_{18}	X_{19}
1	7	1	1	0	1	0	1	0	0	0	0	0	0	0	0	0	0	0	0	0
2	3	1	1	0	1	0	0	1	0	0	0	0	0	0	0	0	0	0	0	0
3	6	1	1	0	1	0	0	0	1	0	0	0	0	0	0	0	0	0	0	0
4	6	1	1	0	1	0	0	0	0	1	0	0	0	0	0	0	0	0	0	0
5	5	1	1	0	1	0	0	0	0	0	1	0	0	0	0	0	0	0	0	0
6	8	1	1	0	1	0	0	0	0	0	0	1	0	0	0	0	0	0	0	0
7	6	1	1	0	1	0	0	0	0	0	0	0	1	0	0	0	0	0	0	0
8	7	1	1	0	1	0	−1	−1	−1	−1	−1	−1	−1	0	0	0	0	0	0	0
1	7	1	0	1	0	1	1	0	0	0	0	0	0	0	0	0	0	0	0	0
2	11	1	0	1	0	1	0	1	0	0	0	0	0	0	0	0	0	0	0	0
3	9	1	0	1	0	1	0	0	1	0	0	0	0	0	0	0	0	0	0	0
4	11	1	0	1	0	1	0	0	0	1	0	0	0	0	0	0	0	0	0	0
5	10	1	0	1	0	1	0	0	0	0	1	0	0	0	0	0	0	0	0	0
6	10	1	0	1	0	1	0	0	0	0	0	1	0	0	0	0	0	0	0	0
7	11	1	0	1	0	1	0	0	0	0	0	0	1	0	0	0	0	0	0	0
8	11	1	0	1	0	1	−1	−1	−1	−1	−1	−1	−1	0	0	0	0	0	0	0
1	8	1	−1	−1	−1	−1	1	0	0	0	0	0	0	0	0	0	0	0	0	0
2	14	1	−1	−1	−1	−1	0	1	0	0	0	0	0	0	0	0	0	0	0	0
3	10	1	−1	−1	−1	−1	0	0	1	0	0	0	0	0	0	0	0	0	0	0
4	11	1	−1	−1	−1	−1	0	0	0	1	0	0	0	0	0	0	0	0	0	0
5	12	1	−1	−1	−1	−1	0	0	0	0	1	0	0	0	0	0	0	0	0	0
6	10	1	−1	−1	−1	−1	0	0	0	0	0	1	0	0	0	0	0	0	0	0
7	11	1	−1	−1	−1	−1	0	0	0	0	0	0	1	0	0	0	0	0	0	0
8	12	1	−1	−1	−1	−1	−1	−1	−1	−1	−1	−1	−1	0	0	0	0	0	0	0

1	16	−1	1	0	−1	0	0	0	0	0	0	0	0	1	0	0	0	0	0	0
2	7	−1	1	0	−1	0	0	0	0	0	0	0	0	0	1	0	0	0	0	0
3	11	−1	1	0	−1	0	0	0	0	0	0	0	0	0	0	1	0	0	0	0
4	9	−1	1	0	−1	0	0	0	0	0	0	0	0	0	0	0	1	0	0	0
5	10	−1	1	0	−1	0	0	0	0	0	0	0	0	0	0	0	0	1	0	0
6	11	−1	1	0	−1	0	0	0	0	0	0	0	0	0	0	0	0	0	1	0
7	8	−1	1	0	−1	0	0	0	0	0	0	0	0	0	0	0	0	0	0	1
8	8	−1	1	0	−1	0	0	0	0	0	0	0	0	−1	−1	−1	−1	−1	−1	−1
1	16	−1	0	1	0	−1	0	0	0	0	0	0	0	1	0	0	0	0	0	0
2	10	−1	0	1	0	−1	0	0	0	0	0	0	0	0	1	0	0	0	0	0
3	13	−1	0	1	0	−1	0	0	0	0	0	0	0	0	0	1	0	0	0	0
4	10	−1	0	1	0	−1	0	0	0	0	0	0	0	0	0	0	1	0	0	0
5	10	−1	0	1	0	−1	0	0	0	0	0	0	0	0	0	0	0	1	0	0
6	14	−1	0	1	0	−1	0	0	0	0	0	0	0	0	0	0	0	0	1	0
7	11	−1	0	1	0	−1	0	0	0	0	0	0	0	0	0	0	0	0	0	1
8	12	−1	0	1	0	−1	0	0	0	0	0	0	0	−1	−1	−1	−1	−1	−1	−1
1	24	−1	−1	−1	1	1	0	0	0	0	0	0	0	1	0	0	0	0	0	0
2	29	−1	−1	−1	1	1	0	0	0	0	0	0	0	0	1	0	0	0	0	0
3	10	−1	−1	−1	1	1	0	0	0	0	0	0	0	0	0	1	0	0	0	0
4	22	−1	−1	−1	1	1	0	0	0	0	0	0	0	0	0	0	1	0	0	0
5	25	−1	−1	−1	1	1	0	0	0	0	0	0	0	0	0	0	0	1	0	0
6	28	−1	−1	−1	1	1	0	0	0	0	0	0	0	0	0	0	0	0	1	0
7	22	−1	−1	−1	1	1	0	0	0	0	0	0	0	0	0	0	0	0	0	1
8	24	−1	−1	−1	1	1	0	0	0	0	0	0	0	−1	−1	−1	−1	−1	−1	−1

Table 8.8 ANOVA Summary Table for the Regression Implementation of the Full Mixed Factorial Experimental Design GLM (Includes Experimental Conditions and Subject Effects)

Source	SS	*df*	MS	*F*	*p*
Regression	1455.333	19	76.596	8.488	
Residual	252.667	28	9.024		

R: 0.923; R^2: 0.852; adjusted R^2: 0. 752.

Table 8.9 ANOVA Summary Table for the Mixed Factorial Experimental Design GLM Omitting the Effect of Subjects Nested Within Factor A

Source	SS	*df*	MS	*F*	*p*
Regression	1328.000	5	265.600	29.356	
Residual	380.000	42	9.048		

R: 0.882; R^2: 0.778; adjusted R^2: 0.751.

Table 8.10 ANOVA Summary Table for the Reduced GLM That Omits the Factor A Experimental Conditions

Source	SS	*df*	MS	*F*	*p*
Regression	1023.333	18	56.852	2.408	0.017
Residual	684.667	29	23.609		

R: 0.774; R^2: 0.599; adjusted R^2: 0.350.

ANOVA presented in Table 8.11 provides the Factor B reduced GLM residual SS. Therefore,

			dfs
Factor A reduced GLM residual SS	=	924.667	30
Full GLM residual SS	=	252.667	28
SS attributable to Factor A	=	672.000	2

The Factor A × Factor B interaction reduced GLM is applied by omitting only the predictor variables representing the Factor A × Factor B interaction (X_4 and X_5). The

Table 8.11 ANOVA Summary Table for the Reduced GLM That Omits the Factor B Experimental Conditions

Source	SS	*df*	MS	*F*	*p*
Regression	783.333	17	46.078	1.495	0.163
Residual	924.667	30	30.822		

R: 0.677; R^2: 0.459; adjusted R^2: 0.152.

Table 8.12 ANOVA Summary Table for the Reduced GLM That Omits the Factor A × Factor B Interaction

Source	SS	*df*	MS	*F*	*p*
Regression	1231.333	17	72.431	4.559	<0.001
Residual	476.667	30	15.889		

R: 0.849; R^2: 0.721; adjusted R^2: 0.563.

summary of this ANOVA presented in Table 8.12 provides the Factor A × Factor B reduced GLM residual SS. Therefore,

			dfs
Factors A × B interaction reduced GLM residual SS	=	476.667	30
Full GLM residual SS	=	252.667	28
SS attributable to Factor A × Factor B interaction	=	224.000	2

Using the SS and *dfs* calculated for each effect by comparing full and reduced GLMs, an ANOVA summary table similar to Table 8.5 can be constructed to present the results of applying the two-factor mixed measures ANOVA GLM.

8.5 EFFECT SIZE ESTIMATION

As mixed measures designs combine independent and related factors, so the approaches developed to estimate independent and related factors are applied separately to the estimation of effects in mixed measures designs. Therefore, for a two-factor mixed design, the procedures described for single factor independent measures designs (Section 4.3) and for single factor related designs (Section 6.6) are most relevant. As the interaction also employs the related factors error term, it is treated as a related measures effect.

8.6 FURTHER ANALYSES

Further analysis of mixed measures designs also applies the approaches developed for independent and related factor main and interaction effects separately to the pertinent factors.

8.6.1 Main Effects: Independent Factor—Encoding Instructions

Further analysis of the independent factor follows exactly the same approach and employs exactly the same calculations described in Section 5.6.1. However, note that the error SS for the independent factor, that is, the S(A), is one-third of the size of the error SS in the fully independent design. This is due to the F-test being applied to "independent" scores obtained by taking the average of the three scores provided by each subject. Although the error SS for Factor A is reduced, the error *dfs* are reduced similarly, so comparable F-values are obtained.

8.6.2 Main Effects: Related Factor—Study Time

Further analysis of the related factor follows exactly the same approach and employs exactly the same calculations described in Sections 6.8 and 7.6.1.

8.6.3 Interaction Effect: Encoding Instructions × Study Time

The combination of independent and related factors in a mixed design can complicate some aspects of the further analysis of the interaction effect. However, as will be explained below, more recent approaches to further analysis of the interaction effect simplify matters by employing specific error terms for the particular comparisons examined.

8.6.3.1 Simple Effects: Comparing Differences Between the Three Levels of Factor B (Study Time) at Each Level of Factor A (Encoding Instructions)

As always, the three levels of Factor B, study time, are examined at each level of Factor A, encoding instructions. In the current two-factor mixed design, each level of the independent Factor A, provides a single factor related ANOVA applied to the three study time conditions (30, 60, and 180 s). The results of these two simple effect analyses already have been presented in Table 8.2 to illustrate the nature of the omnibus error terms. One simple effect single factor related ANOVA is applied to the scores obtained from three study time conditions under memorization instructions and another simple effect single factor related ANOVA is applied to the scores obtained from three study time conditions under story and imagery instructions (see Figure 5.4). Therefore, the approach to the simple effect analysis and the simple effect analysis calculations described in Section 7.6.2.1 also apply here.

8.6.3.2 Simple Effects: Comparing Differences Between the Two Levels of Factor A (Encoding Instructions) at Each Level of Factor B (Study Time)

In the current two-factor mixed measures design, certain comparisons drawn across the levels of the independent factor can be slightly more complicated to deal with than in other designs. Table 8.13 presents a reminder of the two-factor mixed design applied. One group of subjects receives memorization encoding instructions and the other group receives story and imagery encoding instructions, but all subjects apply these encoding instructions under all three study time conditions. When comparisons across study time conditions are applied separately by encoding instructions, as in Section 8.6.3.1, separate S × B error terms were employed. One of these error terms represents the within group variance of the three study time conditions (after removal of the covariation due to related scores) with memorization encoding instructions

Table 8.13 Mean Recall by Encoding Instructions and Study Times in the Mixed Design Applied

	Study Time		
Encoding Instructions	b1	b2	b3
a1 Memorize	6	10	11
a2 Story and imagery	10	12	23

(the a1 row in Table 8.13) and the other S × B error term represents the within group variance of the three study time conditions (after removal of the covariation due to related scores) with story and imagery encoding instructions (the a2 row in Table 8.13). Section 8.2 described and Table 8.2, illustrated that the two Subject SSs and the sum of the two error SSs from the single factor repeated measures ANOVAs, respectively, provide the error SS for the independent factor and the error SS for the related factor in the two factor mixed measures ANOVA. The two MSe terms used in the two factor mixed design are pooled weighted averages of these SSs (accurate averages are provided by the different dfs, but there is equal weighting in a balanced design), but the point of this account is the bases of the mixed design omnibus error estimates are two entirely separate sets of related scores (the a1 row and the a2 row in Table 8.13).

When a comparison is drawn across the levels of the independent factor, for example, condition a1b1 ($\overline{X} = 6$) versus a2b1 ($\overline{X} = 10$), it is drawn across two separate groups of subjects. Usually, the error term employed to assess the interaction would be considered as an appropriate term to assess the difference between these two experimental condition means. However, as described above, the interaction error term is based on two separate repeated measures error terms, but error arising from the differences between the two groups of subjects providing the mean scores is not represented. Some way of accommodating the error arising from the two different groups of subjects is required. One way of accommodating this error term is to simply to combine the between subjects error term [S(A)] with the repeated measures main effect and interaction error term [S×B] to obtain what is known as the *MS within cell* (e.g., Howell, 2010; Kirk, 1995; Winer, Brown, and Michels, 1991). As the MS within cell error term combines the two omnibus error terms from the two factor mixed measures ANOVA it is vital that these error terms comply with their respective variance homogeneity and sphericity assumptions. However, as such assumption compliance is unlikely and assumption violations can have a profound influence on Type 1 error rates, the most frequent recommendation currently is to conduct specific comparisons involving only the experimental conditions of interest. Therefore, the comparison of a1b1 ($\overline{X} = 6$) vs a2b1 ($\overline{X} = 10$) would be implemented by applying a single factor independent ANOVA only to the data obtained from the a1b1 (memorize 30s) experimental condition and the a1b2 (story and imagery 30s) experimental condition, the comparison of a1b2 ($\overline{X} = 10$) vs a2 b2 ($\overline{X} = 12$) would be implemented by applying a separate single factor independent ANOVA only to the data obtained from the a1b2 (memorize 60s) experimental condition and the a1b2 (story and imagery 60s) experimental condition, and the comparison of a1b3 ($\overline{X} = 11$) vs a2b3 ($\overline{X} = 23$) would be implemented by applying another single factor independent ANOVA only to the data obtained from the a1b3 (memorize 180s) experimental condition and the a1b3 (story and imagery 180s) experimental condition. Although conducting specific comparisons as described is a safer approach, when the omnibus error term assumptions are tenable, applying the MS within cell approach will provide more powerful tests, for exactly the same reasons described with respect to the use of the omnibus ANOVA MSe in independent designs (see Section 3.7.1).

8.7 POWER

To conduct power analysis to determine sample size for mixed measures designs using tables, each omnibus main effect and interaction effect, as well as specific comparison effects are assessed separately. Therefore, the procedures described for single factor independent measures designs (Section 4.7) and for single factor related designs (Section 6.8) apply to two-factor mixed measures designs. (The interaction also is treated as a related measures effect.) Nevertheless, more accurate results will be obtained if one of the statistical software packages available for this purpose is employed. For power analysis of two-factor fully related measures designs and all other such designs, the free statistical software package G*Power 3 (Faul et al. 2007) is recommended highly.

CHAPTER 9

The GLM Approach to ANCOVA

9.1 THE NATURE OF ANCOVA

In Chapter 5, the single factor independent measures experimental design presented in Chapters 3 and 4 was extended by establishing a new factor labelled encoding instruction. Memorization and story and imagery instructions defined the two levels of the encoding instruction factor, and these two factor levels were crossed with the three levels of the study time factor, defined by 30 s, 60 s and 180 s, in a factorial independent measures design. The present chapter considers only the 30 s, 60 s and 180 s story and imagery instruction conditions, as if a separate single factor independent measures experiment had been designed.

All of the subjects in this experiment are asked to construct stories using the presented words and to form images of the story events, but story and imagery ability is very likely to differ across subjects. Experimental studies employ random sampling and random allocation of subjects to avoid systematic effects attributable to subjects. If a measure of each subject's story and imagery ability was available, then the most likely observation after random sampling and random allocation would be equal mean story and imagery ability across the three study time groups. Nevertheless, within each of the study time conditions, some subjects will possess greater story and imagery ability, while other subjects will possess lesser story and imagery ability. If constructing stories and images influences memory, then variation in story and imagery ability will result in variation in subjects' recall scores within each study time group and this variation will increment the ANOVA error term. All else being equal, the consequence of a larger error term is a less powerful analysis and so a greater chance that any influence of study time on memory recall will go undetected. A more precise assessment of the influence of study time on free recall when story and imagery encoding strategies are used would be obtained if all of the subjects had the same ability to construct stories and images.

In psychology, when GLMs include a quantitative variable in addition to the categorical coding of experimental conditions, but experimental effects remain the

ANOVA and ANCOVA: A GLM Approach, Second Edition. By Andrew Rutherford.

major concern, the analysis is termed ANCOVA (cf. Cox and McCullagh, 1982). ANCOVA offers a way to obtain a more precise assessment of the effect of the experimental manipulations on the dependent variable. Indeed, ANCOVA offers a statistical means of assessing the influence of study time on free recall when all of the subjects employing the story and imagery encoding strategies have the same ability to construct stories and images

As well as recording the independent and dependent variables, an ANCOVA design requires the measurement of one or more other variables. These variables (variously known as covariates, predictor variables, concomitant variables, or control variables) represent sources of variation that are thought to influence the dependent variable, but have not been controlled by the experimental procedures. In the present example, the covariates would be measures of each subject's ability to construct stories and images. The rationale underlying ANCOVA is that the effect of the independent variable(s) on the dependent variable is revealed more precisely when the relationship between the dependent variable and the covariate(s) is used to adjust the dependent variable scores to those predicted if all subjects had obtained the same covariate score.

9.2 SINGLE FACTOR INDEPENDENT MEASURES ANCOVA DESIGNS

The story and imagery conditions of the memory experiment described in Chapter 5 are presented as if they had been obtained in a new experiment. In this new experiment, prior to their allocation to study time experimental condition, subjects also were required to complete a test that provided a single measure of their story and imagery abilities. Table 6.1 presents the subjects' story and imagery task (covariate) scores and the subjects' free recall scores after story and imagery encoding in the three study time conditions.

The equation

$$Y_{ij} = \mu + \alpha_j + \beta Z_{ij} + \varepsilon_{ij} \tag{9.1}$$

describes an experimental design GLM for the single factor independent measures ANCOVA with one covariate applicable to the data presented in Table 9.1. Y_{ij} is the ith score in the jth treatment, μ is the grand mean of the experimental condition population means, α_j is the effect of the jth treatment level and the error term, ε_{ij}, reflects random variation due to any uncontrolled source. The new term, $\beta_{\mathrm{w}} Z_{ij}$, represents the influence of the covariate on the dependent variable, and is comprised of the regression coefficient parameter, β_{w}, which represents the degree of linear relationship between the covariate and the dependent variable and Z_{ij}, the particular covariate score corresponding to the Y_{ij}. (It is important to appreciate that the degree of the linear relationship between the covariate and the dependent variable is determined empirically from the data.) The ANCOVA GLM combines features of an ANOVA GLM and a regression GLM. The (categorical) experimental condition

Table 9.1 Story and Imagery Test Scores and Recall Scores After Story and Imagery Encoding

Study Time	30 s		60 s		180 s	
	Z	Y	Z	Y	Z	Y
	9	16	8	16	5	24
	5	7	5	10	8	29
	6	11	6	13	3	10
	4	9	5	10	4	22
	6	10	3	10	6	25
	8	11	6	14	9	28
	3	8	4	11	4	22
	5	8	6	12	5	24
Sums	46	80	43	96	44	184
Means	5.750	10.000	5.375	12.000	5.500	23.000
Sum of squares	292	856	247	1186	272	4470
Sum squared	2116	6400	1849	9216	1936	33,856

effects are specified as in ANOVA, while the relationship between the (quantitative) covariate and the dependent variable is specified as in regression.

In Figure 9.1, the regression lines of subjects' dependent variable memory recall scores on their story and imagery ability test scores are plotted for each study time

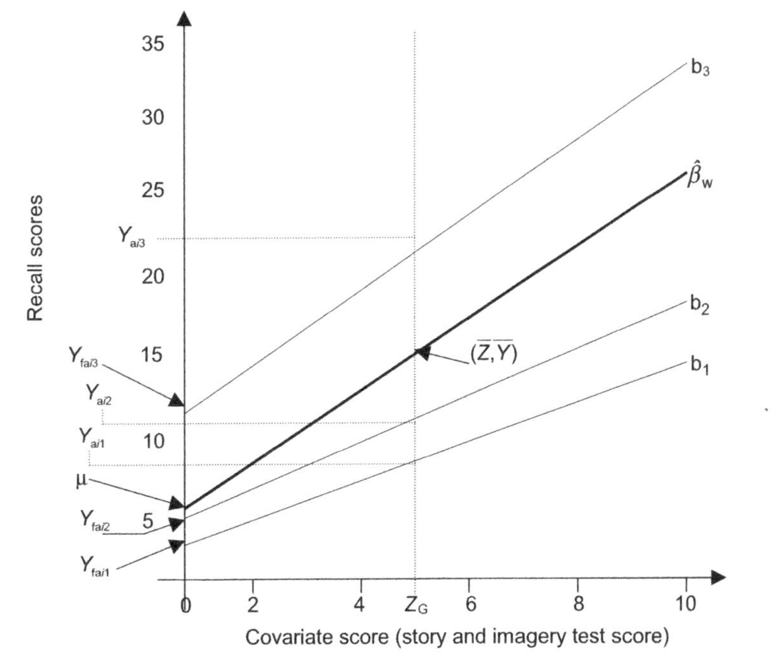

Figure 9.1 Dependent variable memory recall scores plotted on their story and imagery ability test scores for each study time experimental condition(b_1 – 30 s, b_2 –60 s and b_3 – 180 s). The within groups regression line ($\widehat{\beta}_w$) also is plotted.

experimental condition (b_1–30 s, b_2–60 s, and b_3–180 s). Also shown is $\widehat{\beta}_w$, the slope of the regression line employed in the ANCOVA GLM, which is given by

$$\widehat{\beta}_w = \frac{\sum_{j=1}^{p}\sum_{i=1}^{N_j}(Z_{ij} - \overline{Z}_j)(Y_{ij} - \overline{Y}_j)}{\sum_{j=1}^{p}\sum_{i=1}^{N_j}(Z_{ij} - \overline{Z}_j)^2} \tag{9.2}$$

$\widehat{\beta}_w$ also may be calculated from

$$\widehat{\beta}_w = \frac{\sum_{i=1}^{N_j}(Z_{i1} - \overline{Z}_1)^2 b_1 + \sum_{i=1}^{N_j}(Z_{i2} - \overline{Z}_2)^2 b_2 + \sum_{i=1}^{N_j}(Z_{i3} - \overline{Z}_3)^2 b_3}{\sum_{i=1}^{N_j}(Z_{i1} - \overline{Z}_1)^2 + \sum_{i=1}^{N_j}(Z_{i2} - \overline{Z}_2)^2 + \sum_{i=1}^{N_j}(Z_{i3} - \overline{Z}_3)^2} \tag{9.3}$$

Equation (9.3) reveals $\widehat{\beta}_w$ as the weighted average of the separate (within group) regression lines, where each experimental condition regression coefficient (b_1, b_2, and b_3) is weighted by the variation of the covariate scores in that experimental condition. Consequently, $\widehat{\beta}_w$ may be called the within groups regression coefficient. An important point to appreciate is that equations (9.2) and (9.3) provide a regression coefficient that is free of the influence exerted on the dependent variable scores by the experimental conditions. However, in common with the pooled weighted estimate of MSe, is good summary estimate only when the dependent variable on covariate regressions across experimental conditions are homogeneous (see Sections 10.4 and 10.2.3).

A little algebra applied to equation (9.1) reveals

$$Y_{faij} = Y_{ij} - \beta_w Z_{ij} = \mu + \alpha_j + \varepsilon_{ij} \tag{9.4}$$

where Y_{faij} is the fundamental adjusted dependent variable score observed if all influence of the covariate is removed from the dependent variable score. The Y_{faij} correspond to the points on the dependent variable axis intersected by each of the experimental condition regression lines (see Figure 9.1). This is where the value of the covariate equals zero. Traditionally in ANCOVA, however, the dependent variable scores are not adjusted to remove all influence of the covariate. Instead, adjustment is made so it is as if all subjects had obtained a covariate score equal to the general covariate mean ($Z_G = 5.542$). Replacing $\beta_w Z_{ij}$ in equation (9.1) with $\beta_w(Z_{ij} - Z_G)$ provides the single factor independent measures experimental design GLM for traditional ANCOVA with one covariate

$$Y_{ij} = \mu + \alpha_j + \beta_w(Z_{ij} - Z_G) + \varepsilon_{ij} \tag{9.5}$$

Applying the same algebra to equation (9.5) as was applied to equation (9.1) provides

$$Y_{aij} = Y_{ij} - \beta_w(Z_{ij} - Z_G) = \mu + \alpha_j + \varepsilon_{ij} \tag{9.6}$$

where Y_{aij} is the adjusted dependent variable score based on the difference between the recorded covariate score and the general covariate mean scaled by the regression coefficient estimated from the data. The experimental condition means of the Y_{aij} scores correspond to the points on the dependent variable axis where the separate experimental condition regression lines intersect the line representing Z_G (see Figure 9.1).

Although the GLMs described by equations (9.1) and (9.5) employ the same regression coefficient, obviously the adjustments provided by $\beta_w Z_{ij}$ and $\beta_w(Z_{ij} - Z_G)$ do not provide identical adjusted dependent variable scores (see Y_{aij} and Y_{faij} in Figure 9.1). Nevertheless, as the effect of the experimental conditions is represented by the vertical differences between the experimental condition regression lines, and traditional ANCOVA assumes the regression lines of the dependent variable on the covariate in each of the treatment groups are parallel, the experimental condition effect estimates will be constant across all values of the covariate. Therefore, when traditional ANCOVA (which assumes homogeneous regression coefficients/slopes) is applied, the terms Y_{faij} and Y_{aij} provide equivalent estimates of the experimental conditions effect and so accommodate identical variance estimates.

Calculating $\widehat{\beta}_w$ for the data in Table 9.1 provides

30 s		60 s		180 s	
$Z_{ij} - \overline{Z}_j$	$Y_{ij} - \overline{Y}_j$	$Z_{ij} - \overline{Z}_j$	$Y_{ij} - \overline{Y}_j$	$Z_{ij} - \overline{Z}_j$	$Y_{ij} - \overline{Y}_j$
9 − 5.750 = 3.250	16 − 10 = 6	8 − 5.375 = 2.625	16 − 12 = 4	5 − 5.500 = −0.500	24 − 23 = 1
5 − 5.750 = −0.750	7 − 10 = −3	5 − 5.375 = −0.375	10 − 12 = −2	8 − 5.500 = 2.500	29 − 23 = 6
6 − 5.750 = 0.250	11 − 10 = 1	6 − 5.375 = 0.625	13 − 12 = 1	3 − 5.500 = −2.500	10 − 23 = −13
4 − 5.750 = −1.750	9 − 10 = −1	5 − 5.375 = −0.375	10 − 12 = −2	4 − 5.500 = −1.500	22 − 23 = −1
6 − 5.750 = 0.250	10 − 10 = 0	3 − 5.375 = −2.375	10 − 12 = −2	6 − 5.500 = 0.500	25 − 23 = 2
8 − 5.750 = 2.250	11 − 10 = 1	6 − 5.375 = 0.625	14 − 12 = 2	9 − 5.500 = 3.500	28 − 23 = 5
3 − 5.750 = −2.750	8 − 10 = −2	4 − 5.375 = −1.375	11 − 12 = −1	4 − 5.500 = −1.500	22 − 23 = −1
5 − 5.750 = − 0.750	8 − 10 = −2	6 − 5.375 = 0.625	12 − 12 = 0	5 − 5.500 = −0.500	24 − 23 = 1
$\sum = 0$ 27.500		$\sum = 0$ 15.875		$\sum = 0$ 30.000	

30 s	60 s	180 s
$(Z_{ij} - \overline{Z}_j)(Y_{ij} - \overline{Y}_j)$	$(Z_{ij} - \overline{Z}_j)(Y_{ij} - \overline{Y}_j)$	$(Z_{ij} - \overline{Z}_j)(Y_{ij} - \overline{Y}_j)$
3.250(6) = 19.500	2.625(4.00) = 10.500	−0.500(1.00) = −0.500
−0.750(−3) = 2.250	−0.375(−2.00) = 0.750	2.500(6.00) = 15.000
0.250(1) = 0.250	0.625(1.00) = 0.625	−2.500(−13.00) = 32.500
−1.750(−1) = 1.750	−0.375(−2.00) = 0.750	−1.500(−1.00) = 1.500
0.250(0) = 0.000	−2.375(−2.00) = 4.750	0.500(2.00) = 1.000
2.250(1) = 2.250	0.625(2.00) = 1.250	3.500(5.00) = 17.500
−2.750(−2) = 5.500	−1.375(−1.00) = 1.375	−1.500(−1.00) = 1.500
−0.750(−2) = 1.500	0.625(0.00) = 0.000	−0.500(1.00) = −0.500
$\sum = 0$ 33.000	$\sum = 0$ 20.000	$\sum = 0$ 68.000

$$\widehat{\beta}_w = \frac{33 + 20 + 68}{27.500 + 15.875 + 30.000} = 1.649$$

One consequence of including the dependent variable on the covariate regression component in the ANCOVA GLM is the definition of the parameter μ changes slightly. For balanced ANOVA, μ was defined as the mean of the experimental condition means, but this is not the case even for balanced ANCOVA. Nevertheless, in ANCOVA, μ remains the intercept on the Y-axis (see Figures 2.3 and 9.1) of the $\widehat{\beta}_{\mathrm{w}}$ regression line, which passes through the general covariate and dependent variable means and so, it may be calculated simply by estimating the reduction from the dependent variable mean on the basis of the general mean of the covariate. In Figure 9.1, the $\widehat{\beta}_{\mathrm{w}}$ regression line intercepts the Y-axis at a point below the dependent variable general mean. This point is determined by the distance from the general mean of the covariate to the origin, scaled by the $\widehat{\beta}_{\mathrm{w}}$ regression coefficient. Therefore

$$\widehat{\mu} = Y_{\mathrm{G}} - \widehat{\beta}_{\mathrm{w}}(Z_{\mathrm{G}}) \tag{9.7}$$

Applying equation (9.7) to the data in Table 9.1 provides

$$\widehat{\mu} = 15 - 1.649(5.542) = 5.861$$

The adjusted experimental condition means is given by

$$\overline{Y}_{\mathrm{a}j} = \overline{Y}_j - \beta_{\mathrm{w}}(\overline{Z}_j - Z_{\mathrm{G}}) \tag{9.8}$$

Therefore

$$\overline{Y}_{\mathrm{a}1} = \overline{Y}_1 - \beta_{\mathrm{w}}(\overline{Z}_1 - Z_{\mathrm{G}}) = 10 - 1.649(5.750 - 5.542) = 9.657$$
$$\overline{Y}_{\mathrm{a}2} = \overline{Y}_2 - \beta_{\mathrm{w}}(\overline{Z}_2 - Z_{\mathrm{G}}) = 12 - 1.649(5.375 - 5.542) = 12.275$$
$$\overline{Y}_{\mathrm{a}3} = \overline{Y}_3 - \beta_{\mathrm{w}}(\overline{Z}_3 - Z_{\mathrm{G}}) = 23 - 1.649(5.500 - 5.542) = 23.069$$

In common with ANOVA, adjusted experimental condition means in ANCOVA are comprised of the constant μ plus the effect of the experimental condition

$$\overline{Y}_{\mathrm{a}j} = \widehat{\mu} + \widehat{\alpha}_j \tag{9.9}$$

and so it follows

$$\widehat{\alpha}_j = \overline{Y}_{\mathrm{a}j} - \widehat{\mu} \tag{9.10}$$

For the data in Table 9.1 this provides

$$\widehat{\alpha}_1 = 9.657 - 5.861 = 3.796$$
$$\widehat{\alpha}_2 = 12.275 - 5.861 = 6.414$$
$$\widehat{\alpha}_3 = 23.069 - 5.861 = 17.208$$

9.3 ESTIMATING EFFECTS BY COMPARING FULL AND REDUCED ANCOVA GLMs

Although $\widehat{\mu}$ and the experimental condition effects are not needed to calculate the error terms for full and reduced GLMs, their calculation methods are provided for completeness. In fact, these experimental condition effect estimates should not be used to obtain the experimental condition SS, in the manner described for ANOVA GLMs, as this provides an inaccurate estimate (Cochran, 1957; Maxwell, Delaney, and Manheimer, 1985; Rutherford, 1992).

The reduced and full models for the traditional single factor single covariate ANCOVA design are presented below

$$\text{Reduced GLM:} \quad Y_{ij} = \mu + \beta_t(Z_{ij} - Z_G) + \varepsilon_{ij} \tag{9.11}$$

$$\text{Full GLM:} \quad Y_{ij} = \mu + \alpha_j + \beta_w(Z_{ij} - Z_G) + \varepsilon_{ij} \tag{9.5, rptd}$$

There are two differences between these full and reduced GLMs. First, the reduced GLM omits the variable representing the experimental conditions. Second, the reduced GLM employs the regression coefficient for the total set of scores, $\widehat{\beta}_t$, where allocation to experimental condition is ignored and all scores are treated as one large group, whereas the full GLM employs the within groups regression coefficient, $\widehat{\beta}_w$.

Equation (9.11) describes the reduced GLM as a simple linear regression, where the difference between subjects' covariate scores and the general covariate mean is used to predict subjects' dependent variable scores. This GLM may be compared with the reduced GLM that employs Z_{ij} as the predictor

$$Y_{ij} = \mu + \beta_t(Z_{ij}) + \varepsilon_{ij} \tag{9.12}$$

Equation (9.12) is the reduced GLM for the ANCOVA model described by equation (9.1). As was described with respect to the GLMs described by equations (9.1) and (9.5), the GLMs described by equations (9.11) and (9.12) employ a common regression coefficient, and accommodate equal amounts of variation.

The application of a little algebra to the full GLM for the single factor, single-covariate ANCOVA defines the error term in the following way. If

$$Y_{aij} = Y_{ij} - \beta_w(Z_{ij} - Z_G) = \mu + \alpha_j + \varepsilon_{ij} \tag{9.6, rptd}$$

then

$$Y_{ij} - \beta_w(Z_{ij} - Z_G) = Y_{aij} = \mu + \alpha_j + \varepsilon_{ij}$$

Omitting the terms to the left of the first equals sign and employing equation (9.9) provides

$$Y_{aij} = \overline{Y}_{aj} + \varepsilon_{ij}$$

Therefore

$$\varepsilon_{ij} = Y_{aij} - \overline{Y}_{aj} \tag{9.13}$$

In other words, the full GLM errors are equal to each subjects' adjusted score minus the adjusted mean for the experimental condition. Equation (9.6) defines subjects' adjusted scores as

$$Y_{aij} = Y_{ij} - \beta_w(Z_{ij} - Z_G) \qquad \text{(9.6, rptd)}$$

which are calculated as below.

30 s $Y_{i1} - \beta(Z_{i1} - Z_G) = Y_{ai1}$	60 s $Y_{i2} - \beta(Z_{i2} - Z_G) = Y_{ai2}$	180 s $Y_{i3} - \beta(Z_{i3} - Z_G) = Y_{ai3}$
16 − 1.649(9 − 5.542) = 10.298	16 − 1.649(8 − 5.542) = 11.947	24 − 1.649(5 − 5.542) = 24.894
7 − 1.649(5 − 5.542) = 7.894	10 − 1.649(5 − 5.542) = 10.894	29 − 1.649(8 − 5.542) = 24.947
11 − 1.649(6 − 5.542) = 10.245	13 − 1.649(6 − 5.542) = 12.245	10 − 1.649(3 − 5.542) = 14.192
9 − 1.649(4 − 5.542) = 11.543	10 − 1.649(5 − 5.542) = 10.894	22 − 1.649(4 − 5.542) = 24.543
10 − 1.649(6 − 5.542) = 9.245	10 − 1.649(3 − 5.542) = 14.192	25 − 1.649(6 − 5.542) = 24.245
11 − 1.649(8 − 5.542) = 6.947	14 − 1.649(6 − 5.542) = 13.245	28 − 1.649(9 − 5.542) = 22.298
8 − 1.649(3 − 5.542) = 12.192	11 − 1.649(4 − 5.542) = 13.543	22 − 1.649(4 − 5.542) = 24.543
8 − 1.649(5 − 5.542) = 8.894	12 − 1.649(6 − 5.542) = 11.245	24 − 1.649(5 − 5.542) = 24.894
$\overline{Y}_{ai1} = 9.657$	$\overline{Y}_{ai2} = 12.275$	$\overline{Y}_{ai3} = 23.069$

As a check on their accuracy, the adjusted scores can be used to calculate group means for comparison with those calculated from equation (9.8). Error terms are calculated using the adjusted scores and adjusted experimental condition means according to equation (9.13). As always, the sum of the errors per experimental condition and across conditions equals zero (given rounding error). The sum of the squared errors across experimental conditions is the (reduced) error term for the full GLM.

30 s $Y_{ai1} - \overline{Y}_{a1} = \varepsilon_{i1}$	60 s $Y_{ai2} - \overline{Y}_{a2} = \varepsilon_{i2}$	180 s $Y_{ai3} - \overline{Y}_{a3} = \varepsilon_{i3}$
10.298 − 9.657 = 0.641	11.947 − 12.275 = −0.328	24.894 − 23.069 = 1.825
7.894 − 9.657 = −1.763	10.894 − 12.275 = −1.381	24.947 − 23.069 = 1.878
10.245 − 9.657 = 0.588	12.245 − 12.275 = −0.030	14.192 − 23.069 = −8.877
11.543 − 9.657 = 1.886	10.894 − 12.275 = −1.381	24.543 − 23.069 = 1.474
9.245 − 9.657 = −0.412	14.192 − 12.275 = 1.917	24.245 − 23.069 = 1.176
6.947 − 9.657 = −2.710	13.245 − 12.275 = 0.970	22.298 − 23.069 = −0.771
12.192 − 9.657 = 2.535	13.543 − 12.275 = 1.268	24.543 − 23.069 = 1.474
8.894 − 9.657 = − 0.763	11.245 − 12.275 = −1.030	24.894 − 23.069 = 1.825
$\sum_{i=1}^{N_j} \varepsilon_{i1} = 0.002$	$\sum_{i=1}^{N_j} \varepsilon_{i2} = 0.005$	$\sum_{i=1}^{N_j} \varepsilon_{i3} = 0.004$
$\sum_{i=1}^{N_j} \varepsilon_{i1}^2 = 21.944$	$\sum_{i=1}^{N_j} \varepsilon_{i2}^2 = 11.207$	$\sum_{i=1}^{N_j} \varepsilon_{i3}^2 = 95.312$
	$\sum_{j=1}^{p} \sum_{i=1}^{N_j} \varepsilon_{ij}^2 = 128.463$	

The reduced GLM errors for this ANCOVA design may be found in the same way as above. First, reordering the terms in equation (6.9) allows the subjects' adjusted scores to be specified. As

$$Y_{ij} = \mu + \beta_t(Z_{ij} - Z_G) + \varepsilon_{ij} \tag{9.11, rptd}$$

So

$$Y_{ij} - \beta_t(Z_{ij} - Z_G) = Y_{aij} = \mu + \varepsilon_{ij} \tag{9.14}$$

As experimental conditions are ignored, μ is equal to the general mean of the adjusted scores, Y_{aG}. It follows that for the reduced GLM

$$\varepsilon_{ij} = Y_{aij} - Y_{aG} \tag{9.15}$$

A first requirement to calculate the reduced GLM errors is the regression coefficient β_t

$$\hat{\beta}_t = \frac{\sum_{j=1}^{p}\sum_{i=1}^{N}(Z_{ij} - Z_G)(Y_{ij} - Y_G)}{\sum_{j=1}^{p}\sum_{i=1}^{N}(Z_{ij} - Z_G)^2} \tag{9.16}$$

Applied to the data in Table 9.1 provides

30 s $(Z_{ij} - Z_G)$	60 s $(Z_{ij} - Z_G)$	180 s $(Z_{ij} - Z_G)$
$(9 - 5.542) = 3.548$	$(8 - 5.542) = 2.458$	$(5 - 5.542) = -0.542$
$(5 - 5.542) = -0.542$	$(5 - 5.542) = -0.542$	$(8 - 5.542) = 2.458$
$(6 - 5.542) = 0.458$	$(6 - 5.542) = 0.458$	$(3 - 5.542) = -2.542$
$(4 - 5.542) = -1.542$	$(5 - 5.542) = -0.542$	$(4 - 5.542) = -1.542$
$(6 - 5.542) = 0.458$	$(3 - 5.542) = -2.542$	$(6 - 5.542) = 0.458$
$(8 - 5.542) = 2.458$	$(6 - 5.542) = 0.458$	$(9 - 5.542) = 3.458$
$(3 - 5.542) = -2.542$	$(4 - 5.542) = -1.542$	$(4 - 5.542) = -1.542$
$(5 - 5.542) = -0.542$	$(6 - 5.542) = 0.458$	$(5 - 5.542) = -0.542$
$\sum = 27.846$	$\sum = 16.098$	$\sum = 30.014$
	$\sum_{j=1}^{p}\sum_{i=1}^{N_j} = 73.958$	

30 s $(Z_{ij} - Z_G)(Y_{ij} - \overline{Y}_G)$	60 s $(Z_{ij} - Z_G)(Y_{ij} - \overline{Y}_G)$	180 s $(Z_{ij} - Z_G)(Y_{ij} - \overline{Y}_G)$
$(9 - 5.542)(16 - 15) = 3.458$	$(8 - 5.542)(16 - 15) = 2.458$	$(5 - 5.542)(24 - 15) = -4.878$
$(5 - 5.542)(7 - 15) = 4.336$	$(5 - 5.542)(10 - 15) = 2.710$	$(8 - 5.542)(29 - 15) = 34.412$
$(6 - 5.542)(11 - 15) = -1.832$	$(6 - 5.542)(13 - 15) = -0.916$	$(3 - 5.542)(10 - 15) = 12.710$
$(4 - 5.542)(9 - 15) = 9.252$	$(5 - 5.542)(10 - 15) = 2.710$	$(4 - 5.542)(22 - 15) = -10.794$
$(6 - 5.542)(10 - 15) = -2.290$	$(3 - 5.542)(10 - 15) = 12.710$	$(6 - 5.542)(25 - 15) = 4.580$
$(8 - 5.542)(11 - 15) = -9.832$	$(6 - 5.542)(14 - 15) = -0.458$	$(9 - 5.542)(28 - 15) = 44.954$
$(3 - 5.542)(8 - 15) = 17.794$	$(4 - 5.542)(11 - 15) = 6.168$	$(4 - 5.542)(22 - 15) = -10.794$
$(5 - 5.542)(8 - 15) = 3.794$	$(6 - 5.542)(12 - 15) = -1.374$	$(5 - 5.542)(24 - 15) = -4.878$
$\sum = 24.680$	$\sum = 24.008$	$\sum = 65.312$
	$\sum_{j=1}^{p} \sum_{i=1}^{N_j} = 114.000$	

$$\widehat{\beta}_t = \frac{114.000}{73.958} = 1.541$$

This regression coefficient estimate is used to calculate each subjects' adjusted score, according to equation (9.14).

30 s $Y_{ij} - \beta_t(Z_{ij} - Z_G) = Y_{aij}$	60 s $Y_{ij} - \beta_t(Z_{ij} - Z_G) = Y_{aij}$	180 s $Y_{ij} - \beta_t(Z_{ij} - Z_G) = Y_{aij}$
$16 - 1.541(3.548) = 10.671$	$16 - 1.541(2.458) = 12.212$	$24 - 1.541(-0.542) = 24.835$
$7 - 1.541(-0.542) = 7.835$	$10 - 1.541(-0.542) = 10.835$	$29 - 1.541(2.458) = 25.212$
$11 - 1.541(0.458) = 10.294$	$13 - 1.541(0.458) = 12.294$	$10 - 1.541(-2.542) = 13.917$
$9 - 1.541(-1.542) = 11.376$	$10 - 1.541(-0.542) = 10.835$	$22 - 1.541(-1.542) = 24.376$
$10 - 1.541(0.458) = 9.294$	$10 - 1.541(-2.542) = 13.917$	$25 - 1.541(0.458) = 24.294$
$11 - 1.541(2.458) = 7.212$	$14 - 1.541(0.458) = 13.294$	$28 - 1.541(3.458) = 22.671$
$8 - 1.541(-2.542) = 11.917$	$11 - 1.541(-1.542) = 13.376$	$22 - 1.541(-1.542) = 24.376$
$8 - 1.541(-0.542) = 8.835$	$12 - 1.541(0.458) = 11.294$	$24 - 1.541(-0.542) = 24.835$
	$\overline{Y}_i = 15.001$	

With the reduced GLM, the mean of the adjusted scores equals the mean of the unadjusted scores, that is, 15. Given rounding error, this is the value obtained from the adjusted scores. As specified by equation (9.15), the discrepancy between this mean and the subjects' adjusted scores provides the error term estimates.

30 s $Y_{ai1} - \overline{Y}_{a1} = \varepsilon_{i1}$	60 s $Y_{ai2} - \overline{Y}_{a2} = \varepsilon_{i2}$	180 s $Y_{ai3} - \overline{Y}_{a3} = \varepsilon_{i3}$
10.671 − 15 = −4.329	12.212 − 15 = −2.788	24.835 − 15 = 9.835
7.835 − 15 = −7.165	10.835 − 15 = −4.165	25.212 − 15 = 10.212
10.294 − 15 = −4.706	12.294 − 15 = −2.706	13.917 − 15 = −1.083
11.376 − 15 = −3.624	10.835 − 15 = −4.165	24.376 − 15 = 9.376
9.294 − 15 = −5.706	13.917 − 15 = −1.083	24.294 − 15 = 9.294
7.212 − 15 = −7.788	13.294 − 15 = −1.706	22.671 − 15 = 7.671
11.917 − 15 = −3.083	13.376 − 15 = −1.624	24.376 − 15 = 9.376
8.835 − 15 = −6.165	11.294 − 15 = −3.706	24.835 − 15 = 9.835

$$\sum_{j=1}^{p} \sum_{i=1}^{N_j} \varepsilon_{ij} = 0.012$$

$$\sum_{j=1}^{P} \sum_{i=1}^{N_j} \varepsilon_{i2}^{2} = 936.279$$

Given rounding error involved, 0.012 is not too bad an estimate of the correct value of zero. The sum of the squared errors provides the estimate of the reduced GLM error SS. Therefore, the error reduction as a consequence of taking account of the experimental conditions is

$$\text{Reduced GLM SS}_{\text{error}} - \text{Full GLM SS}_{\text{error}} = 936.279 - 128.463 = 807.816$$

Again the full GLM SS error is employed as the error term, but an additional degree of freedom is lost due to the use of the dependent variable on covariate regression line—for every regression line, or equivalently, for every covariate employed, an error *df* is lost. If desired, the SS accommodated by the covariate may be determined by comparing the error SS from the full ANCOVA with the error SS from an equivalent full ANOVA GLM. As before, all of this information can be displayed conveniently in an ANCOVA summary table, as presented in Table 9.2.

The tabled critical F-values presented in Appendix B may be used to determine significance if hand calculation is employed or the statistical software employed does not output the required p-values.

Table 9.2 ANCOVA Summary Table for the Single Factor, Single-Covariate Experiment

Source	SS	*df*	MS	*F*	*p*
Error reduction due to experimental conditions	807.816	2	403.908	62.883	<0.001
Error reduction due to covariate	199.537	1	199.537	31.065	<0.001
Full GLM error	128.463	20	6.423		

9.4 REGRESSION GLMs FOR THE SINGLE FACTOR, SINGLE-COVARIATE ANCOVA

The experimental design GLM equation (9.1) may be compared with the equivalent regression equation

$$Y_{ij} = \beta_0 + \beta_1 X_{i,1} + \beta_2 X_{i,2} + \beta_3 Z_{ij} + \varepsilon_{ij} \tag{9.17}$$

Similarly, the experimental design GLM equation (9.5) may be compared with the equivalent regression equation

$$Y_{ij} = \beta_0 + \beta_1 X_{i,1} + \beta_2 X_{i,2} + \beta_3 (Z_{ij} - Z_G) + \varepsilon_{ij} \tag{9.18}$$

In both equations (9.17) and (9.18), β_0 represents a constant common to all Y scores, β_1 is the regression coefficient for the predictor variable X_1 and β_2 is the regression coefficient for the predictor variable X_2, where the variables X_1 and X_2 code the differences between the three experimental conditions, β_3 is the regression coefficient for the covariate, Z_{ij} is the covariate score for the ith subject in the jth condition and as always, the random variable, e_{ij}, represents error. Table 9.3 presents effect coding for

Table 9.3 Effect Coding and Covariate for a Single Factor ANCOVA With One Covariate

Subject	Z	X_1	X_2	Y
1	9	1	0	16
2	5	1	0	7
3	6	1	0	11
4	4	1	0	9
5	6	1	0	10
6	8	1	0	11
7	3	1	0	8
8	5	1	0	8
9	8	0	1	16
10	5	0	1	10
11	6	0	1	13
12	5	0	1	10
13	3	0	1	10
14	6	0	1	14
15	4	0	1	11
16	6	0	1	12
17	5	−1	−1	24
18	8	−1	−1	29
19	3	−1	−1	10
20	4	−1	−1	22
21	6	−1	−1	25
22	9	−1	−1	28
23	4	−1	−1	22
24	5	−1	−1	24

Subject number and the dependent variable score also are shown.

Table 9.4 Results for the Full Single Factor, Single-Covariate ANCOVA Regression GLM

Variable	Coefficient	Standard Error	Standard Coefficient	t	p (Two-Tailed)
Constant	5.861	1.719	0.000	3.409	0.003
B_1	−5.344	0.734	−0.641	−7.278	<0.001
B_2	−2.725	0.733	−0.327	−3.716	0.001
Z	1.649	0.296	0.425	5.574	<0.001

the single factor, single-covariate regression GLM. It can be seen that apart from the addition of the Z covariate, the setup is identical to the effect coding for a single factor ANOVA with three levels.

Implementing a single factor, single-covariate ANCOVA is a two-stage procedure if only the variance attributable to the experimental conditions is to be assessed, and a three-stage procedure if the covariate regression is to be assessed. Consistent with estimating effects by comparing full and reduced GLMs, the first regression carried out is for the full single factor, single-covariate experimental design GLM, when all experimental condition predictor variables (X_1 and X_2) and the covariate are included. The results of this analysis are presented in Tables 9.4 and 9.5.

Table 9.4 presents the predictor variable regression coefficients and standard deviations, the standardized regression coefficients, and significance tests (t- and p-values) of the regression coefficient. As can be seen, the constant (coefficient) is equivalent to μ. Another useful value in Table 9.4 is the estimate of the full ANCOVA GLM covariate regression coefficient, $\widehat{\beta}_{\mathrm{w}}$. A t-test of this regression coefficient is also provided.

Table 9.5 presents the ANOVA summary table for the regression GLM describing the complete single factor, single-covariate ANCOVA. As the residual SS is that obtained when both covariate and experimental conditions are included in the regression, this is the error term obtained when the single factor, single-covariate ANCOVA GLM is applied.

At the second stage, a regression is applied which omits experimental conditions and employs the covariate (Z) as the only predictor. This regression GLM is equivalent to the reduced GLM for the single factor, single-covariate ANCOVA. The results of this analysis are presented in Tables 9.6 and 9.7. Of most interest is the regression residual/error SS provided in Table 9.7. The difference between the residual/error SS in Table 9.5 and that in Table 9.7 is equivalent to the SS attributable to experimental conditions. (This SS is presented in Table 9.9.) However, the SS attributed to the

Table 9.5 ANOVA Summary Table for Covariate and Experimental Conditions Regression

Source	SS	df	MS	F	p
Regression	983.537	3	327.846	51.041	<0.001
Residual	128.463	20	6.423		

R: 0.940; R^2: 0.884; adjusted R^2: 0.867.

Table 9.6 Results for the Reduced Single Factor, Single-Covariate ANCOVA Regression GLM

Variable	Coefficient	Standard Error	Standard Coefficient	t	p (Two-Tailed)
Constant	6.458	4.410	0.000	1.465	0.157
Z	1.541	0.759	0.398	2.032	0.054

Table 9.7 ANOVA Summary Table for Covariate Regression

Source	SS	df	MS	F	p
Regression	175.721	1	175.721	4.129	0.054
Residual	936.279	22	42.558		

R: 0.398; R^2: 0.158; adjusted R^2: 0.120.

covariate in Table 9.7 is not the covariate SS calculated when the full ANCOVA GLM is applied, as the regression coefficient when experimental conditions are omitted is $\widehat{\beta}_t$ and the full ANCOVA GLM employs the regression coefficient $\widehat{\beta}_w$ to estimate the variation in the dependent variable attributable to the covariate. As mentioned in Section 9.3, the SS for the covariate in the full ANCOVA GLM may be obtained by comparing the error SS from the full ANCOVA GLM with the error SS from an equivalent full ANOVA GLM. A full ANOVA GLM is implemented by a regression that uses only the predictors representing the experimental conditions (X_1 and X_2). Table 9.8 presents the ANOVA summary of this analysis.

Armed with the error term from the regression GLM implementation of the single-factor ANOVA, the error reduction attributable to the covariate can be calculated. This information is summarized in Table 9.9.

Table 9.8 ANOVA Summary Table for Experimental Conditions Regression

Source	SS	df	MS	F	p
Regression predictors for experimental conditions	784.000	2	392.000	25.098	<0.001
Residual	328.000	21	15.619		

R: 0.840; R^2: 0.705; adjusted R^2: 0.677.

Table 9.9 ANOVA Summary Table for Covariate and Experimental Conditions Regression

Source	SS	df	MS	F	p
Error reduction due to experimental conditions	807.816	2	403.908	62.883	<0.001
Error reduction due to covariate	199.537	1	199.537	31.065	<0.001
Full ANCOVA GLM residual	128.463	20	6.423		

R: 0.940; R^2: 0.884; adjusted R^2: 0.867.

Table 9.10 Adjusted Means for the Three Study Time Experimental Conditions

Factor Levels Study Time	a1 30 s	a2 60 s	a3 180 s
Adjusted means	9.66	12.28	23.07

9.5 FURTHER ANALYSES

Further analysis in ANCOVA usually involves comparison between the adjusted experimental condition means. However, due to the inclusion of a covariate in the GLM, a slightly different approach is employed to compare the experimental condition adjusted means. For example, it is assumed that in the hypothetical ANCOVA experiment, there is interest in the comparison between the 30 and 180 s condition adjusted means. The adjusted means were described in Section 9.2 and are presented in Table 9.10. The square of the difference between the two adjusted means divided by the variance of this difference is distributed as F with 1 numerator and $(N-p-1)$ denominator *dfs*.

The first step in obtaining the square of the difference between the two adjusted means is to determine the value of the linear contrast. The linear contrasts between adjusted means are defined as

$$\widehat{\psi} = \sum c_j \overline{Y}_{\mathrm{a}} \tag{9.19}$$

Applying equation (9.19) to the comparison between the 30 and 180 s experimental condition adjusted means provides

$$\widehat{\psi}_{1 \text{ versus } 3} = (-1)9.66 + (0)12.28 + (1)23.07 = 13.41$$

Its square is

$$\widehat{\psi}^2_{\mathrm{a3-a1}} = 13.41^2 = 179.83$$

The variance of the difference between the adjusted means is estimated by

$$s^2_{\psi} = \mathrm{MSe}\left[\sum_j \frac{c_j^2}{N_j} + \frac{(\sum_j c_j \overline{Z}_j)^2}{\sum_j \sum_i (Z_{ij} - \overline{Z}_j)^2}\right] \tag{9.20}$$

The calculation of the full ANCOVA GLM dependent variable on covariate regression coefficient in Section 9.2 provides the components to determine $\sum_j \sum_i (Z_{ij} - \overline{Z}_j)^2$. Applying equation (9.20) provides

$$s^2_{\overline{Y}a3-\overline{Y}a1} = 6.423\left[\frac{-1^2}{8} + \frac{1^2}{8} + \frac{(-1)(5.750)^2 + (1)(5.500)^2}{73.375}\right]$$

$$s^2_{\overline{Y}a3-\overline{Y}a1} = 6.423\left[\frac{2}{8} + \frac{(-5.750 + 5.500)^2}{73.375}\right] = 6.426[0.251]$$

$$s^2_{\overline{Y}a3-\overline{Y}a1} = 1.613$$

Therefore

$$F_{1,20} = \frac{\psi^2_{a3-a1}}{s^2_{\overline{Y}a3-\overline{Y}a1}} = \frac{179.83}{1.613} = 111.488, \quad p < 0.001$$

Assuming this is a planned comparison, no Type 1 error adjustment is necessary and difference between the two groups would be declared significant. Two other pairwise comparisons are available. They are a comparison of the 30 and 60 s (i.e., a1 versus a2) experimental condition means and a comparison of the 60 and 180 s (i.e., a2 versus a3) experimental condition means. Applying equation (9.19) to these comparisons provide

$$\widehat{\psi}_{1 \text{ versus } 2} = (-1)9.66 + (1)12.28 + (0)23.07 = 2.62$$

and

$$\widehat{\psi}_{2 \text{ versus } 3} = (0)9.66 + (-1)12.28 + (1)23.07 = 10.79$$

The squares are

$$\widehat{\psi}^2_{1 \text{ versus } 2} = 6.864$$

$$\widehat{\psi}^2_{2 \text{ versus } 3} = 116.424$$

Therefore, the two F-tests provide

$$F_{1,20} = \frac{\psi^2_{a2-a1}}{s^2_{\overline{Y}a2-\overline{Y}a1}} = \frac{6.864}{1.613} = 4.255, \qquad p = 0.052$$

$$F_{1,20} = \frac{\psi^2_{a3-a2}}{s^2_{\overline{Y}a3-\overline{Y}a2}} = \frac{116.424}{1.613} = 72.179, \quad p < 0.001$$

Although both of these comparisons are unplanned, the significant ANCOVA omnibus F-test rejects the omnibus null hypothesis and Shaffer's (1986) account of logically related hypotheses informs that only one pairwise null hypothesis possibly could be true. Therefore, the classical p-values associated with the adjusted mean pairwise comparison F-tests can be accepted without any Type 1 error adjustment because these values assume only one possibly true null hypothesis. Therefore, the comparison between the 60s and 180s experimental condition adjusted means is declared significant, but the comparison between the 30 and 60 s experimental condition adjusted means is declared not significant. (*Note:* The substantial F-values probably reflect the hypothetical nature of the data—such large F-values are unlikely in real research studies.)

9.6 EFFECT SIZE ESTIMATION

Just as related design effect estimates are modified for comparison with equivalent independent designs, it may be thought that something similar might be done to allow effect size comparisons across ANCOVA designs and their equivalent independent designs. However, while related designs reduce error without influencing the experimental effect estimate(s), ANCOVA reduces error and influences the experimental effect estimate(s). This is a key difference and one that precludes the comparability between ANCOVA and independent designs available between related and independent designs.

9.6.1 A Partial $\widehat{\omega}^2$ SOA for the Omnibus Effect

The most appropriate estimate of effect size for ANCOVA is an estimate that takes into account the full consequences for the experimental effect and error term of the dependent variable on covariate regression. A partial ω^2 can be defined for the single factor ANCOVA omnibus (population) effect. Angle brackets denote the partial effect estimates

$$\omega^2_{\langle\text{effect}\rangle} = \frac{\sigma^2_{\text{effect}}}{\sigma^2_{\text{effect}} + \sigma^2_{\text{error}}} \qquad (5.28, \text{rptd})$$

The partial ω^2 employs the adjusted experimental effect and the adjusted error, but omits from the denominator the variation explained by the regression of the dependent variable on the covariate. The partial $\widehat{\omega}^2$ effect size estimate for the single factor ANCOVA omnibus effect is defined by

$$\widehat{\omega}^2_{\langle\text{effect}\rangle} = \frac{df_{\text{effect}}(F_{\text{effect}} - 1)}{df_{\text{effect}}(F_{\text{effect}} - 1) + N} \qquad (5.29, \text{rptd})$$

Applied to the hypothetical ANCOVA experimental data provides

$$\widehat{\omega}^2_{\langle\text{effect}\rangle} = \frac{2(62.883 - 1)}{2(62.883 - 1) + 24} = 0.838$$

Again, the finding that as much as 84% of the variance associated with the experimental effect is attributable to the experimental effect probably reflects the hypothetical nature of the data.

9.6.2 A Partial ω^2 SOA for Specific Comparisons

As with independent and related designs, there is likely to be considerable interest in the SOA for particular comparisons between experimental conditions in ANCOVA designs. For example, just as it was in the hypothetical independent design, the SOA between the 30 and 180 s conditions in the hypothetical repeated measures study time experiment is of interest.

For single factor ANCOVA designs, the specific comparison partial ω^2 can be defined as

$$\widehat{\omega}^2_{\langle\psi\rangle} = \frac{F_\psi - 1}{F_\psi - 1 + 2N_j} \qquad (4.16, \text{rptd})$$

Applying equation (9.26) to the F-value obtained in Section 9.5 provides

$$\widehat{\omega}^2_{\langle\psi\rangle} = \frac{111.488 - 1}{111.488 - 1 + 2(8)} = \frac{110.488}{126.488} = 0.874$$

Therefore, 87% of the variance in the 30 and 180 s populations is explained by the comparison between these two experimental conditions. As said before, this high ω^2 for the specific comparison probably reflects the hypothetical nature of the data – it is unlikely that experimental manipulations would accommodate such a large proportion of the variance when real data is obtained.

9.7 POWER

Determining the sample size required to achieve a specific level of power in an ANCOVA design is virtually identical to determining the sample size required to achieve a specific level of power in an equivalent single or multi factor independent design (see Sections 4.3 and 5.7). When the planned ANCOVA experiment is based on a previous ANCOVA experiment, the first step is to calculate the omnibus or specific comparison partial $\widehat{\omega}^2$ effect size estimates. Once these values are obtained, the remaining procedure replicates that described for single or multifactor independent designs. When an ANOVA is the basis of the planned ANCOVA, it is necessary to estimate the expected ANCOVA partial ω^2 effect sizes from the ANOVA $\widehat{\omega}^2$ effect size(s). In population parameter terms, the relationship between an ANCOVA partial $\widehat{\omega}^2$ effect size and an ANOVA $\widehat{\omega}^2$ effect size is

$$\widehat{\omega}^2_{\langle\text{effect}\rangle\text{ANCOVA}} = \frac{\sigma^2_{\text{effect ANOVA}}}{\sigma^2_{\text{effect ANOVA}} + (1 - \rho^2_{ZY})\sigma^2_{\text{S(A)}}} \qquad (9.21)$$

where ρ_{ZY}^2 is the correlation between the covariate and the dependent variable. When applied to experimental data, and an estimate of the correlation between the covariate and the dependent variable is available, perhaps from a pilot study, the ANCOVA partial ω^2 effect is estimated by

$$\widehat{\omega}^2_{\langle\text{effect}\rangle\text{ANCOVA}} = \frac{\omega^2_{\text{effect ANOVA}}}{\widehat{\omega}^2_{\text{effect ANOVA}} + (1 - \widehat{\rho}^2_{ZY})(1 - \omega^2_{\text{effect ANOVA}})} \tag{9.22}$$

In factorial designs, each ANOVA main and interaction effect should be modified as described above.

9.8 OTHER ANCOVA DESIGNS

Many texts on experimental design and statistical analysis often are vague about any form of ANCOVA other than single factor independent measures ANCOVA. Although there is insufficient space here to consider other ANCOVA designs in a detailed fashion, the following discussion is presented to provide some appreciation of the different types ANCOVA designs and analyses available.

9.8.1 Single Factor and Fully Repeated Measures Factorial ANCOVA Designs

Single factor repeated measures designs and indeed, all fully repeated measures factorial designs derive no benefit from ANCOVA. In these designs, as all subjects experience and provide a dependent variable score in all of the experimental conditions, there are no group differences to adjust and so no role for ANCOVA. (e.g., Keppel and Zedeck, 1989.)

9.8.2 Mixed Measures Factorial ANCOVA

In contrast to fully related factorial designs, there may be advantage in applying ANCOVA to mixed designs. There are two sorts of mixed ANCOVA design. In the simplest of these designs (Figure 9.2a), each subject provides a single score on the covariate(s). In the more complicated design (Figure 9.2b), each subject provides covariate scores in each experimental condition and it provides covariate scores for each dependent variable score.

For the simpler mixed measures factorial ANCOVA design, the covariate(s) will have consequence for the independent measures factor (Factor A in Figure 9.2a) and, for the same reasons as described for single factor repeated measures designs, the covariate will have no effect on the related factor effect (Factor B in Figure 9.2a), nor will the covariate exert any influence on the interaction between the related and the independent factors. Therefore, the simpler mixed measures factorial ANCOVA design may be analyzed by carrying out two separate analyses: (1) A single independent measures factor ANCOVA and (2) a mixed measures factorial ANOVA.

			Factor B		
			b1	b2	b3
Factor A	Subject	Z	Y	Y	Y
a1	1 2 3				
a2	4 5 6				

(a)

		Factor B					
		b1		b2		b3	
Factor A	Subject	Z	Y	Z	Y	Z	Y
a1	1 2 3						
a2	4 5 6						

(b)

Figure 9.2 Mixed factorial ANCOVA designs. (a) Mixed factorial ANCOVA design with one covariate measure per subject. (b) Mixed factorial ANCOVA design with one covariate measure per repeated measure.

The effect of the independent measures factor is assessed by a single factor ANCOVA applied to the subjects' covariate score(s) and the mean of their repeated measures scores. The related factor main effect and the independent factor and related factor interaction effect are assessed by a mixed measures factorial ANOVA.

In the more complicated mixed measures factorial ANCOVA design (Figure 9.2b), the covariate(s) have consequence for both independent and related factor effects. Consequently and in contrast to the simpler mixed measures factorial ANCOVA design, there are no convenient shortcuts or checks. The traditional ANOVA approach to both mixed factorial ANCOVA designs is presented by Winer, Brown, and Michels (1991), while Huitema (1980) describes the regression GLM for the simpler mixed measures factorial ANCOVA (Figure 9.2a). Considerable care also should be taken when using statistical software to implement these sorts of ANCOVA. Indeed, there is sufficient ambiguity over the form of ANCOVA implemented by some statistical software packages that testing the package by using it to analyze example data (as provided in statistics texts) and inspecting the results output to see if they match with the expected results is a wise strategy.

CHAPTER 10

Assumptions Underlying ANOVA, Traditional ANCOVA, and GLMs

10.1 INTRODUCTION

Ensuring the statistical assumptions underlying analyses are tenable often is considered to be a technical matter to be addressed and resolved prior to the streamlined presentation of the study results in a research journal. Unfortunately this can convey the impression that assumption assessment is not done and so cannot be important. However, the authors of nearly all higher level psychology statistics texts endorse the view that assumption checks should be carried out to ensure the validity of the data analysis (e.g., Cohen et al., 2003; Howell, 2010; Keppel and Wickens, 2004; Kirk, 1995; Maxwell and Delaney, 2004; Myers, Well, and Lorch, 2010; Winer, Brown, and Michels, 1991). Indeed, a good understanding of analysis assumptions enables a researcher to determine the extent to which any of the assumption violations detected jeopardize the validity of the analyses applied.

Given this account, it may be surprising to discover that whether or not assumption checks should be conducted before an analysis is applied has become a matter of debate with some authors unequivocally advising against any assumption checks (e.g., Wells and Hintze, 2007). The argument for abandoning assumption checks is considered at the end of this chapter. However, the prior presentation of ANOVA, traditional ANCOVA, and GLM assumptions and methods of assessing these assumptions anticipates the outcome of this consideration.

10.2 ANOVA AND GLM ASSUMPTIONS

A least squares GLM specification includes more than an equation describing the data in terms of model parameters and error terms. There is also a set of assumptions specifying restrictions on the model parameters and error terms. Some model

ANOVA and ANCOVA: A GLM Approach, Second Edition. By Andrew Rutherford.

parameter assumptions were discussed in Section 2.8, but assumptions about ANOVA model parameters often are expressed implicitly in terms of the experimental design and are made explicit only when the ANOVA is discussed in greater statistical detail. Nevertheless, when GLM statistical assumptions are considered, there is an almost exclusive focus on the assumptions about the error terms – only in ANCOVA does the focus shift to assumptions about model parameters (see Sections 10.2.3 and 10.4.2.1). However, this focus may be regarded as a parsimonious strategy, as examining errors not only provides a direct assessment of the error assumptions, but the nature of any error assumption violations also can reveal problems with assumptions about the model parameters.

Although only one small set of assumptions underlie all GLMs, sometimes it seems there are a number of distinct analyses and all make different statistical assumptions. A major contributor to this impression is the different terminologies that developed within the research areas employing regression and ANOVA (see Section 1.2). Misidentifying the matrix algebra expression as the GLM (see Section 1.3) also encourages the view that GLM analyses differ from regression and ANOVA analyses, while the use of regression terminology by GLMs (due to the regression format being more elemental and so more widely applicable) also suggests a false distinction between GLMs and regression on one side, and ANOVA on the other. However, another contributor to the impression of distinct analyses making different assumptions is the additional assumptions some analyses make to simplify their calculation and interpretation.

The following section on independent measures designs presents the typical expression of ANOVA assumptions and the typical expression of GLM and regression assumptions. It is explained why these two sets of assumptions are equivalent and that all GLMs make this small single set of statistical assumptions. Subsequently, the further assumptions made to simplify parameter estimation and interpretation when related ANOVA and traditional ANCOVA are applied are discussed.

10.2.1 Independent Measures Designs

The set of assumptions underlying all GLM analyses are most apparent in the context of independent measures designs. The ANOVA and GLM expressions of these assumptions are presented in Table 10.1. As the ANOVA and GLM assumptions labelled **a** in Table 10.1 are expressed identically, their equivalence is appreciated easily. However, the assumptions labelled **b**, **c** and **d**, in Table 10.1 show that typically, ANOVA assumptions refer to the dependent variable scores, whereas GLM assumptions are expressed with respect to their error components.

Consideration of the independent measures experimental design GLM reveals the model component provides the predictions of the experimental condition means and the only deviation from these predictions is provided by the error term. As the only variation in scores within an experimental condition is due to the error term, it follows that examining scores by experimental condition is equivalent to examining their errors. Therefore, when the ANOVA assumptions **b** and **c** refer to scores within experimental conditions, really they are referring to the GLM errors and so these

Table 10.1 ANOVA and GLM Assumptions

	ANOVA Assumptions	GLM Assumptions
a	Each condition contains a random sample of the population of such scores	Each condition contains a random sample of the population of such scores
b	The scores in each condition are distributed normally	Errors are distributed normally
c	The scores in each condition are independent of each other	Errors are independent
d	The variances of the scores in each experimental condition are homogeneous	Errors are homoscedastic (errors exhibit common variance across all values of the predictor variables)

ANOVA and GLM assumptions are equivalent. For assumption **d**, the same logic applies with respect to the equivalence of examining errors and dependent variable scores per condition. However, the homoscedasticity assumption is a more general form of the homogeneity assumption. The homoscedasticity assumption is that the error variances observed at any combination of predictor variable values will be equivalent, whereas the ANOVA assumption refers only to equivalent error variance across experimental conditions. For example, in ANCOVA, the homoscedasticity assumption is that error variance will be equivalent, not only across experimental conditions, but also at any covariate value within the experimental conditions.

As the only way to assess error assumptions using full scores is on a condition by condition basis, each assessment is limited to the number of scores per condition. Although errors still can be identified by condition, as they are free of the influence of the experimental conditions, all errors can be examined together. Therefore, assessing error assumptions directly not only facilitates graphical and test based assumption checks by maximizing the data set, but it also allows sophisticated assumption checks developed for regression to be applied to ANOVA and ANCOVA.

Valid significance tests require normally and independently distributed (NID) errors (e.g., Draper and Smith, 1998; Kirk, 1995; Pedhazur, 1997; Snedecor and Cochran, 1980). Although these assumptions are unnecessary when a GLM is used only to describe data, they still would enhance the accuracy of the description. However, as GLMs usually are applied with the intent to test the significance of parameter estimates, usually NID errors are necessary assumptions.

The level of measurement appropriate for the GLM dependent variable is not presented as an ANOVA or GLM assumption. Some authors consider the level of measurement of the dependent variable as determining which statistical analyses are and are not appropriate (e.g., Stevens, 1951; Suppes and Zinnes, 1963). Typically, such authors would consider ANOVA as assuming an interval level of dependent variable measurement. However, there are also authors who consider the level of the dependent variable measurement to be largely irrelevant as far as choosing a statistical technique is concerned (e.g., Townsend and Ashby, 1984; Mitchell, 1986). ANOVA texts have tended, either implicitly (Kirk, 1995) or explicitly (e.g., Howell, 2010;

Winer, Brown, and Michels, 1991) to concord with the latter view. Currently, the general opinion seems to be that the measurement issue falls within the realm of methodology, rather than statistics and it is more important that the numbers representing the dependent variable accurately reflect the entity in which there is interest, than they comply with the requirements of a particular measurement scale. After all, it may be that the entity in which we are interested does not increment in an orderly fashion.

10.2.2 Related Measures

Related measures designs (i.e., repeated measures and other randomized block designs) present a problem for the typical GLM assessment of the statistical assumptions concerning errors. In Chapter 6, it was described how each subject providing only one score per experimental condition precluded separation of the subject by factor interaction estimates from the error estimates and so, error estimates per se cannot be obtained. One consequence of the lack of error term estimates is that unique and self-contained assumption assessment and remedial techniques have developed for related measures designs.

When the covariance matrix of the experimental condition scores is spherical, the biases in the numerator and the denominator mean squares cancel out to provide a valid and accurate *F*-test (Huynh and Feldt, 1970; Rouanet and Lépine, 1970). Therefore, when related ANOVAs are applied, spherical experimental condition score covariance matrices are assumed. Sphericity is a property apparent in the matrix algebra representation of the GLM experimental conditions covariance matrix when there is homogeneity of the variances of the differences between the scores in the related experimental conditions – the related factor levels. For example, given a repeated-measures ANOVA with three experimental conditions, the variance of the differences between the subjects scores in Conditions 1 and 2 should be the same as the variance of the differences between the subjects scores in Conditions 1 and 3, or 2 and 3. Despite homogeneity of variances of differences providing a clear statement of the nature of the assumption, the more obscure sphericity label is more popular, probably due to its brevity.

A more constrained form of sphericity is termed compound symmetry or circularity. The experimental conditions covariance matrix will exhibit compound symmetry if and only if there is homogeneity of score variance across all conditions (1, 2, and 3) and homogeneity of all correlations between scores across all pairs of conditions (i.e., 1 and 2, 1 and 3, 2 and 3). Although sphericity without compound symmetry is possible (but not compound symmetry without sphericity), if real data are spherical, then they almost always exhibit compound symmetry (Howell, 2010; Maxwell and Delaney, 2004). Nevertheless, the need to assume a spherical covariance matrix is a significant restriction for psychological data.

10.2.2.1 Assessing and Dealing with Sphericity Violations

Box (1954) described a parameter that indexes the extent of the sphericity assumption violation. (Unfortunately, as the Greek letter epsilon, ε, is used to denote this parameter, the potential for confusion arises as epsilon also denotes the error term parameter.) Box's ε varies between 0 and 1, with lower values indicating greater violation of sphericity, but it tends to underestimate the parameter ε, so overestimating the extent of the sphericity violation. However, due to the laborious calculations required to obtain an estimate of ε ($\hat{\varepsilon}$), most workers used Geisser and Greenhouse's lower bound adjustment instead. Geisser and Greenhouse (1958) demonstrated that the lowest possible value for the parameter ε in a single factor ANOVA design with p levels provides a numerator *df* of 1 and denominator *dfs* of $(p - 1)$. Of course, this is a very conservative adjustment, as for most data the true value of ε would be much larger than that which would provide the lower bound adjustment. In place of Box's $\hat{\varepsilon}$, Huynh and Feldt (1976) suggested the estimate $\tilde{\varepsilon}$. However, $\tilde{\varepsilon}$ tends to overestimate ε slightly and so, it slightly underestimates the sphericity violation. (Keep in mind that lower values of ε indicate greater sphericity violation.) Laborious calculations are required to estimate $\hat{\varepsilon}$ and $\tilde{\varepsilon}$, but from the late 1970s, these estimates were provided by many statistical packages and began to be used in preference to Geisser and Greenhouse's lower bound adjustment (see Table 10.2). Confusingly, however, the statistical packages usually label $\hat{\varepsilon}$ as the Geisser and Greenhouse adjustment because these programs follow Geisser and Greenhouse's (1958) generalization of $\hat{\varepsilon}$ to more complicated designs.

Greater violation of the sphericity assumption increases the F-test Type 1 error rate. However, reducing the F-test numerator and denominator *dfs* increases the p-values associated with the F-statistic. Therefore, once the extent of the sphericity assumption violation has been assessed, the F-value numerator and denominator *dfs* can be reduced appropriately to rectify the F-test Type 1 error rate and this is exactly what Box's Geisser and Greenhouse epsilon and Huynh and Feldt's epsilon are employed to do.

Table 10.2 presents the SYSTAT repeated measures ANOVA summary table for the experimental data presented in Chapter 6. In common with most statistical software packages, the output includes Geisser and Greenhouse's generalization of Box's $\hat{\varepsilon}$, Huynh and Feldt's (1976) estimate, $\tilde{\varepsilon}$, and the adjusted F-test p-values based on these two epsilon estimates. (SPSS provides the same information, but in a very unhelpful fashion.) There are three levels of the experimental conditions factor in this experiment, but when an experimental factor has only two levels, there can be only one difference between the factor levels and with only one difference, homogeneity of variance of differences (i.e., sphericity) cannot be an issue.

When factorial repeated measures experimental design GLMs are applied, the typical output from a factorial repeated measures ANOVA provides

Table 10.2 SYSTAT Output: Summary Table for the Single Factor Repeated Measures ANOVA with Greenhouse–Geisser (G–G) and Huynh–Feldt (H–F) p-Value Adjustments

Source	SS	*df*	MS	*F*	*p*	G–G	H–F
Experimental conditions	112.000	2	56.000	20.634	<0.001	0.002	0.001
Error	38.000	14	2.714				

Greenhouse–Geisser Epsilon, 0.562; Huynh–Feldt Epsilon, 0.595.

Greenhouse-Geisser and Huynh-Feldt epsilon estimates and p-value adjustments for each factor and each interaction between factors. Similarly, when mixed measures experimental design GLMs are applied, the typical output from a mixed measures ANOVA provides Greenhouse-Geisser and Huynh-Feldt epsilon estimates and p-value adjustments for each repeated measures factor and for each interaction involving a repeated measures factor. This demonstrates the applicability of the sphericity assumption to these factors and factor interactions. The GLM assumptions **a** and **b** also apply to these factors and their interactions. In mixed measures ANOVAs, all of the GLM assumptions apply to the independent factors.

Greater access to statistical computing resources has supported the application of a multivariate approach to the analysis of related measures data (e.g., Hand and Taylor, 1987; Maxwell and Delaney, 2004; O'Brien and Kaiser, 1985). However, empirical evidence indicates that with balanced designs, both the corrected *df* and multivariate approaches provide valid and effective control of Type 1 error (Keselman, Lix, and Keselman, 1996), but generally the univariate approach is more powerful (see Davidson, 1972, or the summary provided by Maxwell and Delaney, 2004). For this reason and also because the univariate approach links more easily to multilevel modeling approaches to the analysis of related measures, the focus here is on the univariate GLM approach to related measures designs.

10.2.3 Traditional ANCOVA

In addition to all of the ANOVA assumptions, traditional ANCOVA also makes the assumptions listed in Table 10.3. These assumptions have no counterparts in GLM terms, as they are made to simplify the interpretation and/or calculation of the ANCOVA. Although orthogonal predictors are preferable and correlations can cause interpretation problems, covariate-treatment correlations do not preclude accurate and informative analysis (Cohen and Cohen, 1983; Rutherford, 1992). GLMs also can accommodate nonlinear regression of the dependent variable on the covariate (see polynomial regression, e.g. Draper and Smith, 1998; Kutner et al, 2005) and heterogeneous regressions (e.g. Rutherford, 1992; Searle, 1979, 1987). Indeed, the popularity of heterogeneous regression ANCOVA seems to be increasing (e.g., Maxwell and Delaney, 2004).

The form of ANCOVA incorporating the assumptions listed in Table 10.3 is termed traditional ANCOVA to distinguish it from less constrained forms of ANCOVA.

Table 10.3 Specific ANCOVA Assumptions

a	The covariate is independent of the treatments
b	In each treatment group the relationship between the covariate and the dependent variable is linear (the covariate and dependent variable are expressed at the first power only)
c	The regression coefficients of the dependent variable on the covariate in each treatment group are homogeneous

Traditional ANCOVA is still the most common form of ANCOVA applied in psychological research and the programs labeled ANCOVA in most commercial statistical packages implement this form of ANCOVA. Beyond the benefits accrued from simplifying ANCOVA interpretation and/or calculation, there are other reasons for choosing traditional ANCOVA. First, the good ANCOVA design practice of measuring the covariate before administering the experimental manipulation(s) usually ensures the experimental conditions cannot influence the covariate and second, most relationships between covariates and dependent variables in psychology appear to be linear, or are approximately linear. Therefore, it is very likely that two of the three traditional assumptions will be tenable for most ANCOVAs. Unfortunately, however the assumption of homogeneous regressions across experimental conditions becomes more tenuous as the number of experimental conditions increases and as the number of experimental factors increase (e.g., Winer, Brown, and Michels, 1991).

10.3 A STRATEGY FOR CHECKING GLM AND TRADITIONAL ANCOVA ASSUMPTIONS

A general strategy for checking GLM and traditional ANCOVA assumptions is presented. Lack of space prevents the detailed description of measures—mainly data transformations—to remedy data that fail to meet ANOVA or ANCOVA assumptions. Nevertheless, excellent accounts are provided by Daniel and Wood (1980), Emerson (1991), Hoaglin, Mosteller, and Tukey (1985), Kirk (1995), Mosteller and Tukey (1977), Kutner, Nachtstein, Neter and Li (2005), Tukey (1977), and Weisberg (1985). An interesting debate as to whether and which transformations should be applied can be found in Games (1983, 1984), Levine and Dunlap (1982, 1983), and Grissom (2000). Moreover, if certain assumptions underlying traditional ANCOVA are violated, some of the alternatives to traditional ANCOVA (Chapter 11) may be applied.

The GLM assumptions provide criteria to judge the statistical validity of any GLM, while the nature of error assumption violations can indicate other problems with the model parameters applied. For example, violations of the extra traditional ANCOVA assumptions typically manifest as violations of one or more GLM error assumptions.

The general strategy advocated to check ANOVA and ANCOVA assumptions employs a stepwise, and, if necessary, an iterative approach. The basic outline of this strategy is presented in Figure 10.1.

First, the analysis is implemented and the GLM residuals (the error term estimates) are obtained (Box 1 and Box 2). These residuals are analyzed in terms of their conformity to GLM assumptions (Box 3). At this point the first branch in the assessment path is reached. If the GLM assumptions are judged to be tenable, nothing more need to be done and the analysis results can be interpreted (Box 4). However, if any of the GLM assumptions is judged to be untenable after an ANOVA was implemented (Box 5), remedial action(s) must be undertaken with respect to the model or data (Box 6). If any of the GLM assumptions are judged untenable after an ANCOVA was implemented (Box 7), then it is possible that the cause of the GLM assumption violation(s) is a failure of one or more of the specific traditional ANCOVA

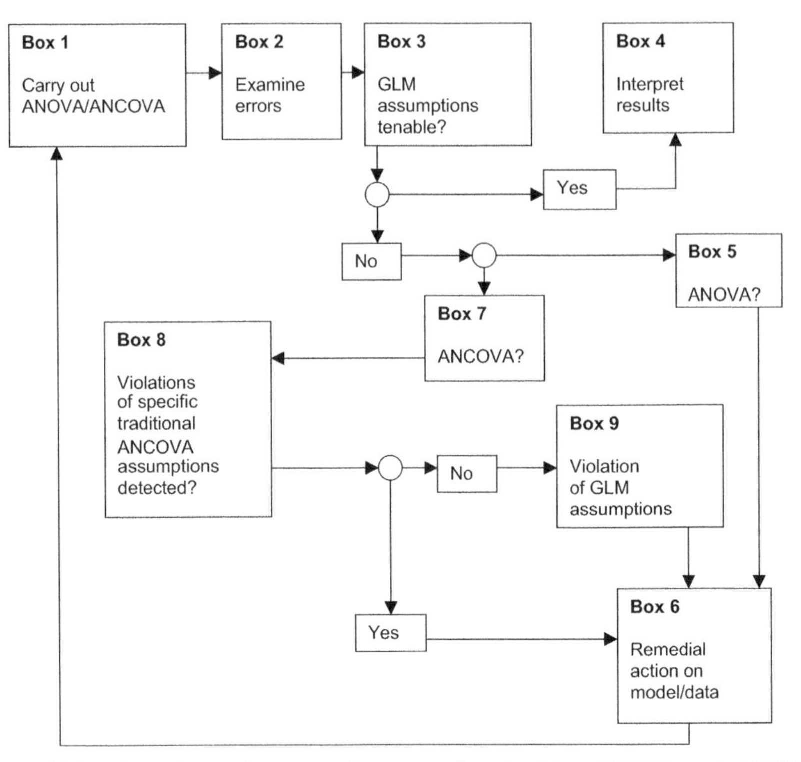

Figure 10.1 Flow chart of a general strategy for checking ANOVA and ANCOVA assumptions.

assumptions (Box 8). If violations of any specific traditional ANCOVA assumptions are detected, then appropriate remedial action on the model or data should be undertaken (Box 6). If there are no specific traditional ANCOVA assumption violations, a failure of one or more of the GLM assumptions is indicated (Box 9). In such circumstances, remedial actions with respect to the model or data should be undertaken (Box 6) and the analysis repeated (Box 1). Nevertheless, after the second analysis, it is necessary to ensure the underlying assumptions are tenable before interpreting the results.

10.4 ASSUMPTION CHECKS AND SOME ASSUMPTION VIOLATION CONSEQUENCES

There are both graphical and significance test methods for assessing assumption conformity. Although the former approach seems more popular (e.g., Cohen et al, 2003; Cook and Weisberg, 1983; Draper and Smith, 1998; Lovie, 1991b; Montgomery and Peck, 1982; Kutner, Nachtsheim, Neter, and Li, 2005; Norusis, 1990; Pedhazur, 1997), it may be difficult for the less experienced to determine assumption violations in this way. With graphical approaches there may be a tendency to ignore or fail to appreciate the spread of scores in relation to the specific aspect of the data considered. In contrast, it is exactly these relations that most significance tests formalize. Nevertheless, as sample size increases, test power increases and with large enough samples, virtually all tests will have sufficient power to reject the null hypothesis. This is a problem because the influence of random processes makes exact conformity with assumptions extremely unlikely and so the issue always is the extent to which assumptions are met. With large data sets, the ability to reject the null hypothesis may not be the best assessment criterion. As well as determining the significance level, the extent of the assumption violation should be considered, perhaps by comparing the size of the test statistic obtained with its expected value under the null hypothesis.

To illustrate graphical and significance test methods, some of the assessment techniques described are applied to the single factor independent measures ANCOVA with one covariate example (see Chapter 9). Most commercially available statistical packages are able to describe ANOVA and ANCOVA in terms of the experimental design GLM and can provide estimates of the model errors (residuals) as part of the output. Once obtained, the residuals can be input to other programs in the statistical package for examination. However, implementing ANOVA and ANCOVA as regression models does offer an advantage with respect to the analysis of errors. As most good quality regression software provides a range of techniques for examining errors regression implementations of ANOVA and ANCOVA can make use of these (sometimes automatic) regression diagnostics programs.

10.4.1 Independent Measures ANOVA and ANCOVA Designs

10.4.1.1 Random Sampling

Ideally, assumption (a) in Table 10.1 should be satisfied by implementing two randomization procedures, one after the other. First, subjects are sampled randomly from the population of interest. The manner of this sampling determines the validity of the inferences from the sample to population of interest. However, very few experiments in psychology invest great effort to ensure a random sample of the population to which inferences will be generalized. Most research conducted in European and American universities employs convenience samples of the undergraduate population. Usually results are generalized to the population of the western world, if not the population of the whole world, on the presumption that with respect to the psychological processes examined in the experiment, there are no real differences between the participating undergraduates, the rest of the undergraduate population, and the western and world populations (see Maxwell and Delaney, 2004; Wright, 1998). Second, the subjects constituting the sample are assigned randomly to the experimental conditions. After such random assignment, it is most likely that any subject characteristics such as academic ability, friendliness, and so on, will be distributed equally across all conditions. In other words, it is most unlikely that differences in subjects' dependent variable scores will be due to differences in subject characteristics across experimental conditions. Consequently, random assignation of subjects to experimental conditions is the basis for attributing any differences observed between the experimental conditions to these experimental conditions. The validity of any ANOVA is severely questioned whenever these sampling procedures are compromised.

10.4.1.2 Independence

The value of one GLM error is assumed not to affect the value of another. However, while error terms are conceived as independent, when h GLM parameters are estimated from N observations, there are only $N - h$ *dfs*, so the residuals, the error term estimators, will covary (e.g., Draper and Smith, 1998). Residuals related only in this way are of little concern. However, residuals may be related in other ways and relatively few statistical texts point out that ANOVA and ANCOVA are not robust with respect to violation of the independent errors assumption (e.g., Maxwell and Delaney, 2004).

As well as being part of the basis for attributing experimental effects to the experimental manipulation, randomization procedures also increase the likelihood of independent scores; there is no reason to believe that scores from subjects randomly selected from the population of interest and randomly allocated to experimental conditions will be related. Generally, appropriate randomization procedures (and the application of a pertinent analysis) provide independent errors. However, contrary to some claims (e.g., Winer, Brown, and Michels, 1991), randomization procedures cannot assure error independence. Sometimes despite random sampling and assignment, relations between scores can arise. In particular, the way in which the dependent variable scores are collected may produce related scores. For example, scores from subjects tested as a group, or scores obtained using a certain piece of equipment, or questionnaire, and so on may be related. Consequently, consideration of the full study methodology, ideally before its implementation, is necessary to ensure independent errors.

Kenny and Judd (1986) describe means of assessing the nonindependence of ANOVA errors due to groups, sequence and space, as well as methods to eliminate the F-ratio numerator and denominator mean square biases caused by nonindependent errors (see Section 10.2.2). Of these three sources, nonindependence due to groups is the most frequently encountered in psychological research. Groups may be defined in a variety of different ways (e.g., see blocking, Hays 1994; Kirk, 1995; Winer, Brown, and Michels, 1991). Most familiar and obvious is the situation where groups and experimental conditions are equivalent. However, data sharing any common feature can be grouped. Ideally, groups should be crossed with experimental conditions. In a single factor ANOVA design, where data are arranged on the basis of the grouping criteria, the nonindependence of errors due to groupings when a spherical covariance matrix is assumed can be estimated by the Within Groups Correlation

$$\text{WGC} = \frac{\text{MS}_\text{b} - \text{MS}_\text{w}}{\text{MS}_\text{b} + \text{MS}_\text{w}(N_j - 1)} \tag{10.1}$$

where MS_b and MS_w are the mean squares between and within groups, and N_j is the number of scores in each group (e.g., Kenny and Judd, 1986). The WGC calculated can be treated as an F-ratio with $(N - 1)$ numerator and $(N - 1)$ denominator *dfs*

(Donner and Koval, 1980). Negative group linkage is indicated by significant WGC values below 1 and positive group linkage is indicated by significant WGC values above 1. Positive values indicate greater similarity of residuals within groups than between, while negative values indicate the converse. Preliminary application of this formula may indicate nonindependence that should be accommodated in the experimental design GLM (see below).

Graphical assessment of errors related through groups also is possible. When errors are plotted against the suspected groupings, dependence should be revealed by errors bunching within groups. This is a graphical analog of equation (10.1). In addition to nonindependence due to grouping, Kenny and Judd (1986) also discuss significance test methods of assessing error dependence due to sequence and space, while Draper and Smith (1998), Montgomery and Peck (1982), and Kutner at al, (2005) present graphical methods of assessing error dependence due to sequence.

Once the related scores are identified, they are treated just as scores from the same subject (or block) would be treated. In other words, the statistical procedures used to analyse related measures (or blocked) designs are applied. Nevertheless, prevention of non-independence is far preferable to cure, not least because, unlike the use of blocks in a planned experiment, post-hoc groupings may not be easily identified nor conveniently nested or crossed.

10.4.1.3 Normality

Wilcox (1998a) raised the profile of the normality assumption. Wilcox argues strongly that even slight deviations from the normal distribution can have substantial consequences for analysis power. However, most psychology statistical texts report ANOVA (and ANCOVA) as being robust with respect to violations of the normality assumption (e.g., Hays, 1994; Kirk, 1995; Maxwell and Delaney, 2004; Winer, Brown, and Michels, 1991), especially when the experimental condition sample distributions are symmetrical and the sample sizes are equal and greater than 12 (Clinch and Keselman, 1982; Tan, 1982). Indeed, Hays (1994) describes the robustness of ANOVA to non-normal distributions to be in proportion to the sample size; greater non-normality exerts less influence on the F-test as the sample size increases. Although data that mildly violates the normality assumption is not uncommon, severe departures from normality are quite rare (but see the differing views of Bradley, 1978; Glass, Peckham, and Sanders, 1972; Micceri's, 1989, critique of the normality assumption for achievement and psychometric variables). Nevertheless, robustness is a matter of degree and greater departures from normality are likely to exert some effect on the F-test Type 1 error rate.

Conformity to a normal distribution can be assessed in a number of ways. Hays (1994) suggests the use of the Kolmogrov–Smirnov test, which assess the discrepancy between hypothetical and sample distributions. This appears to be more powerful than the alternative chi-square test (Siegel and Castellan, 1988). However, the Shapiro–Wilk (1965) test is another popular means of assessing normality, as is the Lilliefors test (Lilliefors, 1967), which is a modification of the Kolmogrov–Smirnov test, specifically for sample data. The Shapiro–Wilk test is more conservative than the Lilliefors test, but as ANOVA and ANCOVA appear to be robust with respect to normality assumption violations and the aim is to screen for large deviations, generally it is accepted that powerful tests are unnecessary.

Most statistical packages can provide skew and kurtosis statistics and also the standard errors of these statistics. (Karl Pearson originally labeled the skew index g_1 and the kurtosis index g_2. Skew and Kurtosis also are known as the third and fourth moments of the normal distribution.) Dividing the skew and kurtosis statistics by their respective standard errors provides an approximation to a *Z*-score that can be used as a significance test. However, standard error is a function of sample size (all else being equal, larger samples have smaller standard errors) and so larger samples tend to provide greater *Z*-scores. Consequently, although *Z*-tables have no obvious markers of sample size, such as *dfs*, it should be appreciated that the power of these *Z*-score tests also increase with sample size.

Normal probability plots of GLM error terms provide a graphical alternative to determine whether the data is likely to have been sampled from a normally distributed population. Normal probability plots are constructed by first ranking the observed errors from smallest to largest. Subsequently, these ranked or ordered observations are plotted against their observed cumulative frequency (centiles) on a graph where the *Y*-axis has been scaled appropriately for the hypothesized distribution. When this is done, the observed cumulative frequency (centiles) should equal the ordered observations, that is, $y = x$ (e.g., Chambers et al., 1983) and so, the plot of the observed cumulative frequency (centiles) on ordered observations should lie along a straight line with unit slope that passes through the graph origin (i.e., 0,0). A SYSTAT normal probability plot of the ANCOVA errors (see Section 9.3) and the results of the Lilliefors test are presented in Figure 10.2. Although it would be very useful if SYSTAT provided an option to include the straight line upon which the plotted points should lie, this line can be drawn on the plot very easily, as described. First, request the default normal probability plot and note the minimum values on the *X*- and *Y*-axes. In the default normal probability plot of the ANCOVA errors (Plot A in Figure 10.2), the minimum *Y*-axis value is -3, while the minimum *X*-axis value is -10. The next step is simply to increase the shorter length axis to match the length of the longer axis. In this case, the minimum *Y*-axis is extended to -10. When this is done and the graph is replotted, a straight line drawn from the bottom left corner (where both *X*- and *Y*-axes now have equivalent minimum values < 0), through the origin toward the top right corner will be the line upon which the observed cumulative frequency (centiles) on ordered observations should lie (Plot B in Figure 10.2).

Both the normal probability plot and the significance test indicate some deviation from a normal distribution but given the robustness of ANOVA and the size and balanced nature of the sample, insufficient to affect the *F*-test interpretation. (Miles and Shevlin, 2001, provide a good introduction to interpreting the normal probability plot deviations from the straight line.) Sometimes, half-normal probability plots of residuals (obtained by plotting absolute residual values) are suggested when there are few data points (e.g., Lovie, 1991). However, these plots often suggest greater deviation from linearity than equivalent "full-normal" probability plots and greater experience may be required for accurate interpretation (Draper and Smith, 1998, also see Judd and McClelland, 1989).

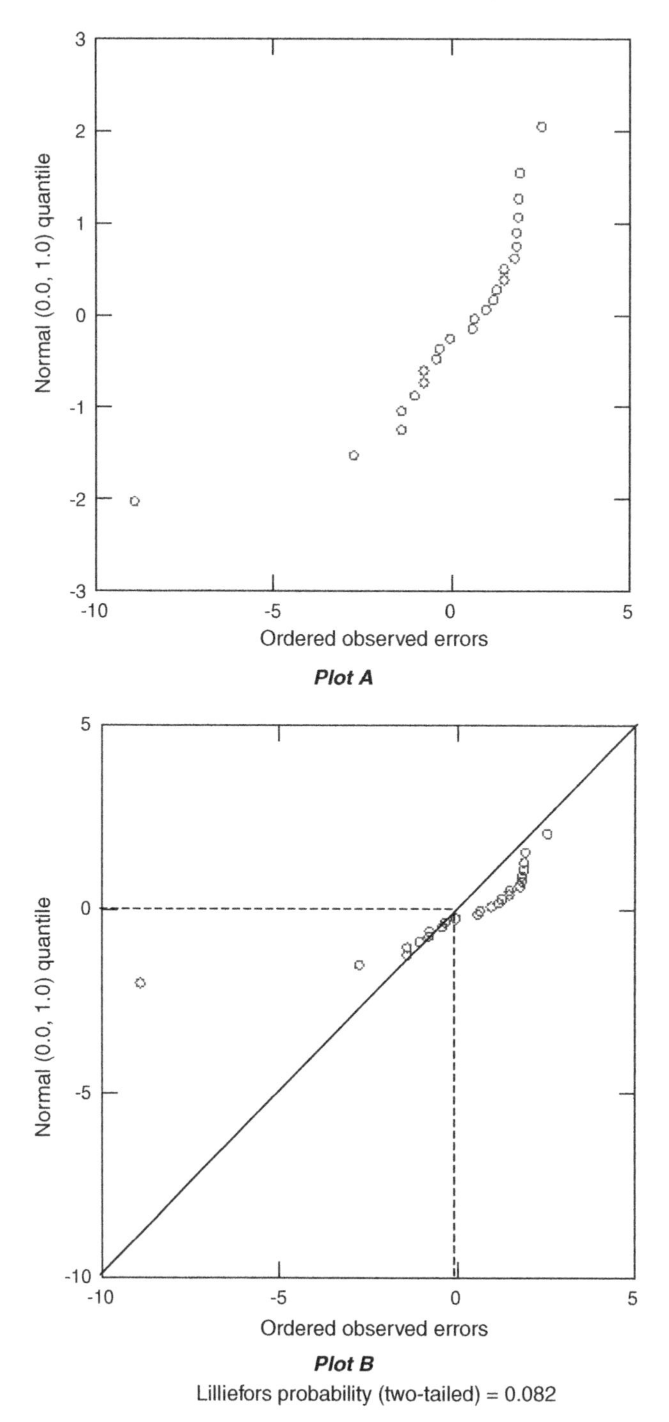

Figure 10.2 Normal probability plots of ANCOVA errors and the Lilliefors significance test result.

10.4.1.4 Homoscedasticity: Homogeneity of Variance

Homoscedasticity can be defined as constant error variance (estimated by the GLM residuals) at every value of the predictor variable, and/or every combination of the predictors (e.g., Kirby, 1993), while some authors define homoscedasticity as constant error variance across all of the predicted values of the dependent variable (e.g., Cohen et al, 2003; Judd and McClelland, 1989). However, the most detailed set of predictor value combinations available from any data set is provided by the predictor combinations associated with each data point. As each data point has a corresponding predicted value, examining variance by predicted values is equivalent to variance examination at every combination of predictor values.

The deviations between the predicted scores (i.e., the experimental condition means) and the observed dependent variable scores are squared and summed for each experimental condition to provide the separate $SS_{\text{within groups}}$. As a pooled average of these separate variance estimates, the MSe should reflect the variation within each of the experimental conditions. However, with heteroscedasticity, the MSe poorly reflects the variation within each experimental condition and so compromises the validity of the F-test.

Most psychology statistical texts (e.g., Hays, 1994; Kirk, 1995; Maxwell and Delaney, 2004; Myers and Well, 2003; Winer, Brown, and Michels, 1991) report ANOVA (and ANCOVA) as being robust with respect to moderate violations of heteroscedasticity, provided the experimental condition sample sizes are equal (i.e., balanced) and greater than five (Clinch and Keselman, 1982; Tomarken and Serlin, 1986). It is worth emphasizing that when sample sizes are unequal, moderate heteroscedasticity can have a substantial effect on F-test Type 1 error rates (e.g., Lix and Keselman, 1998; Scheffe, 1959). However, as said with regard to the normality assumption, robustness is a matter of degree and increasing heteroscedasticity will increase the F-test Type 1 error rate. Therefore, screening to ensure that there is no gross departure from homoscedasticity and that the GLM is appropriate for the data would appear to be sensible.

Errors plotted on the ANOVA or ANCOVA experimental conditions should take the shape of a horizontal band, but this is not the case in Figure 10.3, where the plotted errors exhibit more of a wedge shape (although the 'direction' of the wedge depends on the inclusion or exclusion of the score nearest the X-axis). Discrepancies in the length of the vertical error "stripes" over the experimental conditions indicate heterogeneous error variance as a function of the experimental conditions. However, errors should also be plotted on the predicted values, as in Figure 10.4. Indeed, this may be the easier method of graphical assessment, particularly if more than two experimental conditions have been coded for a regression model implementation of ANOVA or ANCOVA. Shapes other than a horizontal band most likely indicate error variance increasing with the size of the GLM estimate. Although in theory, error variance may decrease as the predicted values increase, rarely is this seen in practice.

Also presented in Figures 10.3 and 10.4 are the results from Cook and Weisberg's (1983) score test. The score statistic is the regression of U, the standardized residual squared, on the predicted values (or any independent variable), divided by 2. The score statistic has an approximate chi-square distribution with one df, which provides a very

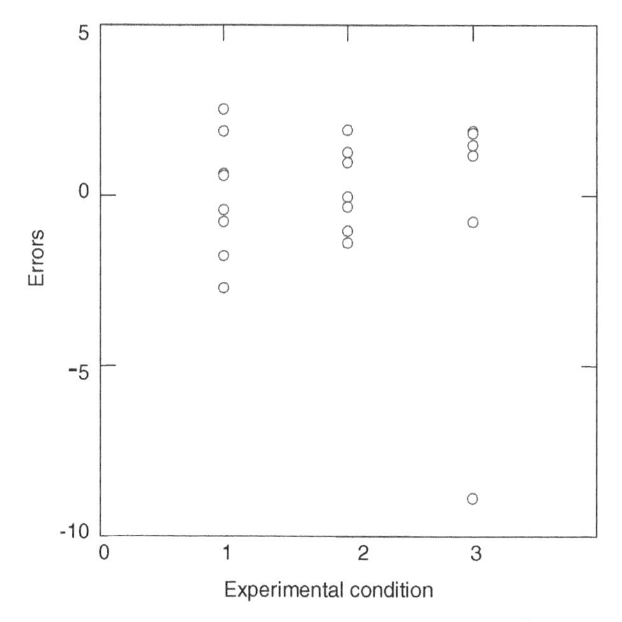

Cook and Weisberg's score test: predicted scores, $\chi^2_{(1)} = 5.871$, $p = 0.015$.

Figure 10.3 Errors plotted on experimental conditions.

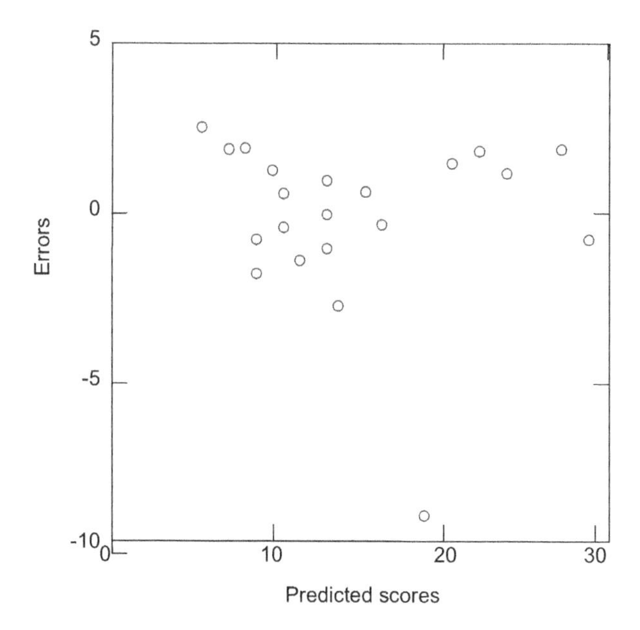

Cook and Weisberg's score test: predicted scores, $\chi^2_{(1)} = 1.246$, $p = 0.264$

Figure 10.4 Errors on predicted scores and Cook and Weisberg's score test for predicted scores.

useful significance test of homoscedasticity. Step-by-step accounts of how to calculate the score test are presented by Weisberg (1985) and Kirby (1993). This test is also incorporated into the BMDP regression program P2R. (Unfortunately, however, the BMDP score test is very poorly documented and easily could be missed. The score test result appears as a single line directly below the diagnostic error variance plot to which it refers. Moreover, there is no account in the *BMDP Manuals* of the fact that separate score test statistics are calculated for each predictor and for the predicted values, but only the plot and score test for the variable associated with the largest score test and so, greatest heteroscedasticity, are presented. Indeed, nowhere in the *BMDP Manuals* is there mention of this extremely useful significance test of homoscedasticity, see *BMDP Communications*, 1983, Vol. 16, No. 2, p. 2). Whereas the graphical and significance test assessments of errors by predicted scores suggest conformity to the homoscedasticity assumption (Figure 10.4), this is not the case when the graphical and significance test assessments examine errors by experimental condition (Figure 10.3). This difference emphasizes the value of examining errors as a function of a number of GLM components. The error attributable to an outlier (the data point nearest to the X-axis in both graphs) is very noticeable also and deleting this outlier may benefit assumption conformity substantially. A good introduction to dealing with outliers is provided by Pedhazur (1997). Finally, it should be mentioned that heteroscedasticity resisting all remedial measures may indicate the omission of a predictor variable.

10.4.2 Traditional ANCOVA Designs

10.4.2.1 Covariate Independent of Experimental Conditions

Although it is inevitable that the covariate(s) and experimental conditions will have some sort of conceptual link, traditional ANCOVA assumes that the covariate(s) and experimental conditions are statistically independent: Experimental conditions should not affect the distribution of covariate scores, nor should the covariate(s) influence the nature of the experimental conditions. When the covariate(s) and experimental conditions are related, traditional ANCOVA adjustments on the basis of the general covariate mean(s) are equivalent to modifying the experimental conditions, so that adjustments to the dependent variable can remove part of the experimental effect or produce an artefactual experimental effect (Cochran, 1957; Elashoff, 1969; Keppel, 1991; Kirk, 1995; Smith, 1957). Good design practice in ANCOVA involves measuring the covariate(s) before the experimental manipulation. Logically, this makes it impossible for the experimental conditions to influence the covariate (e.g., Howell, 2010; Keppel, 1991; Kirk, 1995; Winer, Michels, and Brown, 1991). Measuring the covariate(s) after the experimental manipulation affords the opportunity for the experimental conditions to exert an influence on the covariate(s) and is one way in which a relation between covariate(s) and experimental conditions can arise. However in ANCOVA, a relation between covariate(s) and experimental conditions can also arise as a consequence of the procedures employed to assign subjects to experimental conditions.

The least serious relation between covariate(s) and experimental conditions has been termed fluke assignment (Maxwell and Delaney, 2004). Fluke assignment is when, despite random assignment, different experimental condition covariate distributions are obtained. With fluke assignment, the differences between experimental condition covariate distributions that produce the relation between covariate(s) and experimental conditions reflect a Type 1 error. However, in these circumstances, ANCOVA is the appropriate analysis to control any resulting bias (Permutt, 1990; Senn, 1989; Shirley and Newnham, 1984).

A more serious relation between covariate(s) and experimental conditions caused by assignment procedures is known as biased assignment. Here, the covariate scores are used as an assignment criterion. Two types of assignment may be used. In one, only subjects with particular covariate scores (e.g., below the mean of all covariate scores recorded) are assigned to the experimental conditions. With the other type of assignment, subjects scoring high on the covariate are placed in one experimental condition and low scoring subjects are placed in another experimental condition (Huitema, 1980). Even when biased assignment has been used, traditional ANCOVA will adjust for the differences between the covariate distributions and will provide an unbiased test of the experimental effects (Rubin, 1977). Nevertheless, as experimental condition covariate distributions become more distinct, so the tenability of the traditional model assumptions becomes more important for the interpretation of the analysis (see Cochran, 1957; Huitema, 1980; Maxwell and Delaney, 2004). Taylor and Innocenti (1993) also refer to this issue, when they assert that if the general covariate mean is not logical for a variable, it should not be used as a covariate.

The most serious covariate-experimental condition relation caused by assignment procedures occurs when intact groups serve as the subjects in each of the experimental conditions. A typical example is the use of two classes of school children to compare two types of teaching method. When intact groups constitute the experimental conditions, interpretation of the ANCOVA results should proceed with considerable caution. With biased assignment, the basis for the difference between experimental conditions is known. In contrast, intact groups may be distinguished on the basis of a whole range of unknown variables. If the covariate scores of the intact experimental conditions differ, this can be conceived as an effect of the experimental conditions on the covariate(s). However, as the relationship between the covariate(s) and any set of (unknown) variables distinguishing the experimental conditions cannot be determined, there is a model specification error and the nature and consequences of the ANCOVA adjustment cannot be known (e.g., Overall and Woodward, 1977). (Intact groups also violate the random sampling and independence assumptions and so creates a number of problems for ANOVA.)

The issue of covariate measurement error is also pertinent here. Strictly, the covariate (in common with all of the independent variables) is assumed to be measured without error. However, provided random assignment or even biased assignment to experimental conditions is employed and all other assumptions are tenable, the consequence of covariate measurement error (cf. no covariate measurement error) is only a slight reduction in the power of the ANCOVA. However, when intact experimental conditions are used, covariate measurement error is expected to provide biased ANCOVA experimental effects. (For further discussion of assignment to experimental conditions procedures, and the consequences of covariate measurement errors and methods of repair, see Maxwell and Delaney, 2004; Huitema 1980; Bollen, 1989.)

In the GLM context, correlations among predictor variables, such as the covariate and the experimental conditions, is termed multicolinearity (e.g., Cohen et al, 2003; Kutner et al, 2005; Pedhazur, 1997). Previously, this term was used to describe predictors that were exact linear combinations of other model predictors (e.g., Draper and Smith, 1998), but now it tends to be applied more generally. Data exhibiting multicolinearity can be analyzed, but this should be done in a structured manner (e.g., Cohen et al, 2003, also see Rutherford, 1992, and Section 10.7.1 regarding the use of heterogeneous regression ANCOVA to attenuate the problems caused by covariate-experimental conditions dependence). Multicolinearity may also arise through correlations between two or more covariates. However, because there is seldom concern about the relative composition of the extraneous variance removed by covariates, this is much less problematic than a relation between the covariate(s) and the experimental conditions.

One symptom of a covariate-experimental conditions relation is that the ANCOVA regression homogeneity assumption may not be tenable (Evans and Anastasio, 1968). Elashoff (1969), Maxwell and Delaney (2004), and Winer, Brown, and Michels (1991) claim that carrying out an ANOVA on the treatment group covariate scores can be a useful indicator of experimental conditions influencing the covariate. However, with random assignment, the expectation is equal covariate treatment means and given that the covariate was measured before any experimental manipulation, what is to be made of a significant F-test? In such circumstances, covariate imbalance should reflect just less likely covariate distributions and, as argued by Senn (1989) and Permutt (1990), ANCOVA is the appropriate analysis to control any resulting bias. Consequently, an ANOVA on the treatment group covariate scores is appropriate only when there are theoretical or empirical reasons to believe that something more serious than fluke assignment has occurred. Adopting good ANCOVA design practice and applying all knowledge about the relationships between the study variables seem the only ways to avoid violating this assumption.

10.4.2.2 Linear Regression

When a linear regression is applied to describe a relationship between two variables that is not linear, the regression line will not only provide a poorer overall fit to the data but it will also fit the data better at some points than at others. At the well fitting points there will be smaller deviations between the actual and adjusted scores than at the ill fitting points. Consequently, the residual variance is likely to be heterogeneous and it is possible that the residuals may not be distributed normally (Elashoff, 1969). Moreover, as the points through which the regression line passes provide the predicted scores, a regression line that does not track the data properly provides predicted scores of questionable meaning.

Atiqullah (1964) examined in purely mathematical terms the traditional ANCOVA F-test when the real relationship was quadratic. With just two experimental conditions Atiqullah found that the F-test was biased unless there was random assignment of subjects to experimental conditions. The other situation considered by Atiqullah was a single factor ANCOVA where the number of experimental conditions approached infinity. In such circumstances, despite random assignment, the traditional ANCOVA F-test was gravely biased, with the amount of bias depending upon the size of the quadratic component. However, precisely because the number of experimental conditions approached infinity, there is considerable dubiety concerning the relevance of these conclusions for any real ANCOVA study (Glass, Peckham, and Saunders, 1972). Moreover, precisely because Atiqullah's examination was in purely mathematical terms, dealing with expectations over many experiments, these reservations may extend to the whole study.

Huitema (1980) states that the assumption of linear regression is less important than the traditional ANCOVA assumptions of random assignment, covariate-experimental conditions independence, and homogeneity of regression. This claim is made on the grounds that linear regression provides an approximate fit to most behavioral data and that nonlinearity reduces the power of the ANCOVA F-test by only a small amount. Nevertheless, as Huitema illustrates, in the face of substantial nonlinearity, ANOVA can provide a more powerful analysis than traditional ANCOVA. Moreover, with nonlinearity, as experimental condition covariate distribution imbalance increases, ANCOVA adjustments become extremely dubious.

Although it may interfere with the smooth execution of the planned data analysis, nonlinearity should not be considered as a statistical nuisance preventing proper analysis of the data. Nonlinearity is a pertinent finding in its own right, as well as a feature of the data that should be accommodated in the model in order to allow its proper analysis.

Many psychology statistical texts deliver a rather enigmatic presentation of regression linearity assessment. Most of these texts merely state the regression linearity assumption (Hays, 1994), or like Keppel (1991), they refer readers to Kirk (1995), who cites Kendall (1948), or Winer, Brown, and Michels (1991). Winer, Brown, and Michels (1991) distinguish between ANCOVA assumption tests and tests of other properties of the data. However, Winer (1962, 1971) includes tests of the regression linearity of the dependent variable experimental condition means on the covariate experimental condition means ($\widehat{\beta}_{\text{ECM}}$) and the linearity of the regression line based on the total set of scores ($\widehat{\beta}_{\text{t}}$), plus a test of $\widehat{\beta}_{\text{t}} = \widehat{\beta}_{\text{w}}$ (see Chapter 9). Winer (1962) states,

> if this regression ··($\widehat{\beta}_{\text{ECM}}$)·· does not prove to be linear, interpretation of the adjusted treatment means becomes difficult (Winer, 1962, p. 588).

Winer (1971) omits this sentence, but as the presentation is identical otherwise, the same meaning is conveyed. Similarly, in his section on ANCOVA assumptions, Kirk (1968, 1982, 1995) includes tests of the regression linearity of $\widehat{\beta}_{\text{ECM}}$, and the linearity of $\widehat{\beta}_{\text{t}}$, but no distinction is made in the text between these tests and those that assess the specific ANCOVA assumptions. Unfortunately, considerable misunderstanding and confusion about the nature of the traditional ANCOVA regression linearity assumption and how it should be tested can be caused by these accounts.

The traditional ANCOVA linearity assumption is that the regression of the dependent variable on the covariate(s) in each of the experimental conditions is linear (i.e., the $\widehat{\beta}_j$ are linear). No other tests of regression linearity are pertinent. Linearity of $\widehat{\beta}_t$ and $\widehat{\beta}_{ECM}$ is expected only when there are no experimental effects. Given that regression linearity should be determined prior to the assessment and in the presence of experimental effects, indirect tests of $\widehat{\beta}_j$ linearity are not satisfactory.

Probably the most obvious way to assess the linearity of the separate groups regressions is to plot the dependent variable against the covariate (or each covariate) for each experimental condition. Another popular, but much less direct approach is to ignore experimental conditions and to plot residuals against the predicted scores. This approach has the advantage of generalizing over covariates. However, nonlinearity within one condition may be masked by the linearity within the other conditions, particularly when there are many conditions. Moreover, when any nonlinearity is detected, it will need to be traced to source and so eventually, checks per covariate per condition will be required. Further discussion of graphic checks of regression linearity is provided by Draper and Smith (1998), Montgomery and Peck (1982), and Kutner et al, (2005). Assessing linearity by inspecting data plots may be more difficult than the graphic assessments of normality and homoscedasticity, particularly when the linearity assessments are carried out per experimental condition, where there are fewer data upon which to form an opinion. Consequently, significance test methods may have a larger role to play, particularly for those less experienced in graphical assessment.

Regression linearity also may be assessed by applying a significance test for the reduction in errors due to the inclusion of nonlinear components (e.g., Maxwell and Delaney, 2004). A nonlinear relationship between the dependent variable and a covariate can be modeled by including the covariate raised above the first power as a predictor. For example, the ANCOVA GLM equation

$$Y_{ij} = \mu + \alpha_j + \beta_w(Z_{ij} - Z_G) + \beta_w(Z_{ij} - Z_G)^2 + \varepsilon_{ij} \tag{10.2}$$

is termed as second-order polynomial model and describes a quadratic curve. The ANCOVA GLM equation

$$Y_{ij} = \mu + \alpha_j + \beta_w(Z_{ij} - Z_G) + \beta_w(Z_{ij} - Z_G)^2 + \beta_w(Z_{ij} - Z_G)^3 + \varepsilon_{ij} \tag{10.3}$$

is termed as third-order polynomial model and describes a cubic curve. Further components (e.g., quartic, quintic, etc.) can be added to increase the order of a GLM, but the highest order any equation may take is equal to the number of experimental conditions minus 1 (here, $p - 1$). However, it is exceptional for more than a third-order polynomial model to be needed to describe psychological data.

To apply a polynomial model as described by equation (10.3), two new predictor variables must be created: Z^2 and Z^3. However, these variables will be correlated with Z, so the problem of multicolinearity arises. To deal with this, the data should be analyzed in a structured manner (e.g., Cohen et al., 2003). For example, the traditional ANCOVA model

$$Y_{ij} = \mu + \alpha_j + \beta_w(Z_{ij} - Z_G) + \varepsilon_{ij} \quad (9.5, \text{rptd})$$

should be compared with the traditional ANCOVA model described by equation (10.2). Any decrement in the error estimate of GLM equation (10.2) in comparison to the same estimate in equation (9.5) can be attributed to the $\beta_w(Z_{ij} - Z_G)^2$ component. The component is retained if an F-test of the variance attributed to the component is significant. However, if error examination suggests that further curvilinearity exists, a third-order GLM may be compared with the second-order GLM. (For further information on curvilinear ANCOVA, see Cohen et al., 2003; Huitema, 1980; Maxwell and Delaney, 2004. For further information on polynomial models, see Cohen et al., 2003; Draper and Smith, 1998; Neter, Wasserman, and Kutner, 1990; Pedhazur, 1997.) One advantage of this approach is that when all significant curvilinear components are included, the description of the curvilinear ANCOVA GLM is complete and the ANCOVA results can be interpreted.

Figure 10.5 presents a plot of the dependent variable on the covariate with the linear regression lines depicted for each experimental condition. The ANCOVA Summary Table for the GLM described by equation (9.5) is presented in Chapter 9 (Table 9.2), while the ANCOVA summaries of the second- and third-order ANCOVA GLMs are presented in Table 10.4. As can be seen, assessing linearity per experimental condition graphically with only a few data can be a difficult task. However, the insignificant reduction in error variance attributable to the inclusion of the quadratic and cubic

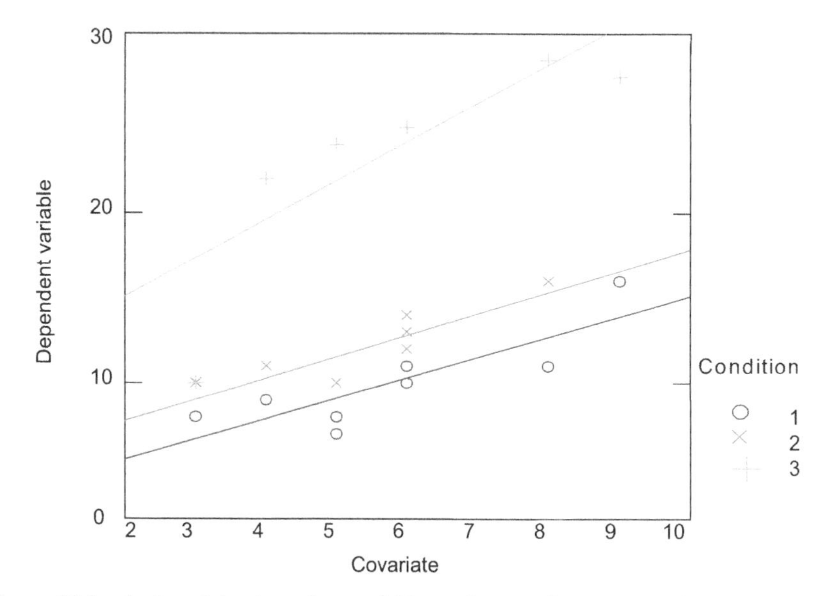

Figure 10.5 A plot of the dependent variable on the covariate per experimental condition.

Table 10.4 Summaries of Error Reduction Due to the Second- and Third-Order ANCOVA GLMs

Source	SS Increment	*df*	MS	*F*	*p*
Covariate (Z2) GLM equation (10.2)	2.086	1	2.086	0.308	0.586
Covariate (Z3) GLM equation (10.3)	4.464	1	4.464	0.659	0.428
Error from GLM equation (10.3)	121.913	18	6.773		

R: 0.941; R^2: 0.886; adjusted R^2: 0.862.

components suggests the tenablity of the dependent variable on covariate regression linearity assumption.

10.4.2.3 Homogeneous Regression

The final traditional ANCOVA assumption is that the regression slopes, described by the regression coefficients, are homogeneous across the experimental conditions. As the regression coefficient employed in ANCOVA is a weighted average of the separate experimental conditions' regression coefficients ($\hat{\beta}_w$), two problems occur if there is heterogeneity of experimental condition regression coefficients. The first problem concerns the effect on the F-test. Monte Carlo investigations employing random assignment (e.g., Hamilton, 1976; also see Huitema, 1980) indicate that provided sample sizes are equal and exhibit homogeneous variance, heterogeneity of regression coefficients tends to result in conservative F-values, reducing the sensitivity or power of the analysis. This is because averaging over heterogeneous regression coefficients introduces error into $\hat{\beta}_w$, with the result that it is a poor estimate of the dependent variable on covariate regression slopes in all of the experimental conditions. Therefore, in comparison to the homogeneous regression slopes situation, where $\hat{\beta}_w$ is a good descriptor of all of the experimental condition regression slopes, there will be larger discrepancies between the actual and predicted scores. Consequently, the error variance will be larger and so the power of the analysis will be lower. However, this applies to ANCOVA employing random assignment, where the differences between experimental condition covariate means are expected to be zero. Hollingsworth (1976) (cited by Huitema, 1980) found Type 1 error increased with regression heterogeneity when nonzero differences between experimental condition covariate means were provided by nonrandom assignment. The second problem posed by heterogeneous regression ANCOVA is that treatment effects vary as a function of the covariate. As a result, and in contrast to the homogeneous regression situation, an assessment of experimental effects at any one measure of the covariate cannot be taken to reflect experimental effects at any other measures of the covariate (see Section 11.2).

As with the regression linearity of the dependent variable on the covariate, heterogeneous regression coefficients across experimental conditions should not be considered as statistical nuisance interfering with the proper analysis of the data. Heterogeneous regression across experimental conditions is an important finding.

Regression heterogeneity indicates that the dependent variable on covariate relationship differs between experimental conditions. This is a result that should be considered on a par with differences observed between experimental condition dependent variable means and it should be accommodated in the GLM to allow proper analysis of the data. However, it is worth repeating that heterogeneous regression coefficients may be symptomatic of a relationship between the covariate and the experimental conditions (Evans and Anastasio, 1968).

Regression homogeneity may be assessed graphically by judging the relative slopes of experimental condition regression lines (see Figure 10.5). Alternatively, a significance test approach can be applied by examining the reduction in errors due to the inclusion in the GLM of a term representing the interaction between the covariate and the experimental conditions. The predictor variables representing the interaction between the covariate and the experimental conditions are constructed in exactly the same manner as these variables representing factor interactions were constructed. (For further details see Cohen et al, 2003; Howell, 2010; Kutner et al, 2005; Pedhazur, 1997.) However, as with the tests of regression linearity, the problem of multicolinearity arises; the predictor variables representing the interaction between the covariate and the experimental conditions will be correlated with Z. Therefore, the data analysis should proceed in a structured manner (e.g., Cohen et al, 2003). This emphasizes the point that the significance test approach examines the (further) reduction in errors due to the inclusion in the GLM of a term representing the interaction between the covariate and the experimental conditions, after those terms representing experimental conditions and the single regression line have been included in the GLM. A significant reduction in error (i.e., a significant interaction term) indicates regression heterogeneity. This means the model fit to data can be improved by employing a different regression coefficient in at least one of the experimental conditions.

Table 10.5 presents the ANCOVA summary of the error reduction when separate regression slopes are employed in the different experimental conditions. As no significant improvement is observed, the tenability of the assumption of homogeneous regression coefficients is accepted. However, if a significant interaction between the covariate and the experimental conditions had been detected, the next question should be, which of the experimental conditions require distinct regression

Table 10.5 Summary of Additional Error Reduction Due to Heterogeneous Regression ANCOVA

Source	SS Increment	*df*	MS	*F*	*p*
Additional error reduction due to covariate x experimental conditions	19.394	2	9.697	1.600	0.229
Full GLM error	109.070	18	6.059		

R: 0.941; R^2: 0.886; adjusted R^2: 0.862.

coefficients? It may be that only one or two of the experimental conditions require a unique regression slope. As a *df* is lost from the error variance estimate with every distinct regression line employed, for this and other reasons applying only the minimum number of terms required is a guiding principle of linear modeling (e.g., Draper and Smith, 1998). Which experimental conditions require distinct regression lines can be determined by comparing different models that employ a common regression line for all but one of the experimental conditions. Rather than obtaining an estimate of the error when all experimental conditions employ distinct regression lines, an error estimate is obtained when only one experimental condition employs a distinct regression line. The reduction in residuals due to this one distinct regression line can then be assessed in comparison to the residual estimate obtained when a common regression line is applied in all treatment groups. Successive estimations can be made and each time a distinct regression line is employed in a different experimental condition, any significant reduction in errors indicates a significant improvement in the fit of the model to the data.

The significance test method described above is equivalent to the standard test of regression homogeneity presented by Kendall (1948), reproduced by Hays (1994), Keppel (1991), Kirk (1995), and Winer, Brown, and Michels (1991). For the single factor, single-covariate independent sample design, this is

$$F[(p-1), p(N-2)] = \frac{S_2/(p-1)}{S_1/(p(N-2))} \tag{10.4}$$

where S_1 is the residual variation when separate group regressions have been employed and S_2 is the variation of the separate experimental condition regressions about the weighted average regression line (see texts above for computational formulas). The sum of squares S_2 estimates the variation not accommodated when the weighted average regression line (rather than separate experimental condition regression lines) is used and is equivalent to the estimate of the interaction effect or the reduction in residuals due to the separate experimental condition regression lines. For all of the significance test methods, the same F-test denominator estimate is employed and a significant F-value indicates that the homogeneity of regression coefficients assumption is untenable. In order to avoid Type 2 errors, Kirk (1995) and Hays (1994) recommend the use of a liberal level of significance (about 0.25) with the regression homogeneity test. However, test power increases with large data sets, so more conservative significance levels should be set when the test is applied to large data sets.

Regression homogeneity also must be checked after polynomial components have been added to the traditional ANCOVA model to accommodate curvilinear regression. This is achieved most easily by applying a significance test in a manner similar to that described above. Another set of predictors is created to represent the interaction between the polynomial components (added to accommodate the curvilinearity) and the experimental conditions. For example, had it been decided that the GLM described by equation (10.2) was most appropriate for the data, incorporating an additional term to represent an experimental conditions–curvilinear interaction would result in the model

$$Y_{ij} = \mu + \alpha_j + \beta_w(Z_{ij} - Z_G) + \beta_w(Z_{ij} - Z_G)^2 + \left[(\alpha_j)(\beta_w(Z_{ij} - Z_G) + \beta_w(Z_{ij} - Z_G)^2)\right] + \varepsilon_{ij} \tag{10.5}$$

An F-test is applied to the reduction in error variance attributed to the interaction term. A significant F-test indicates that better prediction is provided when at least one of the experimental conditions employs a different curvilinear regression line. Therefore, the next step is to determine which experimental conditions actually require distinct regression lines. As always, when any new GLM is considered, it is necessary to check that it conforms to the set of GLM assumptions. As with curvilinear regression, an advantage of this approach is that when all significant heterogeneous regression components are included, the description of the heterogeneous regression ANCOVA GLM is complete and the ANCOVA results can be interpreted.

10.5 SHOULD ASSUMPTIONS BE CHECKED?

Over the past 50 years, a number of research studies have questioned the use of preliminary analyses to check statistical test assumptions to identify the most appropriate test (Arnold, 1970; Bancroft, 1964; Moser and Stevens, 1992; Moser, Stevens, and Watts, 1989; Rao and Saxena, 1981; Saleh and Sen, 1983; Zimmerman, 1996, 2004). These arguments are summarized and considered by Wells and Hintze (2007). Five problems arise when statistical or graphical assumption checks are applied in this two-stage procedure. The first problem is that when a preliminary test determines the main test, the test statistic distribution becomes conditional on the preliminary test outcome. As Wells and Hintze explain, there would be no problem if the preliminary test outcome was always correct, but, unfortunately, the preliminary test Type 1 and Type 2 error rates affect the significance level (and error rate) of the main test. Research has focused on the issue of test choice following preliminary analyses, but the order of application of assumption checks and main analysis is irrelevant. Main test statistics become conditional on assumption checks because the outcome of the assumption checks determines the analysis applied. Perhaps some difficulty is created by the use of the term "applied." Many analyses can be applied, but any use of assumption check information to decide which of these main analyses is appropriate makes all of these main analyses conditional on the assumption check(s). This is clearest when the checks indicate assumptions are untenable and the initial main analysis choice has to be changed, or a data transformation is applied, and the original main analysis (and assumption checks) reapplied. However, the main analysis is also conditional on assumption checks when information from the checks supports assumption tenability and the planned main analysis is applied or the planned and already implemented main analysis is deemed appropriate, reported, and interpreted. The second problem arising when statistical or graphical assumption checks are applied is that the detrimental consequences for the main test accumulate with each assumption that is checked. The third problem is that the usual assumption check null hypothesis is that the assumption (e.g., equal variances) is valid. However, failure to reject the null hypothesis does not allow simple acceptance of the null hypothesis. Therefore, assumption checks are limited in their ability to provide conclusive information about the validity of the assumption assessed. The fourth problem is that preliminary analyses frequently make their own statistical assumptions and so there is a danger of an infinite regress regarding assumption checks. The fifth and last problem is that even when an assumption check identifies a serious assumption violation, it is still possible for the particular violation to exert little effect on the outcome of the subsequent test.

Of these five problems, the focus will be the main difficulty that the distribution of the test statistic becomes conditional on the preliminary test outcome. Problem two is a generalization of the first and main problem, problem three applies to all hypothesis tests and can be mitigated by power analysis (i.e., checking that sufficient power to detect the effects was available in principle—if the hypothesized effect was present, see Section 4.7.5). However, most researchers appreciate that assumption checks do not provide conclusive information about assumption compatibility, but instead are indicators of assumption tenability. Problem four is serious in principle, but less serious in reality, particularly when assumption checks are regarded as indicators of assumption tenability. Finally, the assessment inherent in problem five seems either crude or inaccurate. One of the benefits of larger samples and balanced designs is a robustness to some assumption violations, as assessed by the F-test Type 1 error rate. Therefore, Wells and Hintze may have a point if the exclusive concern of the study is to accept or reject a particular null hypothesis. However, it is to be hoped that psychologists are moving away from this barren approach (see Section 4.1), with specific effect p-values, effect sizes, confidence limits and power considerations all regarded as evidence contributing to an understanding of a study outcome. As many of these estimates also rely on the GLM assumptions, it is extremely unlikely that a serious assumption violation would have little effect on the accuracy and coherence of all of these measures.

Wells and Hintze (2007) suggest that the way to address the problems caused by preliminary analyses is to abandon the use of such analyses and instead employ theory, empirical evidence, and reason to identify an appropriate test. Relevant theory includes psychological theory, measurement theory, and statistical theory. For example, psychological theory may suggest not only higher, but more varied scores in one condition than another and so tests that are unaffected by, or robust with respect to, variance heterogeneity would be preferred. Measurement theory can also assist by identifying the dependent variable measure with the most appropriate properties. Statistical theory can offer tests that are robust to particular assumption violations and where normality violations are a concern, it can inform the sample size chosen so as to ensure central limit theorem applies. Pertinent empirical evidence includes previously conducted studies and pilot studies. Both of these sources provide a great deal of information and pilot studies can offer the advantage of exactly the same experimental design and procedures. Finally, Wells and Hintze suggest reason that provides a synthesis of theory and empirical evidence that allows researchers to determine which assumptions are most, less, and least likely to be tenable. Wells and Hintze's suggestions really are an emphatic iteration of the vital role of careful planning in the design of any study. However, while statistical theory can provide tests that are robust to particular assumption violations, it is misleading to suggest that many equivalent tests are available – most robust tests are quite limited compared to tests typically employed to analyze psychological data. Hopefully, researchers apply ANOVA and ANCOVA, and other tests, because they are most appropriate for their research purposes and this means other tests will not be as appropriate for their research purposes.

Another consequence of applying the preventative approach described by Wells and Hintze (2007) is it precludes the type of optimized experimental design advocated by McClelland (1997, see Section 2.2). Nevertheless, some resolution may be achieved if Wells and Hintze's recommendations are applied primarily to research topics about which relatively little is known, or to research topics which are known to provide problematic data, while McClelland's recommendations are applied primarily to research topics which are known to provide data compliant with the required statistical assumptions.

An analysis of the cost and benefits of checking or not checking assumptions is also informative. For example, Zimmerman (2004) described the effect on the Welch test applied only when Levene's (1960) test declared significant variance heterogeneity. When the group with the greater sample size (i.e., 40 vs. 20) also exhibited greater variance, the Welch test Type 1 error rate actually diminishes from 0.050 to a minimum of 0.037. However, when the group with the smaller sample size (i.e., 20 vs. 40) exhibited greater variance, the Welch test Type 1 error rate increased to a maximum of 0.065. (t-test Type 1 error rates followed the same patterns as Welch test Type 1 error rates, but, as would be expected as variance heterogeneity increased, due to increasing t-test assumption violation, the effect on t-test Type 1 error rates was much greater.) Therefore, with variance heterogeneity and unequal sample sizes designed to exert substantial effects on the subsequent test, the maximum variation around 0.05 level was approximately ± 0.015. In contrast, if analyses are conducted without any checks on assumptions and one or more statistical assumptions underlying an analysis are violated in a substantial fashion, it is possible that a completely invalid analysis may be reported. Given that the two extremes are to tolerate a maximum Type 1 error rate deviation of ± 0.015 or report an invalid analysis, the former situation seems preferable.

Unfortunately, the recommended method of assumption checking by residual examination suffers all of the problems summarized by Wells and Hintze's (2007). Nevertheless, residual examination provides researchers with a vital and necessary appreciation of the statistical validity of their analyses. For these reasons, it is unlikely that data analysts will stop recommending the assessment of the tenability of the statistical assumptions underlying analyses. However, strategies are likely to be developed to minimize the accumulation of assumption check error. For example, it should be possible to reduce the set of assumptions to be checked by taking into account what is known about the data obtained in such research situations and the robustness of the main analysis.

CHAPTER 11

Some Alternatives to Traditional ANCOVA

11.1 ALTERNATIVES TO TRADITIONAL ANCOVA

As the good design practice of recording the covariate measure before the experimental manipulation prevents the experimental conditions affecting the covariate directly, and because most relationships between dependent variables and covariates in psychology are linear or approximately linear, heterogeneous regression coefficients across experimental conditions is the traditional ANCOVA assumption most likely to be violated. As it also is a problem that becomes more likely as the number of factors and conditions in a study increase, the majority of traditional ANCOVA alternatives were presented to address this issue. However, rather than being a statistical assumption required for a valid and accurate analysis, regression homogeneity is an assumption made only to simplify the calculation and interpretation of traditional ANCOVA. In fact, the ANCOVA GLM can be modified easily to accommodate heterogeneous regressions across conditions and most contemporary approaches deal with regression heterogeneity in this fashion.

Heterogeneous regression ANCOVA is able to provide superior representation of variable relationships and influence, and so provide greater prediction accuracy. Indeed, the potential benefits and low cost of applying heterogeneous regression ANCOVA compared with traditional ANCOVA led Maxwell and Delaney (2004) to suggest there should be a bias in favour of heterogeneous regression ANCOVA applications. Consequently, the present chapter focuses on heterogeneous ANCOVA GLMs, but two further alternatives to traditional ANCOVA are presented, as is the ability of heterogeneous regression ANCOVA to ameliorate the problems caused by a relationship between the covariate(s) and the experimental conditions.

ANOVA and ANCOVA: A GLM Approach, Second Edition. By Andrew Rutherford.

11.2 THE HETEROGENEOUS REGRESSION PROBLEM

As the term ANCOVA is applied, the main concern is the determination of experimental effects (see Section 1.3.3). The basic problem posed by heterogeneous regression ANCOVA is that experimental effects vary as a function of the covariate. This is illustrated in Figure 11.1. In order that the issues are understood clearly, a simple independent measures, single-covariate, single factor (with two levels) experimental design is depicted.

From Figure 11.1, it can be appreciated that the experimental effect, represented by the vertical distance between the regression lines, is not constant across the range of the covariate values, as is the case in traditional ANCOVA, but instead varies as a function of the covariate values. As a result and in contrast to the homogeneous regression situation, an assessment of experimental effect at any one measure of the covariate cannot be taken to reflect the experimental effect at any other measures of the covariate. In essence therefore, heterogeneous regression ANCOVA presents a problem of experimental effect determination and description.

If the process of model selection determines that distinct regressions (in graphical terms, lines which pass through all fitted values) are required to accommodate the data, there seems little reason to consider the null hypothesis that the separate regressions predict the same dependent variable values. Yet by definition, nonparallel regression lines intersect at some point. Here, the same values will be predicted by the two distinct regression lines and importantly, the experimental effect observed below this covariate value will be the converse of the experimental effect observed above this value. (A plot of the regression lines is an obvious way to determine if regressions intersect within the range of observed covariate values.) Depending on the location of this intersection point, it may be necessary to determine the covariate values at which significant experimental effects are exerted and to specify the nature of these experimental effects. As the intersection point in Figure 11.1 is below $Z = 0$, although the effect size may vary, the nature of the experimental effect is constant across the range of positive covariate values.

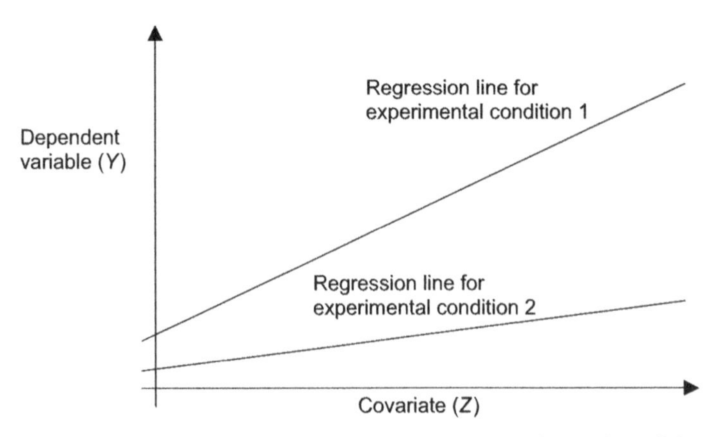

Figure 11.1 Heterogeneous regression across two experimental conditions.

11.3 THE HETEROGENEOUS REGRESSION ANCOVA GLM

In Chapter 9, two versions of the ANCOVA GLM were described. The typical traditional ANCOVA GLM expresses the covariate scores in terms of their deviation from the overall covariate mean, while the more general ANCOVA GLM simply expresses the covariate score. The latter version of the independent measures, single factor, single-covariate experimental design GLM is

$$Y_{ij} = \mu + \alpha_j + \beta_w Z_{ij} + \varepsilon_{ij} \qquad (9.1, \text{rptd})$$

As described in Chapter 9, subtracting the term $\beta_w Z_{ij}$ from the dependent variable score removes all influence of the covariate, leaving what was labelled as the fundamental adjusted score (Y_{faij}). This is the predicted dependent variable score when $Z = 0$

$$Y_{faij} = Y_{ij} - \beta_w Z_{ij} = \mu + \alpha_j + \varepsilon_{ij} \qquad (9.4, \text{rptd})$$

If the general covariate mean (Z_G) is substituted for Z_{ij}, the predicted dependent variable scores are the values Y_{aij}, as in traditional ANCOVA. Both of these predictions are specific instances of the general prediction on the basis of Z, where Z is any measure of the covariate.

To accommodate the heterogeneous regressions depicted in Figure 11.1, equation (9.1) may be rewritten as

$$Y_{ij} = \mu + \alpha_j + \beta_j Z_{ij} + \varepsilon_{ij} \qquad (11.1)$$

where the regression coefficient β_w of equation (9.1) is replaced by β_j, which, by virtue of the j subscript, denotes a different regression coefficient per experimental condition. An important point to appreciate about the separate regression lines is that they are statistically independent (e.g., Searle, 1987).

As heterogeneous regression ANCOVA GLMs simply incorporate terms to accommodate the different slopes, they are able to provide tests comparable with the traditional ANCOVA hypotheses, as well as tests of the covariate effect and factor interactions. In traditional ANCOVA, the omnibus F-test of experimental conditions compares adjusted means, which are those scores predicted on the basis of the general covariate mean (Z_G). However, when heterogeneous regression ANCOVA is applied, the omnibus F-test of experimental conditions compares the predicted scores when $Z = 0$ (e.g., Searle, 1987). In other words, the Y-intercepts of the separate regression lines are compared (see Figure 11.1). Nevertheless, it is unlikely there will be much interest in comparing treatment groups at the zero value of the covariate, not least because a zero-covariate score may be impossible to observe in the real world. Nevertheless once the heterogeneous regression ANCOVA GLM is selected and parameter estimates obtained, it is possible to predict a dependent variable score based on any covariate value for each experimental condition (including $Z = 0$). Moreover, as the standard errors associated with these predicted scores are also determinable, it is possible to carry out F-tests of the effect of experimental conditions at any covariate value(s). For example, F-tests of the experimental effects might be carried out at Z_G, or at the separate experimental condition covariate means ($\overline{Z}_j$).

In ANCOVA, the accuracy of the predicted dependent variable scores and so the power of the F-test of experimental effects is greatest when the covariate value employed lies at the center of the covariate distribution. This is known as the center of accuracy (C_a). Interestingly, Rogosa (1980) revealed that the heterogeneous regression ANCOVA experimental effect at C_a is identical to the traditional homogeneous regression ANCOVA experimental effect at C_a. Moreover, with balanced designs, estimating experimental effects on the basis of the separate $\overline{Z}_j$ values also provides an F-test at C_a. Paradoxically, therefore, estimating experimental effects at $\overline{Z}_j$ provides a simple heterogeneous regression alternative to the traditional ANCOVA experimental effect estimate, which is identical to the traditional ANCOVA experimental effect estimate. There may be theoretical reasons for comparing all subjects across the experimental conditions at the same covariate value, but as prediction accuracy drops with distance from the experimental condition covariate means, the power cost inherent in these comparisons depends on the difference between Z_G and each of the $\overline{Z}_j$.

When statistical software is used to implement heterogeneous regression ANCOVA, it is advisable to check the software documentation or to check empirically (with a known data set) which covariate value(s) are the basis for any adjusted experimental condition means presented. Many statistical packages able to implement heterogeneous regressions will apply GLMs in the form of equation (11.1) and so the ANCOVA summary table will present experimental effects assessed when all influence of the covariate has been removed (i.e., at $Z=0$). If experimental effects are to be assessed at any other covariate values, then further analysis will be necessary.

11.4 SINGLE FACTOR INDEPENDENT MEASURES HETEROGENEOUS REGRESSION ANCOVA

In the following example, two of the conditions reported in Chapter 9 will be presented as if they constituted a separate experiment. Table 11.1 presents the subjects' story and imagery task (covariate) scores and the subjects' memory recall scores after story and imagery encoding in two study time conditions, and a heterogeneous regression ANCOVA is applied (also see Figure 10.5).

The slope of the regression lines ($\widehat{\beta}_j$) for each of the experimental conditions employed in the ANCOVA GLM is given by

$$\widehat{\beta}_j = \frac{\sum_{i=1}^{N}(Z_{ij} - \overline{Z}_j)(Y_{ij} - \overline{Y}_j)}{\sum_{i=1}^{N}(Z_{ij} - \overline{Z}_j)^2} \tag{11.2}$$

Table 11.1 Story and Imagery Test Scores and Recall Scores After Story and Imagery Encoding

	30 s		180 s	
Study Time	Z	Y	Z	Y
	9	16	5	24
	5	7	8	29
	6	11	3	10
	4	9	4	22
	6	10	6	25
	8	11	9	28
	3	8	4	22
	5	8	5	24
$\sum Z/Y$	46	80	44	184
$\overline{Z}/\overline{Y}$	5.750	10.000	5.500	23.000
$(\sum Z/Y)^2$	292	856	272	4470
$\sum Z^2/Y^2$	2116	6400	1936	33,856

For $\widehat{\beta}_1$ this provides

$(Z_{ij} - \overline{Z}_j)^2$	$Y_{ij} - \overline{Y}_j$	$(Z_{ij} - \overline{Z}_j)(Y_{ij} - \overline{Y}_j)$
$9 - 5.750 = 3.250$	$16 - 10 = 6$	19.500
$5 - 5.750 = -0.750$	$7 - 10 = -3$	2.250
$6 - 5.750 = 0.250$	$11 - 10 = 1$	0.250
$4 - 5.750 = -1.750$	$9 - 10 = -1$	1.750
$6 - 5.750 = 0.250$	$10 - 10 = 0$	0.000
$8 - 5.750 = 2.250$	$11 - 10 = 1$	2.250
$3 - 5.750 = -2.750$	$8 - 10 = -2$	5.500
$5 - 5.750 = -0.750$	$8 - 10 = -2$	1.500
$\sum = 27.500$		$\sum = 33.000$

$$\widehat{\beta}_1 = \frac{33.000}{27.500} = 1.200$$

and for $\widehat{\beta}_2$ this provides

$(Z_{ij} - \overline{Z}_j)^2$	$Y_{ij} - \overline{Y}_j$	$(Z_{ij} - \overline{Z}_j)(Y_{ij} - \overline{Y}_j)$
$5 - 5.500 = -0.500$	$24 - 23 = 1$	−0.500
$8 - 5.500 = 2.500$	$29 - 23 = 6$	15.000
$3 - 5.500 = -2.500$	$10 - 23 = -13$	32.500
$4 - 5.500 = -1.500$	$22 - 23 = -1$	1.500
$6 - 5.500 = 0.500$	$25 - 23 = 2$	1.000
$9 - 5.500 = 3.500$	$28 - 23 = 5$	17.500
$4 - 5.500 = -1.500$	$22 - 23 = -1$	1.500
$5 - 5.500 = -0.500$	$24 - 23 = 1$	−0.500
$\sum = 30.000$		$\sum = 68.000$

$$\widehat{\beta}_2 = \frac{68.000}{30.000} = 2.267$$

The formula for calculating adjusted means presented in Chapter 9 is repeated below

$$\overline{Y}_{aj} = \overline{Y}_j - \beta(\overline{Z}_j - Z_G) \qquad (9.8, \text{rptd})$$

Applying this to the heterogeneous regression situation where each experimental condition has a distinct regression line and employs the experimental condition covariate mean as a predictor reveals the adjusted experimental condition means to be equal to the unadjusted experimental means

$$\begin{aligned} \overline{Y}_{aj} &= \overline{Y}_j - \beta(\overline{Z}_j - \overline{Z}_j) \\ \overline{Y}_{aj} &= \overline{Y}_j - \beta(\overline{Z}_j - \overline{Z}_j) \\ \overline{Y}_{aj} &= \overline{Y}_j - \beta(0) \\ \overline{Y}_{aj} &= \overline{Y}_j \end{aligned} \qquad (11.3)$$

However, it is worth noting that the equivalence of adjusted and unadjusted means is a consequence of employing distinct regression lines in each experimental condition and employing the respective experimental condition covariate means as predictors. Consequently, when heterogeneous regression ANCOVA is applied that does not fit separate regression lines per experimental condition (in more complex designs, heterogeneous regressions may be applied over all levels of different factors, rather than per experimental condition) or does not employ experimental condition covariate means as predictors, adjusted and unadjusted means may not be equivalent.

11.5 ESTIMATING HETEROGENEOUS REGRESSION ANCOVA EFFECTS

The full GLM for the single factor, single-covariate heterogeneous regression ANCOVA design was described in equation (11.1). The reduced GLM for this design omits the variable representing experimental conditions and is described by the equation

$$Y_{ij} = \mu + \beta_j Z_{ij} + \varepsilon_{ij} \qquad (11.4)$$

The GLM equation (11.4) describes p dependent variable on covariate regression lines all with a common Y-intercept (μ). However, the estimates of the $\widehat{\beta}_j$ in equation (11.4) and those estimated for equation (11.1) are not equal (see Searle, 1987). Therefore, to minimize the amount of calculation required, an alternative approach to estimating heterogeneous regression ANCOVA effects will be described.

This approach simply applies and extends the check of the regression homogeneity assumption.

First, the full traditional ANCOVA error term is required. As these calculations were described in Chapter 6, albeit for all three experimental conditions, the traditional ANCOVA error sum of squares (SS) will be accepted as, 116.591, with *dfs* = 13. Next, the full heterogeneous regression ANCOVA GLM predicted scores must be calculated. A little algebra applied to equation (11.1) reveals

$$Y_{ij} - \beta_j Z_{ij} = \mu + \alpha_j + \varepsilon_{ij} \tag{11.5}$$

Simply rewriting this equation in terms of the parameter estimates provides

$$\widehat{Y}_{ij} - \widehat{\beta}_j Z_{ij} = \widehat{\mu} + \widehat{\alpha}_j \tag{11.6}$$

which states that when all influence of the covariate has been removed, the predicted dependent variable score is equivalent to the constant plus the effect of the particular experimental condition. Of course, when all influence of the covariate is removed, $Z = 0$. Therefore, $(\widehat{\mu} + \widehat{\alpha}_j)$ must be equivalent to the *Y*-intercept of each experimental condition regression line (see Figure 11.1). Indeed, with balanced data

$$\mu = \frac{\sum_{j=1}^{p} (\mu + \alpha_j)}{p} \tag{11.7}$$

As well as passing through the $(\widehat{\mu} + \widehat{\alpha}_j)$ intercepts, each regression line passes through the point defined by the experimental condition dependent variable mean and the experimental condition covariate mean $(\overline{Z}_j \overline{Y}_j)$. Substituting these mean values into equation (11.7), along with the pertinent regression coefficient estimates, allows calculation of the regression line intercepts. For experimental condition 1,

$$(\widehat{\mu} + \widehat{\alpha}_1) = 10.000 - 1.200(5.750) = 3.100$$

and for experimental condition 2

$$(\widehat{\mu} + \widehat{\alpha}_3) = 23.000 - 2.267(5.500) = 10.532$$

In fact, the $(\widehat{\mu} + \widehat{\alpha}_j)$ values are actually the means of the predicted scores, as can be seen by adding the Y_{faij} term to equation (11.5)

$$Y_{faij} = Y_{ij} - \beta_j Z_{ij} = \mu + \alpha_j + \varepsilon_{ij} \tag{11.8}$$

Therefore, substituting each subjects' dependent variable and covariate scores into the first half of equation (11.8)

$$Y_{faij} = Y_{ij} - \beta_j Z_{ij} \tag{11.9}$$

provides the $\widehat{Y}_{faij}$ scores.

30 s $Y_{ij} - \widehat{\beta}_1(Z_{ij}) = \widehat{Y}_{faij}$	180 s $Y_{ij} - \widehat{\beta}_3(Z_{ij}) = \widehat{Y}_{faij}$
16 – 1.200 (9) = 5.200	24 – 2.267 (5) = 12.665
7 – 1.200 (5) = 1.000	29 – 2.267 (8) = 10.864
11 – 1.200 (6) = 3.800	10 – 2.267 (3) = 3.199
9 – 1.200 (4) = 4.200	22 – 2.267 (4) = 12.932
10 – 1.200 (6) = 2.800	25 – 2.267 (6) = 11.398
11 – 1.200 (8) = 1.400	28 – 2.267 (9) = 7.597
8 – 1.200 (3) = 4.400	22 – 2.267 (4) = 12.932
8 – 1.200 (5) = 2.000	24 – 2.267 (5) = 12.665

As the $\widehat{Y}_{faij}$ are the scores distributed around the $(\widehat{\mu} + \widehat{\alpha}_j)$ means the discrepancy between the $\widehat{Y}_{faij}$ scores and the $(\widehat{\mu} + \widehat{\alpha}_j)$ intercepts provide the error term estimates, $\widehat{\varepsilon}_{ij}$. This may be appreciated if a little algebra is applied to equation (11.8)

$$Y_{faij} = Y_{ij} - \beta_j Z_{ij} = \mu + \alpha_j + \varepsilon_{ij} \qquad (11.8, \text{rptd})$$

so

$$Y_{faij} = (\mu + \alpha_j) + \varepsilon_{ij}$$

and

$$Y_{faij} - (\mu + \alpha_j) = \varepsilon_{ij} \qquad (11.10)$$

30 s $\widehat{Y}_{fai1} - (\widehat{\mu} + \widehat{\alpha}_1) = \widehat{\varepsilon}_{i1}$	180 s $\widehat{Y}_{fai2} - (\widehat{\mu} + \widehat{\alpha}_2) = \widehat{\varepsilon}_{i2}$
5.200 – 3.100 = 2.100	12.665 – 10.532 = 2.133
1.000 – 3.100 = – 2.100	10.864 – 10.532 = 0.334
3.800 – 3.100 = 0.699	3.199 – 10.532 = – 7.333
4.200 – 3.100 = 1.100	12.932 – 10.532 = 2.400
2.800 – 3.100 = – 0.300	11.398 – 10.532 = 0.867
1.400 – 3.100 = – 1.699	7.597 – 10.532 = – 2.933
4.400 – 3.100 = 1.300	12.932 – 10.532 = 2.400
2.000 – 3.100 = – 1.100	12.665 – 10.532 = 2.133
$\sum \varepsilon_{i1}^2 = 16.395$	$\sum \varepsilon_{i1}^2 = 83.858$
$\sum_{i=1}^{N} \sum_{j=1}^{p} \varepsilon_{ij}^2 = 100.253$	

Table 11.2 summarizes the SS error and the *df*s obtained when homogeneous and heterogeneous regression ANCOVA GLMs are applied to the data.

Table 11.2 SS Error and *dfs* with Homogeneous and Heterogeneous Regression ANCOVA GLMs

	Homogeneous Regression ANCOVA GLM	Heterogeneous Regression ANCOVA GLM	Reduction
SS	116.591	100.253	16.338
df	13	12	1

As before, an F-test of the reduction in the error SS, attributed to heterogeneous regressions is given by

$$F = \frac{\text{SS Error reduction}/dfs\ \text{Reduction}}{\text{SS Heterogeneous regression}/df\ \text{Heterogeneous regression}}$$

$$F = \frac{16.338/1}{100.253/12} = 1.956$$

As $F(1,12) = 1.956$ is not significant at the 0.05 level, the traditional ANCOVA homogeneity of regression assumption is tenable. Nevertheless, as Maxwell and Delaney (2004) state and will also be seen later in this chapter, there may be good reasons for applying heterogeneous regression ANCOVA even when the homogeneity assumption is tenable. Therefore, analysis of the data presented in Table 11.1 will continue on the basis of the heterogeneous regression ANCOVA GLM.

Earlier it was said that the omnibus F-test of the effect of experimental conditions, when a heterogeneous regression ANCOVA GLM is applied, compares the separate experimental condition regression line Y-intercepts—the predicted differences between the experimental conditions when $Z = 0$. Therefore, a simple comparison of these Y-intercepts provides the same test of the effect of the experimental conditions.

In general, the experimental effect is denoted by the vertical difference between the experimental condition regression lines at any experimental condition covariate value, that is, the experimental effect can be predicted using different covariate values in the different experimental conditions. Equation (11.11) provides a formula for calculating such experimental effects when there are two experimental conditions

$$F = \frac{(\widehat{Y}_{Zpj} - \widehat{Y}_{Zpj})^2}{\text{MSe}\left[1/N_j + 1/N_j + \dfrac{(Z_{pj} - \overline{Z}_j)^2}{\sum_{i=1}^{N}(Z_{ij} - \overline{Z}_j)^2} + \dfrac{(Z_{pj} - \overline{Z}_j)^2}{\sum_{i=1}^{N}(Z_{ij} - \overline{Z}_j)^2}\right]^2} \tag{11.11}$$

where MSe is the heterogeneous regression ANCOVA mean square error, the $\widehat{Y}_{Zpj}$ are the predicted means given the covariate values for the particular experimental condition, the N_j are the number of subjects per experimental condition, the Z_{pj} are the experimental condition covariate values upon which the dependent variable means are predicted and the $\overline{Z}_j$ are the experimental condition covariate means. Substituting the values for the current example provides

$$F = \frac{(3.100 - 10.532)^2}{8.354\left[1/8 + 1/8 + \frac{(0 - 5.750)^2}{27.500} + \frac{(0 - 5.500)^2}{30.000}\right]}$$

$$F = \frac{55.235}{20.552}$$

$$F_{(1,12)} = 2.688$$

As $F(1,12) = 2.688$ is not significant at the 0.05 level, the null hypothesis that the two experimental condition regression line Y-intercepts are equivalent cannot be rejected. However, examination of Figure 11.1 shows that the difference between the regression lines is at its minimum (ignoring negative covariate values) when $Z = 0$. For this and other theoretical reasons, there may be relatively little interest in comparing the predicted experimental condition means when the covariate is zero. Probably of much more interest is the predicted effect of experimental conditions when subjects obtain covariate scores equal to the experimental condition covariate means. Given balanced data, the convenient fact that the predicted experimental effect at the respective experimental condition covariate means obtained with heterogeneous regression ANCOVA is equal to that obtained with traditional homogeneous regression ANCOVA may be employed. If the strategy outlined here has been applied (i.e., fitting a traditional homogeneous regression ANCOVA GLM and then testing for error reduction after fitting heterogeneous regressions) then the traditional ANCOVA and the heterogeneous regression ANCOVA experimental effects will have been calculated already. Table 11.3 summarizes the traditional ANCOVA GLM applied to the data presented in Table 11.1.

Table 11.3 reveals a significant effect of experimental conditions predicted on the basis of the experimental condition covariate means. The critical F-values presented in Appendix B may be used to determine significance if the calculations were carried out by hand, or the statistical software employed does not output the required p-values.

Table 11.3 Summary of the Traditional ANCOVA of the Data Presented in Table 11.1

Source	SS	*df*	MS	*F*	*p*
Error reduction due to experimental conditions	719.313	1	719.313	80.204	<0.001
Error reduction due to covariate	177.409	1	177.409	19.781	0.001
Full GLM error	116.591	13	8.969		

11.6 REGRESSION GLMs FOR HETEROGENEOUS REGRESSION ANCOVA

The experimental design GLM equation (11.1) may be compared with the equivalent regression equation

$$Y_{ij} = \beta_0 + \beta_1 X_{i,1} + \beta_2 Z_{ij} + \beta_3 (XZ) + \varepsilon_{ij} \qquad (11.12)$$

where β_0 represents a constant common to all Y scores, β_1 is the regression coefficient for the predictor variable X_1, which distinguishes between the two experimental conditions, and β_2 is the regression coefficient for the covariate, Z_{ij} is the covariate score for the ith subject in the jth condition, β_3 is the regression coefficient for the (XZ) interaction, which represents the heterogeneous regression, and, as always, the random variable, e_{ij}, represents error. Table 11.4 presents effect coding for the single factor, single-covariate heterogeneous regression ANCOVA GLM.

As with other design analyses, implementing a single factor, single-covariate heterogeneous regression ANCOVA is a two-stage procedure, if only the variance attributable to the experimental conditions is to be assessed, and a three-stage procedure if the variance attributable to the covariate regression is to be assessed. Consistent with estimating effects by comparing full and reduced GLMs, the first regression carried out is for the full single factor, single-covariate heterogeneous regression experimental design GLM, when all experimental condition predictor variables (X_1), the covariate (Z), and the experimental condition—covariate interaction (XZ) are included. The results of this analysis are presented in Tables 11.5 and 11.6.

Table 11.4 Effect Coding and Covariate for a Single Factor, Single-Covariate Heterogeneous Regression ANCOVA. Subject Number and the Dependent Variable Score Are Also Shown

Subject	Z	X_1	XZ	Y
1	9	1	9	16
2	5	1	5	7
3	6	1	6	11
4	4	1	4	9
5	6	1	6	10
6	8	1	8	11
7	3	1	3	8
8	5	1	5	8
17	5	−1	−5	24
18	8	−1	−8	29
19	3	−1	−3	10
20	4	−1	−4	22
21	6	−1	−6	25
22	9	−1	−9	28
23	4	−1	−4	22
24	5	−1	−5	24

Table 11.5 Results for the Full Single Factor, Single-Covariate Heterogeneous Regression ANCOVA Regression GLM

Variable	Coefficient	Standard Error	Standard Coefficient	t	p (two-tailed)
Constant	6.817	2.267	0.000	3.007	0.011
X_1	−3.717	2.267	0.477	−1.639	0.127
Z	1.733	0.382	0.423	4.543	0.001
XZ	−0.533	0.382	−0.407	−1.398	0.187

Table 11.6 ANOVA Summary Table for Experimental Conditions, Covariate, and Heterogeneous Regressions

Source	SS	df	Mean Square	F	p
Regression	869.733	3	289.911	34.697	<0.001
Residual	100.267	12	8.356		

R: 0.947; R^2: 0.897; adjusted R^2: 0.871.

Table 11.5 presents the predictor variable regression coefficients and standard deviations, the standardized regression coefficients, and significance tests (t- and p-values) of the regression coefficients. Table 11.5 is also interesting in that the Constant is the value of μ free of α_j. This confirms

$$\mu = \frac{\sum_{j=1}^{p}(\mu + \alpha_j)}{p} \qquad (11.7, \text{rptd})$$

Table 11.6 presents the ANOVA summary table for the regression GLM describing the complete single factor, single-covariate ANCOVA. As the residual SS is obtained when both covariate and experimental conditions are included in the regression, this is the error term obtained when the single factor, single-covariate ANCOVA GLM is applied.

The second stage is to carry out a regression where the experimental conditions are omitted, but all other regression predictors are included. This regression GLM is equivalent to the reduced GLM for the single factor, single-covariate heterogeneous

Table 11.7 Results for the Heterogeneous Regression ANCOVA GLM Omitting Experimental Conditions

Variable	Coefficient	Standard Error	Standard Coefficient	t	p (two-tailed)
Constant	7.110	2.402	0.000	2.959	0.011
Z	1.694	0.405	0.413	4.186	0.001
XZ	−1.126	0.130	−0.858	−8.692	<0.001

Table 11.8 ANOVA Summary Table for the Heterogeneous Regression GLM Omitting Experimental Conditions

Source	SS	*df*	MS	*F*	*p*
Regression	847.278	2	423.639	44.876	<0.001
Residual	122.722	13	9.440		

R: 0.935; R^2: 0.873; adjusted R^2: 0.854.

regression ANCOVA. The results of this analysis are presented in Tables 11.7 and 11.8.

The results presented in Table 11.7 are of little interest, but they do demonstrate that the reduced GLM estimates of the Constant (μ) and the dependent variable on covariate regressions per experimental condition differ from those of the full GLM estimate, the additional calculation of which was the reason given earlier for taking an alternative approach to calculating the effects of the experimental conditions. Of most interest is the residual/error term from the heterogeneous regression presented in Table 11.8.

The difference between the residual/error SS in Table 11.6 and that in Table 11.8 is equivalent to the SS attributable to experimental conditions. However, the SS attributed to the regressions in Table 11.8 is not equivalent to the covariate SS calculated when the full ANCOVA GLM is applied. The SS for the covariate in the full ANCOVA GLM may be obtained by comparing the error SS from the full ANCOVA with the error SS from an equivalent full ANOVA GLM. A full ANOVA GLM is implemented by a regression that uses only the predictors representing the experimental conditions (X_1). Table 11.9 presents the ANOVA summary of this analysis.

The error reduction attributable to the covariate can be calculated using the error term from the regression GLM implementation of the single factor ANOVA. This information is summarized in Table 11.10. In common with the full experimental design heterogeneous regression ANCOVA, this regression ANCOVA GLM assess the experimental effect when $Z=0$. The tabled critical F-values presented in Appendix B may be used to determine significance if hand calculation is employed or the statistical software employed does not output the required p-values.

Table 11.9 ANOVA Summary Table for Experimental Conditions Regression

Source	SS	*df*	MS	*F*	*p*
Experimental condition regression predictors	676.000	1	676.000	32.190	<0.001
Residual	294.000	14	21.000		

R: 0.835; R^2: 0.697; adjusted R^2: 0.675.

Table 11.10 ANOVA Summary Table for Experimental Conditions and Heterogeneous Covariate Regressions with Regression Implementation

Source	SS	*df*	MS	*F*	*p*
Error reduction due to experimental conditions	22.456	1	22.456	2.688	0.127
Error reduction due to covariate	188.754	2	94.377	20.636	<0.001
Full ANCOVA GLM residual	100.267	12	8.356	1.954	

R: 0.940; R^2: 0.884; adjusted R^2: 0.867.

11.7 COVARIATE–EXPERIMENTAL CONDITION RELATIONS

Applying traditional ANCOVA when a relationship exists between the covariate and the experimental conditions will provide an analysis which is statistically correct, but it is unlikely to apply easily or usefully to the real world (see Huitema, 1980, for the traditional ANCOVA approach to this issue). In these circumstances, heterogeneous regression provides the most appropriate analysis.

In what probably remains the most detailed consideration, Smith (1957) identified three situations that result in a relationship between the covariate and the experimental conditions (also see Huitema, 1980; Maxwell and Delaney, 2004). The first situation is when a variable not included in the GLM exerts an effect on both the dependent variable and the covariate. The best way of dealing with this source of systematic bias is to include the pertinent variable in the GLM. The second situation, where the covariate and dependent variable are both measures of the one entity is addressed in Section 11.8.2. The final situation is when the experimental conditions influence the covariate measure.

The good ANCOVA design practice of recording the covariate before the experimental manipulation prevents the experimental conditions affecting the covariate directly. (However, this will have no consequence for the relationship between the covariate and the experimental manipulation if both are measures of the same entity.) However, there may be practical or theoretical reasons for recording the covariate after the experimental manipulation. For example, practical reasons may preclude covariate recording before the experimental manipulation has been implemented or for theoretical reasons separate and independent statistical control within experimental conditions may be deemed appropriate. In these situations, the covariate measure is likely to be affected by the experimental manipulation and this creates problems for traditional ANCOVA. Two problems arise for traditional ANCOVA when the experimental conditions affect the covariate. These problems are adjustments based on the general covariate mean and multicolinearity. Each of these problems and how they are addressed by heterogeneous regression ANCOVA is described below.

11.7.1 Adjustments Based on the General Covariate Mean

Traditional ANCOVA is the assessment of an experimental effect at the general covariate mean. This can be seen by examining equation (9.6), which reveals that the traditional ANCOVA adjustment is based on the within groups regression coefficient (β_{w}) and the deviation of subjects' covariate scores (Z_{ij}) from the general covariate mean (Z_{G})

$$Y_{aij} = Y_{ij} - \beta_{\mathrm{w}}(Z_{ij} - Z_{\mathrm{G}}) = \mu + \alpha_j + \varepsilon_{ij} \tag{9.6, rptd}$$

However, when experimental conditions influence the covariate, the experimental conditions are related to (i.e., correlated with) the covariate and this has two important consequences. First, differences between the experimental condition covariate means are expected and second, covariate values correspond with experimental conditions, with the experimental condition covariate mean values reflecting the specific experimental conditions and the general covariate mean value reflecting an intermediate experimental condition, but one which may have no counterpart in reality (Huitema, 1980; Smith, 1957). Therefore, when experimental conditions affect the covariate, traditional ANCOVA can be described as assessing the experimental effect in a fictitious experimental condition. Interpreting the general mean as an experimental condition is one difficulty, but another is interpreting the experimental effect in an experimental condition when it should involve the comparison of experimental conditions.

When experimental conditions influence the covariate, heterogeneous regression ANCOVA is able to fit separate and independent regression lines (Searle, 1987) to the different experimental conditions with the different covariate score distributions. As adjustments can be based on the individual experimental condition covariate means rather than the general covariate mean, the problem of the general covariate mean representing an intermediate experimental condition does not arise (see Urquhart, 1982, for a similar assessment and conclusion). However, while heterogeneous regression ANCOVA maintains the integrity of the experimental condition covariate score distributions, the relationship between the experimental conditions and the covariate remains to be addressed. Nevertheless, by avoiding covariate distribution conflation, the relation between the experimental conditions and the covariate in heterogeneous regression ANCOVA changes from a condition that seriously complicated the interpretation of adjusted experimental effects and questioned the real world validity of the analysis, to the familiar regression topic of multicolinearity.

11.7.2 Multicolinearity

Multicolinearity has been discussed already in Sections 1.4 and 10.4.2.1, where the use of incremental analysis and heterogeneous regression to address multicolinearity problems also was mentioned. Multicolinearity refers to correlated predictor variables. As a consequence of their relationship, the covariate and experimental condition predictors will be associated with the same variation in the dependent variable. Unless predictors are correlated perfectly, each predictor also will accommodate variation in the dependent variable that no other predictor accommodates, but it is the association of more than one predictor with the same variation in the dependent variable that creates the multicolinearity problems described in Section 1.4.

Incremental analysis attributes variance in the dependent variable to predictors that are entered into the GLM cumulatively in a principled order. When the heterogeneous regressions are across experimental conditions, the principled order requires the predictor(s) representing the experimental conditions to enter the GLM first – separate regressions cannot be applied before the experimental condition predictors separate the data by experimental condition. In these circumstances, the estimate of the effect of the experimental conditions will equal that which would be obtained with a conventional ANOVA. At the next step, variance uniquely attributable to the covariate is accommodated with heterogeneous regression across the experimental conditions. This heterogeneous regression ANCOVA adjustment is based on separate and independent experimental condition regression coefficients (β_j) and the deviation of the subjects' covariate scores (Z_{ij}) from their experimental condition covariate mean ($\overline{Z}_j$). Although the variance accommodated by these regressions does not influence the variance attributed to the experimental conditions, it is removed from the error term and so provides more powerful tests of the effects assessed.

When the variance in the dependent variable due to the covariate is extracted first, variance that could have been attributed to the experimental conditions will be removed and so, the variance subsequently attributed to the experimental conditions will be found to be less than that obtained when the experimental condition variance is extracted first, with the difference being a function of the correlation between the experimental conditions and the covariate. For most experiments this form of analysis would not be appropriate, due to a primary interest in the effect of the experimental conditions. However, it does suggest another way of analysing the experimental data (see 11.8.2).

11.8 OTHER ALTERNATIVES

11.8.1 Stratification (Blocking)

Rather than a statistical operation, stratification is a modification to the design of the study, which necessitates a change in the experimental design GLM. The strategy employed is to allocate subjects to groups defined by certain ranges of the covariate scores. This creates another factor in the study design, with the same number of levels as the number of newly defined groups. This modification also changes the ANCOVA into an ANOVA: the dependent variable scores are input to a conventional ANOVA on the basis of the new experimental design GLM. For example, the GLM equation for the independent measures, single-covariate, single factor design, described by equation (11.1), after stratification would be described by equation (11.3)

$$Y_{ijjk} = \mu + \alpha_j + \beta_k + (\alpha\beta)_{jk} + \varepsilon_{ijk} \tag{11.13}$$

where β_k is the new factor with q levels, from 1 to k. The q levels of β_k represent the defined ranges of the covariate values (see Cochran, 1957; Elashoff, 1969; Kirk,

1995; Maxwell and Delaney, 2004; Winer et al., 1991; Winer, Brown, and Michels, 1991).

Ideally, the decision to employ this sort of analysis would be taken before subjects are recruited to the experiment. This would enable the appropriate allocation procedures to be implemented (Maxwell and Delaney, 2004). The major advantage conferred by stratification is that no assumptions are made about the form of the relationship between the treatments and the covariate. Consequently, all of the problems unique to ANCOVA are avoided.

However, Maxwell and Delaney (2004) describe the disadvantages of stratification compared with ANCOVA when ANCOVA assumptions are tenable. First, information is lost in the change from the covariate measurement scale to the nominal stratification measurement scale. The consequence is that variance accommodated by the covariate in ANCOVA cannot be accommodated by the stratified covariate. Second, while ANCOVA accommodates only linear trend, with stratification all possible trends, such as linear, quadratic, cubic, and so on, are accommodated: another trend component with each level of the new factor. The consequence is that where ANCOVA devotes only one *df* to the covariate, stratification devotes $(q - 1)$ *dfs*. Unfortunately, this is not economical, as the linear trend component accommodates the vast majority of the variance in most psychological data. Both of these stratification features result in a loss of analysis power in comparison with ANCOVA. Third, stratification will increase the number of experimental conditions and so reduce the *dfs* associated with the error term. The largest reduction is due to the *dfs* associated with the most complex or highest order interaction involving the covariate stratification factor. This interaction approximates the ANCOVA capability of assessing experimental effects at any value of the covariate. Generally, the reduction in error term *dfs* results in higher error term estimates and so again, less powerful F-tests in comparison with ANCOVA.

Given the preceding points, a stratified analysis is most likely to be applied when one or more traditional ANCOVA assumptions are untenable. As the untenability of most of the traditional ANCOVA assumptions is likely to be determined only after the experiment has been completed and the data analyzed, a major difficulty with stratification is that the distribution of subjects' covariate scores may not allow convenient allocation to useful covariate range groups to conform to conventional ANOVA design requirements. In other words, without discarding selected data, which raises problematic issues, it is likely that this approach will require the analysis of unbalanced designs. Therefore, when planning an ANCOVA experiment it would seem wise to consider assumption failures and, as far as possible, design the experiment so the data obtained would be compatible with stratification analysis requirements.

11.8.2 Replacing the Experimental Conditions with the Covariate

When multicolinearity is caused by the covariate and the experimental conditions expressing the same entity, it may be beneficial to modify the experimental conception by dropping the terms representing the correlated experimental conditions from the experimental design GLM and employ only the covariate to predict dependent variable scores. Certainly, one advantage enjoyed by the covariate is measurement on a ratio or interval scale (although sometimes this is stretched to an ordinal scale), in contrast to the nominal scale on which the category coded experimental manipulation is measured. An analysis based on a new linear model may be carried out by dropping the correlated experimental manipulation and introducing a term to represent the covariate. (In factorial designs, the covariate may be correlated with only one factor and so, one factor and all other terms involving this factor would be replaced by the covariate and a new set of terms representing the main effect and interactions.) With a new different GLM fitted to the data, consideration would need to be given to the suitability of the new hypotheses tested and the conclusions that could be drawn from their rejection or support. (See Cohen and Cohen, 1983; McCullagh and Nelder, 1989; Pedhazur, 1997, regarding the interpretation of categorical and quantitative variable interactions.)

11.9 THE ROLE OF HETEROGENEOUS REGRESSION ANCOVA

Although heterogeneous regression ANCOVA is only an extension of traditional ANCOVA, its application to real problems pushes to the foreground a particularly important issue: the nature of the relationship between the covariate and the dependent variable. Smith (1957) pointed out that a direct causal link between covariate and dependent variable is not a necessary requirement in traditional ANCOVA, but without knowledge of the causal effects, the interpretation of adjusted means is hazardous. In heterogeneous regression this state of affairs would appear to be even more pronounced, due to the potential increase in causal routes provided by the separate regressions. Therefore, the price of achieving an accurate interpretation of effects in heterogeneous regression ANCOVA is a more extensive theoretical consideration of the relationship between the covariate and the dependent variable under the different experimental conditions, than needs to be undertaken when traditional ANCOVA is employed.

With the emphasis on a GLM approach to heterogeneous ANCOVA, the similarities between the theoretical description of causality required of the linear model and the causality, which usually is examined with structural equation models (Bentler, 1980), such as LISREL (e.g., Jöreskog and Sorbom, 1993), becomes more apparent (also see Cohen et al, 2003; Pedhazur, 1997). It is for these reasons that heterogeneous regression should be regarded as a means by which the validity of theoretical accounts can be further assessed and not as a cheap way to circumvent research effort or repair faulty research designs.

CHAPTER 12

Multilevel Analysis for the Single Factor Repeated Measures Design

12.1 INTRODUCTION

Snijders and Bosker (1999) describe multilevel analysis as the coming together of two streams of statistical research: one addressing contextual analysis and the other addressing mixed effects analysis. Contextual analysis is concerned with the way in which the social context influences the individual's behavior. Mixed effects analysis in regression and analysis of variance (ANOVA) concerns analyses in which some variables (or factors) are fixed, while others are random. Multilevel analysis arose in the 1980s from the appreciation that the social context and the individual are separate and random sources of variance. This appreciation linked with statistical developments in mixed effects analysis (and within contextual analysis) that enabled regression models to accommodate nested random variables and their coefficients. (For more detailed historical accounts, see Hüttener and van den Eeden, 1995; Longford, 1993.) One consequence of its varied origins is the number of different names applied to a set of similar, but not identical analyses. These include, hierarchical linear analysis (cf. Section 5.4), random coefficient analysis, or general linear mixed model analysis (e.g., Laird and Ware, 1982). Of these analyses, only hierarchical linear analysis is equivalent to multilevel analysis. Both hierarchical linear analysis and multilevel analysis impose a hierarchical or multilevel structure on the data (see Section 12.3). However, it is also possible to combine the equations describing the separate levels into a single full model equation. When this is done, the full model equation will include both fixed and random components and will employ fixed and random coefficients. Consequently, the model may be referred to as a random coefficient analysis (or model) or a (general) linear mixed model.

ANOVA and ANCOVA: A GLM Approach, Second Edition. By Andrew Rutherford.

Nevertheless, despite these differences, multilevel analysis now seems to be the most commonly applied name. Multilevel analysis is the title employed throughout this chapter not only because it is applied commonly but also because there will be a focus on a particular multilevel structure.

Since the 1980s, multilevel analysis applications have become increasingly popular in the medical sciences, in the social sciences and in developmental, educational, and organizational psychology. In these research areas, data frequently is conceived as being grouped at different levels. For example, children from the same family or nursery may be more alike in their responses than they are to other children, while individual pupils within a class, within a school, or within an education authority area may exhibit greater similarity of response to a new teaching technique than pupils from other classes, schools, or education authority areas. Responses on other measures may also show similar patterns for individuals within different parts of an organization and perhaps at different levels of seniority.

Although it may not be obvious immediately from the account just provided, multilevel analysis also offers a number of benefits for the examination of experimental data (e.g., Hoffman and Rovine, 2007). For example, data from groups tend to be related and often violates the sphericity assumption (see Sections 10.2.2 and 10.4.1.3). Multilevel models are able to deal with such relations by their accommodation of random factors. Multilevel analysis also offers a means of conducting more sophisticated analyses of experimental data. For example, Wright and London (2009) describe how signal detection theory (Macmillan and Creelman, 2005) measures can be estimated and employed economically within a multilevel analysis, while Hoffman and Rovine (2007) describe how multilevel analysis can model reaction times and error rate simultaneously, so allowing direct investigation of speed-accuracy trade-offs. Nevertheless, such sophisticated multilevel analysis will not be considered here. The current chapter has the much simpler aim of introducing readers to the multilevel analysis approach to the single factor repeated measures experimental design described in Chapter 6.

12.2 REVIEW OF THE SINGLE FACTOR REPEATED MEASURES EXPERIMENTAL DESIGN GLM AND ANOVA

In Chapter 6, data from eight subjects, all observed under each of the three experimental conditions was presented in Table 6.6. All but the subject marginal means are presented again in Table 12.1. It also was explained in Chapter 6 that the general linear model (GLM) underlying a single factor related or repeated measures design ANOVA can be described by the equation

$$Y_{ij} = \mu + \pi_i + \alpha_j + \varepsilon_{ij} \qquad (6.3, \text{rptd})$$

where Y_{ij} is the ith subject's dependent variable score in the jth experimental condition, μ is the grand mean of the experimental condition population means,

Table 12.1 The Number of Words Recalled by Subjects After Study Periods of 30, 60, and 180 s Obtained in a Single Factor Repeated Measures Design

Subjects	30 s (Condition 1)	60 s (Condition 2)	180 s (Condition 3)
s1	7	7	8
s2	3	11	14
s3	6	9	10
s4	6	11	11
s5	5	10	12
s6	8	10	10
s7	6	11	11
s8	7	11	12
Means	6	10	11

π_i represents the random effect of the *i*th subject, and α_j is the effect of the *j*th experimental condition. It was also explained that rather than simply reflecting variation due to any uncontrolled source, the error term, ε_{ij}, also incorporates the interaction effect $(\pi\alpha)_{ij}$. However, it is only when the error covariance matrix is spherical that a valid and accurate *F*-test is obtained (see Sections 6.3 and 10.2.2).

In addition to the assumption of a spherical experimental conditions covariance matrix, ANOVA repeated measures designs also require all subjects to provide data across all of the repeated measures conditions. If a subject fails to provide any repeated measures dependent variable score, all of this subject's data must be dropped from the ANOVA of the experimental data.

12.3 THE MULTILEVEL APPROACH TO THE SINGLE FACTOR REPEATED MEASURES EXPERIMENTAL DESIGN

The multilevel analysis approach grew out of regression analysis, so expressing multilevel models in regression terms is to be expected (see Section 2.7.3). Therefore, in common with GLMs, when the independent variable has a quantitative or continuous nature, multilevel analysis provides an appropriate form of analysis (Cohen, 1983; Maxwell and Delaney, 1993; Vargha et al., 1996). Nevertheless, when multilevel models are applied to experimental data, it is just as easy to express them as experimental design GLMs. When a multilevel approach is applied to repeated measures data, the repeated measures are considered to occur "within" the subject. In multilevel parlance, the subject is the context for the repeated measures. Consequently, a Level 1 model describes the repeated measures "within" each subject as

$$Y_i = \beta_0 X_{0i} + \beta_1 X_{1i} + \beta_2 X_{2i} + \varepsilon_i \tag{12.1}$$

where Y_i represents the ith subject's dependent variable score, $\beta_0 X_{0i}$ represents the intercept on the Y-axis for the ith subject, X_{1i} and X_{2i} are indicator variables coding experimental condition for the ith subject, and ε_i the error term reflecting variation due to any uncontrolled source for the ith subject. Therefore, equation (12.1) describes a separate regression line for each subject.

The relations across subjects are expressed by the (higher) Level 2 multilevel regression GLMs

$$\beta_0 = \beta_0 + u_{0i} \tag{12.2a}$$

$$\beta_1 = \beta_1 + u_{1i} \tag{12.2b}$$

$$\beta_2 = \beta_2 + u_{2i} \tag{12.2c}$$

It is worth drawing attention to the unfortunate, but established, use of the ordinary letter "u" to represent the parameter (i.e., the coefficient) for a random factor. Clearly, this has the potential to create confusion with μ, the Greek letter representing the general mean in experimental design GLMs. Nevertheless, examination of equation (12.2a) reveals that β_0 is defined as being composed of fixed (β_0) and random (u_0) effects. As fixed effect parameters are defined frequently as population means and are estimated as the average over the subjects in the sample, they are not associated with individual subjects so the i subscript is not appropriate. In contrast, random effect parameters, such as the parameter u, represent variance. Here, each u_i represents deviation attributable to the ith subject. In other words, equation (12.2a) informs us that β_0 describes a separate intercept for each subject, where each subject varies from the general intercept (β_0) by u_{0i}. Similarly, the first and the second regression coefficients, β_1 and β_2, include fixed components estimated across subjects (β_1 and β_2) and individual deviations from these fixed components given by u_{1i} and u_{2i}.

Substituting the Level 1 components for their Level 2 descriptions provides the full multilevel regression GLM for the repeated measures design presented in Table 12.1

$$Y_i = \beta_0 + u_{0i} + \beta_1 X_{1i} + \beta_2 X_{2i} + (u_i X_{1i} + u_2 X_{2i} + \varepsilon_i) \tag{12.3}$$

(*Note*: X_1 is multiplied by β_1, so when β_1 is replaced by $\beta_1 + u_{1i}$, X_1 is multiplied by $\beta_1 + u_{1i}$ to give $\beta_1 X_{1i} + u_i X_{1i}$. The same applies to β_2.) Examination of the full model (12.3) reveals an equivalence with the single factor repeated measures experimental design GLM when effect coding is employed. With respect to the fixed effects, the β_0 component from equation (12.3) is equal to the μ component of equation (6.3, rptd) and the ($\beta_1 X_{1i} + \beta_2 X_{2i}$) component is equal to the α_j component of equation (6.3, rptd). With respect to the random effects, the u_{0i} component from equation (12.3) is equal to the π_i component of equation (6.3, rptd), while the

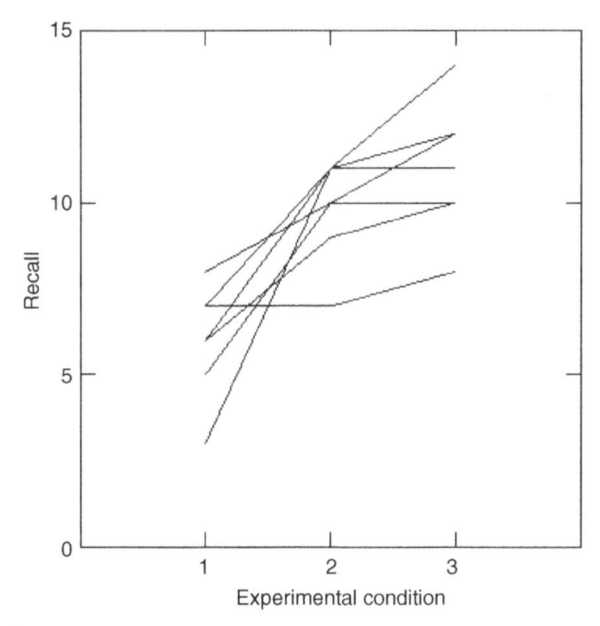

Figure 12.1 Recall on experimental condition for each subject.

$(u_i X_{1i} + u_2 X_{2i} + \varepsilon_i)$ from equation (12.3) is equal to the ε_i, or, more accurately, the $[(\pi\alpha)_{ij} + \varepsilon_{ij}]$ component of equation (6.2).

In multilevel analyses, the dependent variable scores are conceived as randomly sampled measures nested within subjects. Although missing repeated measures is less of a problem for laboratory-based experimental research, a major benefit of this conception for applied research is that it allows the accommodation and use of data from subjects with incomplete sets of repeated measures. This contrasts with the experimental design GLM approach to repeated measures mentioned earlier, where any subject with any missing repeated measures data must be omitted from the analysis. However, in common with the experimental design GLM approach to unbalanced data with independent data, multilevel analyses implicitly assume that missing data is missing at random (i.e., there is no systematic reason for data to be missing).

In Figure 12.1, each line represents each subjects' recall plotted as a function of the experimental condition. As can be seen, most subjects' recalls increase from experimental condition 1 (30 s study time) through experimental condition 2 (60 s study time) to experimental condition 3 (180 s study time), but there are exceptions.

The multilevel model described for this repeated measures design employs a separate regression intercept of scores on experimental conditions per subject. More often than not, "factor" is dropped from the name of the random factor intercept model, it being called a random intercept model for short. (This is mentioned just in

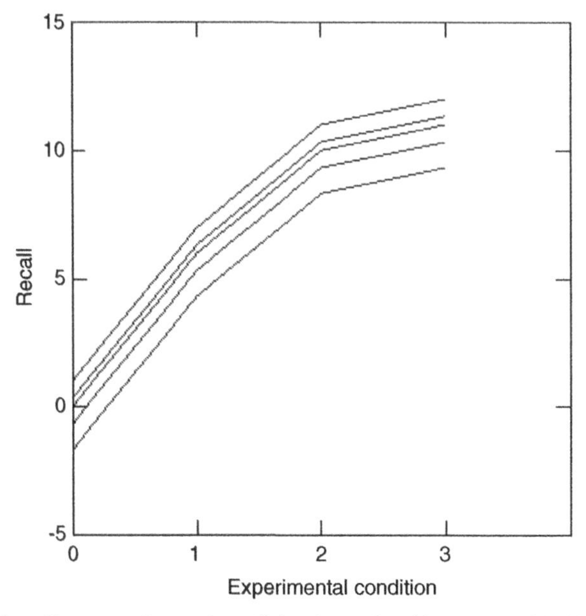

Figure 12.2 Recall on experimental condition for each subject as provided by the multilevel model described by equation (12.3). Each subject provides a regression line, but as four subjects share an intercept, only five lines are apparent (i.e., one regression line represents four subjects).

case any reader thinks random intercept models locate the intercepts on a random basis!) However, although the current multilevel model for the repeated measures design accommodates the influence of experimental conditions on subjects' scores, it also employs a single regression coefficient to represent the subjects' scores across the experimental conditions. Figure 12.2 provides a graphical representation of the multilevel model (and the repeated measures experimental design GLM) account of the repeated measures data presented in Figure 12.1. Hopefully, the graphical presentation demonstrates that employing a single regression coefficient for all subjects results in a set of parallel regression lines (one line per subject) across experimental conditions. All of the GLMs assume independent errors, so in repeated measures designs additional variables are introduced to accommodate relations between the repeated measures to provide independent errors. As discussed, both the repeated measures experimental design GLM and the random intercept multilevel GLM applied to the same repeated measures data accommodate certain relations, but not others. For example, the equivalent repeated measures experimental design and random intercept multilevel GLMs allow subjects scores to be related across experimental conditions, but, as Figure 12.2 shows, they do not allow any other relations between the repeated measures, or between the repeated measures and the experimental condition, as might be suggested by Figure 12.1. Of course, the issue is whether there is sufficient (i.e., significant) departure from compound symmetry expectations to warrant rejection of this assumption.

If the repeated measures experimental data was consistent with the repeated measures experimental design GLM and the random intercept multilevel GLM, then the matrix representing the experimental conditions covariance structure will exhibit compound symmetry (see Section 10.2.2). In other words, when you apply a random intercept multilevel model, you assume it is appropriate for the data and so, you assume the experimental conditions covariance structure will exhibit compound symmetry. In short, you assume compound symmetry when you apply a random intercept multilevel model.

As suggested above, different variables can be introduced into the repeated measures experimental design or multilevel GLMs to accommodate different relations across the repeated measures and experimental conditions. However, accommodating different aspects of the related factors may change the assumed nature of the covariance structure. Therefore, by including particular variables in the repeated measures experimental design or multilevel GLMs, different covariance structures can be accommodated providing more realistic models and more accurate estimates. However, as said in Chapter 6, when repeated measures design GLMs are applied, there is rarely interest in the random factor (i.e., the subject variations over the repeated measures). Usually, this factor is included simply to increase the power of the test of the experimental factor (manipulation). Given this perspective, there will be no further attempt to describe the specific nature of the multilevel GLMs that represent the different correlational structures. Instead, in Section 12.5, attention will focus simply on fitting multilevel models that vary in terms of the covariance structures they assume.

In the first part of this section the equivalence of the multilevel GLM and the repeated measures experimental design GLM for the repeated measures data presented in Table 12.1 was demonstrated and subsequently, it was described how the multilevel approach allows for different covariance structures. However, irrespective of the specific way in which the multilevel GLMs represent the different correlational structures, the effect of the experimental manipulations still may be expressed in terms of a comparison of full and reduced repeated measures GLMs. For example, the full repeated measures experimental design GLM

$$Y_{ij} = \mu + \pi_i + \alpha_j + \varepsilon_{ij} \qquad (6.3, \text{rptd})$$

was compared with the reduced repeated measures experimental design GLM

$$Y_{ij} = \mu + \pi_i + \varepsilon_{ij} \qquad (6.15, \text{rptd})$$

The same comparison expressed in multilevel GLM terms involves the full GLM

$$Y_i = \beta_0 + u_{0i} + \beta_1 X_{1i} + \beta_2 X_{2i} + (u_i X_{1i} + u_2 X_{2i} + \varepsilon_i) \qquad (12.3, \text{rptd})$$

and the reduced GLM

$$Y_i = \beta_0 + u_{0i} + (u_i X_{1i} + u_2 X_{2i} + \varepsilon_i) \qquad (12.3, \text{rptd})$$

In short, the comparison of full and reduced GLMs compares a GLM that includes terms representing the experimental conditions with a GLM that excludes the terms representing the experimental conditions. It follows that irrespective of the way in which the multilevel GLMs represent the different correlation structures, it still will be possible to compare full and reduced models that include and exclude the specific terms representing the experimental conditions. Later, when the effect of the experimental manipulations (i.e., the experimental factor) is presented, it can be assumed that such a comparison of full and reduced models will have been carried out.

12.4 PARAMETER ESTIMATION IN MULTILEVEL ANALYSIS

All of the GLMs considered in previous chapters employed least squares estimation (see Section 1.6) and to allow the least squares based normal equations to be solved to provide estimates of random variable effects requires the imposition of simplifying but restrictive constraints (i.e., assumptions). One way to escape from some of these constraints is to employ a different method of parameter estimation. The ability of multilevel models to provide superior account of random factor effects is tied very closely to their use of different methods of parameter estimation. However, all of the parameter estimation methods used by multilevel models are very demanding computationally and so, generally, multilevel analyses rely on computer implementations via appropriate software. One consequence is that various aspects of the multilevel analysis applied can be attributed to the particular capabilities of the statistical software employed. The same is true of this presentation, which employs the multilevel analyses available in SYSTAT.

SYSTAT offers four estimation methods for random factor and fixed factor effect estimation: ANOVA methods (i.e., method of moments estimation, which can provide Type I, II, and III sums of squares), minimum variance quadratic unbiased estimation (MIVQUE0), maximum likelihood estimation (ML), and restricted (or residual) maximum likelihood estimation (REML). However, all four estimation methods are available only when variance components models are applied to examine fixed and random effects. When (general) linear mixed models and hierarchical linear mixed models (see Section 12.1) are used to examine fixed and random effects, only ML and RML estimation methods are available. Most other statistical packages also offer ML and REML estimation methods for (general) linear mixed models and hierarchical linear mixed models, so ML and REML estimation methods will be the focus here. Although a review of parameter estimation methods is well beyond the scope of this text, some aspects of ML and REML, and other estimation methods will be mentioned as the use of multilevel models to analyze repeated measures designs is presented.

The least squares estimation of random factor effects hardly differs from the least squares estimation of fixed factor effects, whereas ML and REML take quite different approaches to the estimation of fixed and random factors. As the name "maximum likelihood" suggests, both ML and REML provide parameter estimates that are most likely in specific, statistically defined situations. Features of the statistically defined situation include the nature of the model, which defines the factors the factor types (fixed or random), the factor interactions, and parameter estimate distributions, as well as the nature of the covariance structures. However, ML and REML estimation procedures are computationally demanding partly because equations defining the parameter estimates with maximum likelihood are not available for all situations and iterative parameter estimation has to be adopted to identify the most likely parameter estimate. Although the power of contemporary computers makes this a fairly trivial problem, ML and REML exhibit several other features that would make them attractive parameter estimation procedures even if their computational demands were still problematic. Specifically, ML and REML provide consistent, asymptotically efficient (i.e., the estimates are precise—they have low variance), and asymptotically normal parameter estimates that always fall within the theoretically defined parameter space under a range of assumptions. Unfortunately, ML estimates also tend to be biased, due to using residuals as if they were data, when, in fact, as a consequence of the process that generates them, residuals are more correlated than real data. REML addresses the residual correlation by reducing the residual *df*s appropriately to provide unbiased parameter estimates. (For further information on ML and REML, see Searle, Casella, and McCulloch, 1992.)

12.5 APPLYING MULTILEVEL MODELS WITH DIFFERENT COVARIANCE STRUCTURES

In this section, a step-by-step description of how to implement the multilevel GLM described by equation (12.4) in SYSTAT is presented and some aspects of the SYSTAT output is discussed. Subsequently, some other multilevel GLMs for the repeated measures experiment under consideration are presented and discussed. However, this section has a limited set of specific aims and should not be regarded as providing a full account of linear mixed models in SYSTAT. (For further information on linear mixed models in SYSTAT readers should refer to the appropriate SYSTAT manuals.)

12.5.1 Using SYSTAT to Apply the Multilevel GLM of the Repeated Measures Experimental Design GLM

One of the first things required for a multilevel analysis, is an appropriately constructed data file. In SYSTAT and most other statistical packages, this means that each (repeated measures) dependent variable score should be on a separate line and the other column variables should specify the nature of each (repeated measures) dependent variable score. For the data from the hypothetical experiment presented in

Chapter 6, three column variables are required. The first variable specifies the subject, with subjects identified by the numbers 1–8. The next variable specifies the experimental condition, with the 30 s experimental condition identified by the number 30, the 60 s experimental condition identified by the number 60, and the 180 s experimental condition identified by the number 180. The (repeated measures) dependent variable scores occupy the last column. For Subject 1 in experimental condition 30 s, the dependent variable score provided was 7. For Subject 2 in experimental condition 30 s, the dependent variable score provided was 3, and so on. However, all of the subjects also provided scores in the other two experimental conditions and these data are recorded too. A copy of the data file containing this data is presented in Figure 12.3.

	SUBJECTS	TIME	RECALL
1	1.000	30.000	7.000
2	2.000	30.000	3.000
3	3.000	30.000	6.000
4	4.000	30.000	6.000
5	5.000	30.000	5.000
6	6.000	30.000	8.000
7	7.000	30.000	6.000
8	8.000	30.000	7.000
9	1.000	60.000	7.000
10	2.000	60.000	11.000
11	3.000	60.000	9.000
12	4.000	60.000	11.000
13	5.000	60.000	10.000
14	6.000	60.000	10.000
15	7.000	60.000	11.000
16	8.000	60.000	11.000
17	1.000	180.000	8.000
18	2.000	180.000	14.000
19	3.000	180.000	10.000
20	4.000	180.000	11.000
21	5.000	180.000	12.000
22	6.000	180.000	10.000
23	7.000	180.000	11.000
24	8.000	180.000	12.000

Figure 12.3 The typical format of data file required for multilevel modeling.

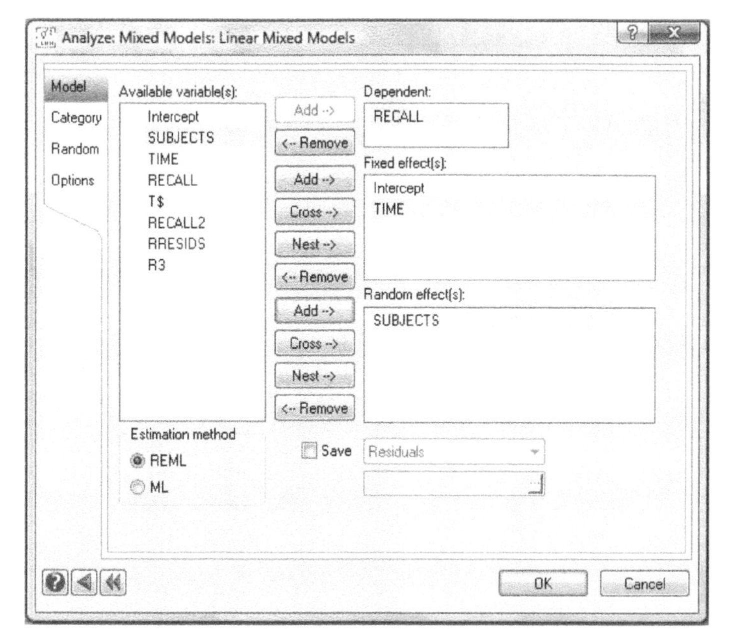

Figure 12.4 The *Linear Mixed Models* dialog box.

While the data editor is open, click on *Data* to obtain the drop down menu and click on *Categorical Variables*, and select, SUBJECTS and TIME, and click OK. (This surmounts a problem setting categorical variables under *Mixed Models*.)

12.5.1.1 The Linear Mixed Model

After completing the steps described in Section 12.5.1, click on *Analyze* to obtain the drop down menu and then select *Mixed Models* and then *Linear Mixed Models*. This opens the *Linear Mixed Models* dialog box. Select the column containing all of the (repeated measures) dependent variable scores (RECALL) as dependent and the experimental conditions column (TIME) as the fixed effect (see Figure 12.4). Notice that ML and REML estimation methods are available and that REML is the default.

Next, click on the *Random* tab. This opens the *Random* dialog box. At the top of this dialog box, identify the section titled, Random effects covariance structure. SUBJECTS should be the only effect listed in the Available effect(s) box on the left. Highlight SUBJECTS and click on the Add button. This will move SUBJECTS into the selected effect(s) box on the right. SUBJECTS also will be accompanied by an indication that a variance components covariance structure will accommodate the SUBJECT effect. The variance components covariance structure is one in which all diagon al entries are equal (i.e., an identity matrix). Unless there are particular features of the SUBJECT effect to be accommodated, this simple covariance structure usually is sufficient. Next, in the section below, labeled error covariance structure, click on the Structure menu (the box with the downward pointing arrow)

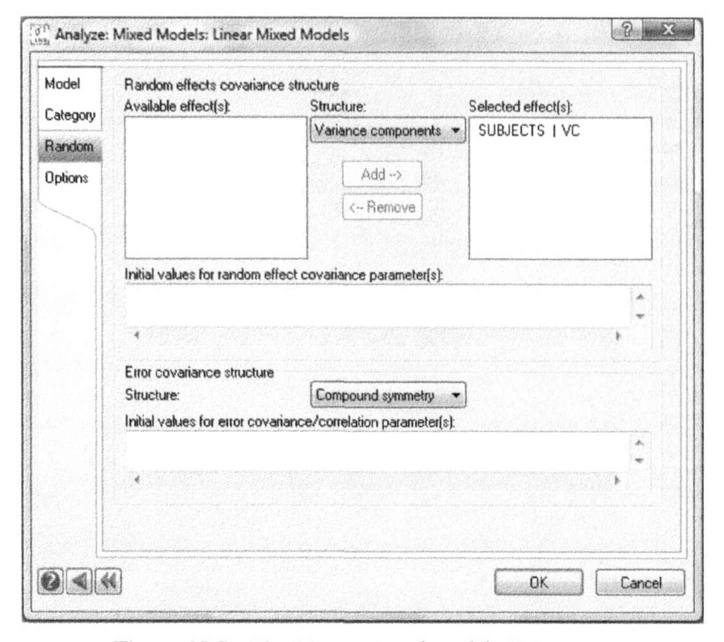

Figure 12.5 The *Linear Mixed Models* dialog box.

and select Compound symmetry, and click OK, which implements the analysis (see Figure 12.5).

The SYSTAT output is presented below.

```
The categorical values encountered during processing are

Variables           ¦                 Levels
--------------------+--------------------------------------
TIME (3 levels)     ¦ 30.000  60.000   180.000
SUBJECTS (8 levels) ¦  1.000   2.000     3.000  4.000  5.000
                    ¦  6.000   7.000     8.000


Dependent Variable  : RECALL
Fixed Factor(s)     : TIME
Fixed Covariate(s)  : Intercept
Random Factor(s)    : SUBJECTS
Estimation Method   : Residual or Restricted Maximum
                      Likelihood (REML)
```

Dimensions

```
Covariance Parameters :  2
Columns in X          :  4
Columns in Z          :  8
No. of Observations   : 24
```

Fit Statistics

Final L-L	:	-42.973
-2L-L	:	85.946
AIC	:	91.946
AIC(Corrected)	:	93.358
BIC	:	95.080

Estimates of Covariance Components

Random Effect	Description	Estimate
SUBJECTS	Variance Parameter	0.451
Error variance	Variance Parameter	2.224
	Error Correlation (CS)	0.000

Estimates of Fixed Effects

Effect	Level	Estimate	Standard Error	*df*	t	p-Value
Intercept		11.000	0.578	7	19.020	0.000
TIME	30	-5.000	0.746	14	-6.706	0.000
	60	-1.000	0.746	14	-1.341	0.201
	180	0.000	0.000	.	.	.

Confidence Intervals of Fixed Effects Estimates

			95.00% Confidence Interval	
Effect	Level	Estimate	Lower	Upper
Intercept		11.000	9.760	12.240
TIME	30	-5.000	-6.599	-3.401
	60	-1.000	-2.599	0.599
	180	0.000	.	.

Type III Tests for Fixed Effects

Effect	Numerator *df*	Denominator *df*	F-Ratio	p-Value
TIME	2	14	25.180	0.000

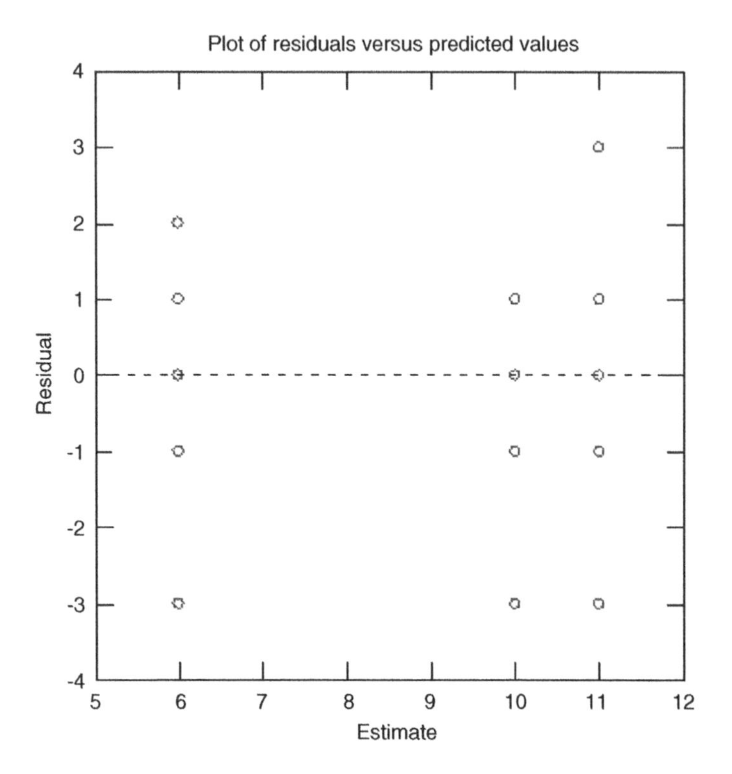

As the multilevel GLM described by equation (12.4) is equivalent to the repeated measures experimental design GLM, the analysis applied here should provide exactly the same results as the repeated measures ANOVA (based on the repeated measures experimental design GLM) presented in Tables 6.10 and 6.13. Above the residual plot (which suggests that there are no homoscedasticity problems—see Section 10.4.1.4), is a table providing the ANOVA (Type III SS) test for the fixed effects of the experimental factor, Time. However, the F-value presented, $F(2,14) = 25.180$, is greater than the value reported in Tables 6.10, 6.13, and 6.19 $F(2,14) = 20.634$. The reason for this is outlined below.

Previously, it was said that ML and REML provide estimates that always fall within the theoretically defined parameter space. The theoretically defined parameter space does not include negative variance, but, due to sampling error, negative variance estimates can arise in exactly the same way that slightly inflated variance estimates can arise. Unfortunately, the repeated measures experimental data in Table 12.1 provides a negative variance estimate for the subject effect. As REML assumes that all variance estimates are greater than or equal to 0, the negative estimate is set to 0. This results in an underestimate of the F-test denominator for the effect of the experimental factor, which, in turn, provides the inflated F-value. Fortunately, however, this problem can be surmounted by implementing a hierarchical linear mixed model analysis.

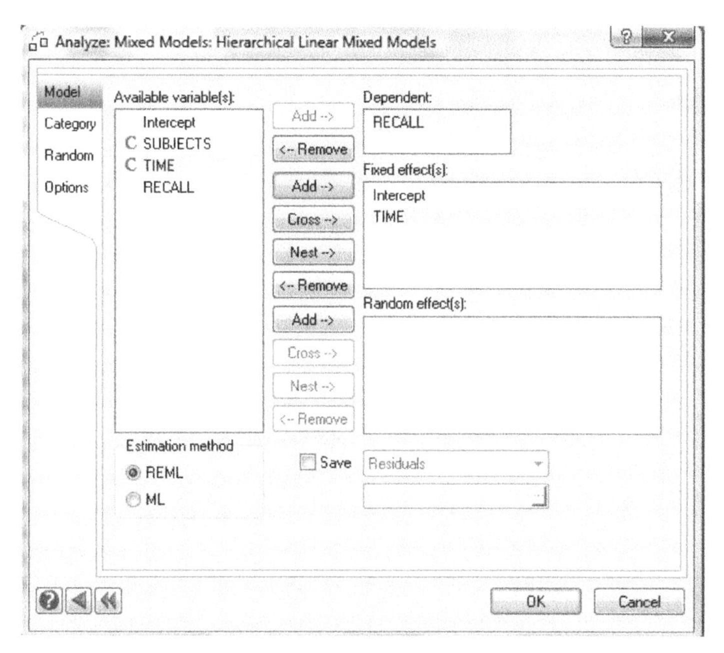

Figure 12.6 The *Hierarchical Linear Mixed Models* dialog box.

12.5.1.2 The Hierarchical Linear Mixed Model

A hierarchical linear mixed model analysis imposes the multilevel structure on the data described in Section 12.3. One consequence of this approach is an alteration in the parameters estimated. Although the two models are essentially equivalent, what was a negative variance estimate for the subject effect in the linear mixed model manifests as (REML acceptable) negative correlations between subjects' repeated measures scores in the hierarchical linear mixed model. Instructions to implement a hierarchical linear mixed model analysis are provided below.

After completing the steps described in Section 12.5.1, click on *Analyze* to obtain the drop down menu and then select *Mixed Models* and then *Hierarchical Linear Mixed Models*. This opens the *Hierarchical Linear Mixed Models* dialog box. Select the column containing all of the (repeated measures) dependent variable scores (RECALL) as dependent and the experimental conditions column (TIME) as the fixed effect (see Figure 12.6).

Next, click on the *Random* tab. This opens the *Random* dialog box. At the top of this dialog box, identify the section titled, Random effects covariance structure, and under *Subject*, from the drop down list, select the variable, SUBJECTS, and then moving to the left of this dialog box, under *Covariance structure*, select Compound symmetry, and click OK, which implements the analysis (see Figure 12.7).

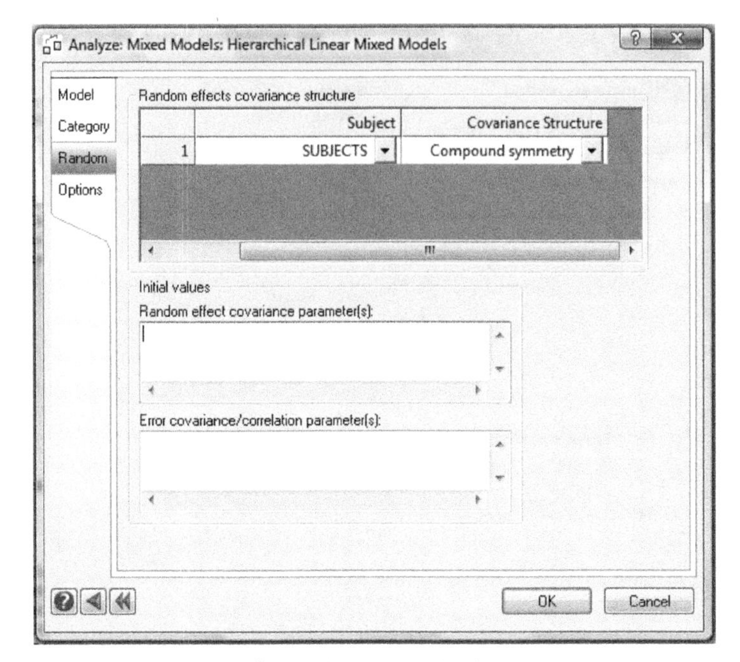

Figure 12.7 The *Random* dialog box—selecting Compound symmetry.

The SYSTAT output is presented below.

```
The categorical values encountered during processing are

Variables            ¦                Levels
---------------------+-----------------------------------
TIME (3 levels)      ¦ 30.000  60.000  180.000
SUBJECTS (8 levels)  ¦  1.000   2.000    3.000  4.000  5.000
                     ¦  6.000   7.000    8.000


Dependent Variable : RECALL
Fixed Factor(s)    : TIME
Fixed Covariate(s) : Intercept
Estimation Method  : Residual or Restricted
                     Maximum Likelihood (REML)


Dimensions
Covariance Parameters         :  1
Columns in X                  :  4
Columns in Z                  :  0
No. of Observations           : 24
Number of subjects            :  8
Max. observations per subject :  3
```

Fit Statistics

Final L-L	:	-42.333
-2L-L	:	84.665
AIC	:	88.665
AIC(Corrected)	:	89.332
BIC	:	90.754

Estimates of Covariance Components

Random Effect	Description	Estimate
Error variance	Variance Parameter	2.476
	Error Correlation (CS)	-0.096

Estimates of Fixed Effects

Effect	Level	Estimate	Standard Error	*df*	t	p-Value
Intercept		11.000	0.556	21	19.772	0.000
TIME	30	-5.000	0.824	21	-6.070	0.000
	60	-1.000	0.824	21	-1.214	0.238
	180	0.000	0.000	.	.	.

Confidence Intervals of Fixed Effects Estimates

Effect	Level	Estimate	95.00% Confidence Interval Lower	Upper
Intercept		11.000	9.843	12.157
TIME	30	-5.000	-6.713	-3.287
	60	-1.000	-2.713	0.713
	180	0.000	.	.

Type III Tests for Fixed Effects

Effect	Numerator *df*	Denominator *df*	F-Ratio	p-Value
TIME	2	21	20.632	0.000

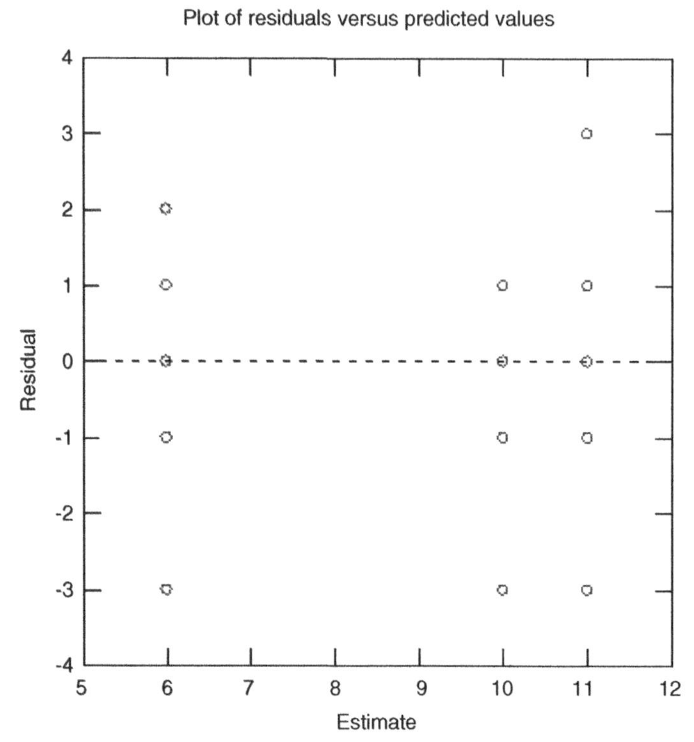

The F-ratio in the table above is virtually identical to the F-value reported in Tables 6.10, 6.13 and 6.19 (i.e., 20.632 cf. 20.634). Nevertheless, there is a discrepancy with respect to the denominator *dfs*. In common with (IBM) SPSS implementations of hierarchical linear mixed models, SYSTAT shows the F-test denominator $df = 21$. However, the repeated measures ANOVA shows the F-test denominator $df = 14$. This discrepancy is due to neither SYSTAT nor SPSS removing the 7 *dfs* contributed by Subjects from the F-test denominator *dfs*. However, no such *dfs* discrepancy arose when the linear mixed model was implemented (Section 12.5.1.1). This and the explanation for the F-estimate discrepancy observed with the linear mixed model (Section 12.5.1.1), reveals that although the linear mixed model and the hierarchical linear mixed model can be expressed equivalently, the two implementations differ and different issues are matters of debate for the different implementations, such as appropriate *dfs* for particular model terms in the hierarchical linear mixed model (e.g., Bolker et al, 2009).

12.5.2 Applying Alternative Multilevel GLMs to the Repeated Measures Data

Of course, the purpose of adopting a multilevel analysis approach to repeated measures designs is not to replicate ANOVAs, but rather to escape from some of the constraints placed on repeated measures ANOVA, so more accurate estimates of experimental effects can be obtained. The requirement that the experimental conditions covariance structure should exhibit sphericity (compound symmetry for current purposes, see Section 10.2.2) has been identified as particularly problematic.

When a linear mixed model was implemented (Section 12.5.1.1), it was assumed that the matrix representing the error covariance structure exhibited compound symmetry, but when a hierarchical linear mixed model was implemented (Section 12.5.1.2), it was assumed that the matrix representing the random Subject effect covariance exhibited compound symmetry. In Section 10.2.2, it was explained how a spherical matrix representing the experimental conditions covariance structure provided independent errors. Similarly, the assumptions about the covariance structures of the error and Subject effect covariance structures suit the different implementation methods but follow from the assumption that the matrix representing the experimental conditions covariance structure exhibits compound symmetry.

SYSTAT offers two alternative covariance structures: variance components and AR(1). The variance components option applies an identity matrix to accommodate the covariance structure, while the AR(1) option applies an autoregressive covariance structure, which assumes that adjacent experimental conditions will be more correlated than experimental conditions that are further apart. (It is also assumed that each correlation depends only upon the previous correlation plus error.) For example, the 30 s experimental condition will be more correlated with the 60 s experimental condition than with the 180 s experimental condition, while the 180 s experimental condition will be more correlated with the 60 s experimental condition than with the 30 s experimental condition. The AR(1) covariance structure applies most coherently to repeated measures data collected over time, where all subjects' experience of the same order of study "conditions" (i.e., with longitudinal data). Nevertheless, this conception also fits with the current experimental factor, study time, as although it is treated as a categorical variable, its origins on a quantitative scale are apparent. (*Note*: As each subject would have experienced a different random order of experimental conditions, it is the theoretical "distance" between the 30, 60, and 180 s experimental conditions and not the actual order of experimental conditions experienced that is important here.) There will be other experimental factor levels that also fit with the AR(1) conception, but some care should be taken to ensure that the AR(1) assumptions have theoretical correspondence with the experimental factor levels. An unstructured covariance matrix is often a serious contender for repeated measures data and this certainly would be a useful addition to the set of covariance structures offered by SYSTAT.

To implement a multilevel GLM with a covariance structure defined by AR(1) to the repeated measures data presented in Table 12.1, follow all of the instructions in Section , up to the opening of the *Random* dialog box. However, rather than selecting Compound symmetry for the SUBJECTS error covariance structure at this point, select AR(1) and then click OK (see Figure 12.8).

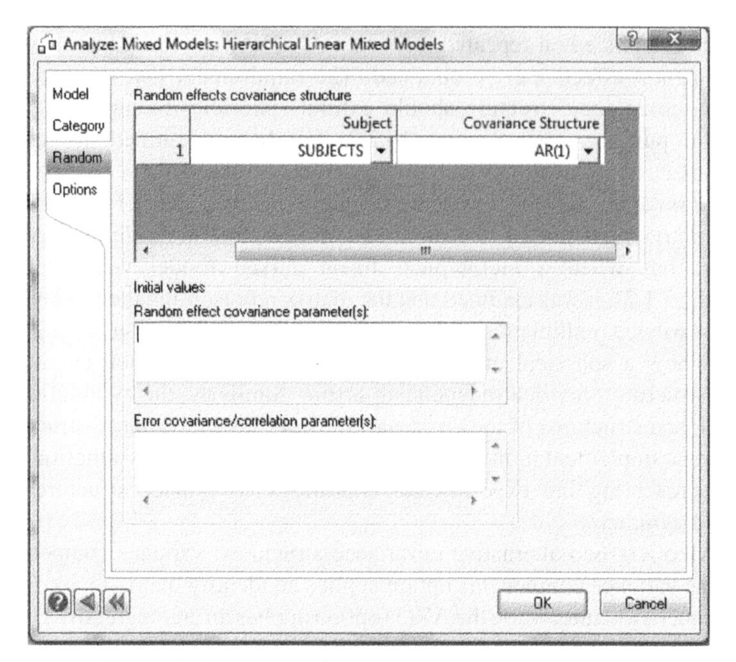

Figure 12.8 The *Random* dialog box—selecting AR(1).

However, when this was done initially, the following appeared in the output:

```
**WARNING**: Maximum number of iterations exceeded. Displaying
             results at last iteration.
 You may: a) Increase the number of ECME iterations using
             NEM=number to improve initial estimates
          b) Increase the number of Newton-Raphson iterations
             using NNR = number
```

Therefore, before running the analysis again, the number of ECME and Newton–Raphson iterations was increased. This is done by clicking on the Options tab in the *Hierarchical Linear Mixed Models* dialog box and simply increasing the pertinent values, as shown in Figure 12.9, and clicking OK.

The SYSTAT output obtained is presented below.

```
The categorical values encountered during processing are

Variables            |                 Levels
---------------------+--------------------------------------
SUBJECTS (8 levels)  |  1.000    2.000    3.000   4.000  5.000
                     |  6.000    7.000    8.000
TIME (3 levels)      | 30.000   60.000  180.000
```

```
Dependent Variable : RECALL
Fixed Factor(s)    : TIME
Fixed Covariate(s) : Intercept
Estimation Method  : Residual or Restricted
                     Maximum Likelihood (REML)
```

Dimensions

```
Covariance Parameters          :  1
Columns in X                   :  4
Columns in Z                   :  0
No. of Observations            : 24
Number of subjects             :  8
Max. observations per subject  :  3
```

Fit Statistics

```
Final L-L        : -41.895
-2L-L            :  83.790
AIC              :  87.790
AIC(Corrected)   :  88.457
BIC              :  89.879
```

Estimates of Covariance Components

Random Effect	Description	Estimate
Error variance	Variance Parameter	2.490
	Error Correlation (AR(1))	0.286

Estimates of Fixed Effects

Effect	Level	Estimate	Standard Error	*df*	t	p-Value
Intercept		11.000	0.558	21	19.719	0.000
TIME	30	-5.000	0.756	21	-6.615	0.000
	60	-1.000	0.666	21	-1.500	0.148
	180	0.000	0.000	.	.	.

Confidence Intervals of Fixed Effects Estimates

Effect	Level	Estimate	95.00% Confidence Interval Lower	Upper
Intercept		11.000	9.840	12.160
TIME	30	-5.000	-6.572	-3.428
	60	-1.000	-2.386	0.386
	180	0.000	.	.

Type III Tests for Fixed Effects

Effect	Numerator *df*	Denominator *df*	F-Ratio	p-Value
TIME	2	21	25.612	0.000

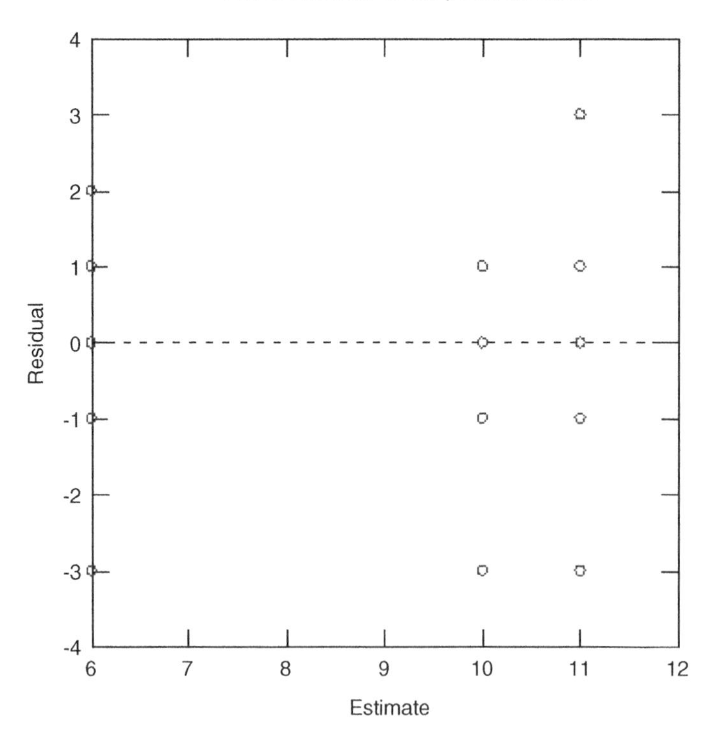

When a covariance structure defined by AR(1) was applied, the ANOVA (Type III SS) test for the fixed effects of the experimental factor, Time, was, $F = 25.612$. This is greater than, $F = 20.632$, which was obtained when Compound symmetry was assumed to provide an adequate description of the covariance structure. So which

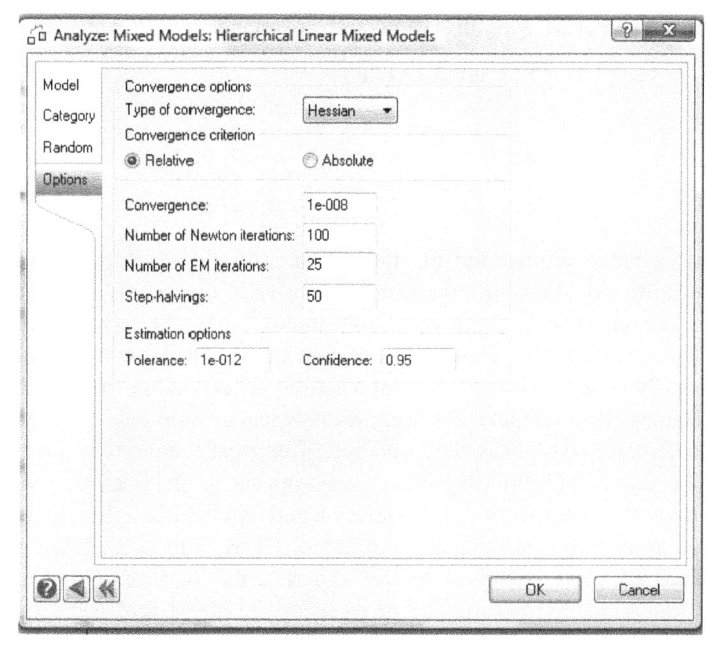

Figure 12.9 The *Options* dialog box to specify the number of ECME and Newton–Raphson iterations.

of these two F-tests is most appropriate to report? The answer is the most appropriate F-tests is the one from the model providing the best fit to the experimental data. How to determine which model fits the experimental data best is addressed in the next section.

12.6 EMPIRICALLY ASSESSING DIFFERENT MULTILEVEL MODELS

In common with most multilevel statistical packages, SYSTAT provides a number of model Fit Statistics and comparing the Fit Statistics for each model provides an empirical assessment of the model fit to the experimental data.

Fit Statistics are presented near the top of the data output. For the multilevel GLM with a compound symmetry error covariance, the fit statistics were

```
Fit Statistics
Final L-L        : -42.333
-2L-L            :  84.665
AIC              :  88.665
AIC(Corrected)   :  89.332
BIC              :  90.754.
```

For the multilevel GLM with an AR(1) error covariance, the fit statistics were

```
Fit Statistics
Final L-L        : -41.895
-2L-L            :  83.790
AIC              :  87.790
AIC(Corrected)   :  88.457
BIC              :  89.879.
```

Generally, the recommended fit statistics are Akiake's information criterion (AIC), the corrected Akiake information criterion (AIC corrected) and the Bayesian information criterion (BIC), with smaller fit statistics indicating a better fit between the model and the data. Unfortunately, however, Keselman, Algina, Kowalchuk and Wolfinger (1998) demonstrated that information criteria alone frequently do not identify the best fitting model and so simple empirical assessments of the best fitting model, particularly when fit statistics do not differ greatly, cannot be guaranteed to select the best model. Identifying the most appropriate model is another matter that requires theoretical perspective, consideration and input. Although a comparison of the indices above suggests that the multilevel GLM with an AR(1) covariance structure provides the better fit to the experimental data than the compound symmetry covariance structure, the differences are slight and some theoretical perspective, consideration and input would be very useful. Nevertheless, as both models indicate a large and significant effect of Time, the choice of the most appropriate covariance structure is unlikely to affect the conclusions drawn.

There is much to recommend linear mixed models and hierarchical linear mixed models. However, a number of issues, such as appropriate *dfs* for some terms and even whether some effects can be assessed, remain to be resolved (e.g., Bolker et al, 2009). Until these matters are resolved fully, it is unlikely linear mixed models and hierarchical linear mixed models will be adopted as widely as least squares GLMs.

Appendix A

SYSTAT

Information about SYSTAT is available on the SYSTAT home page: http://www.systat.com

MYSTAT is a free reduced version of SYSTAT that can be downloaded via the SYSTAT home page.

Although MYSTAT has much less capability than SYSTAT, it does include independent measures ANOVA and ANCOVA.

ANOVA and ANCOVA: A GLM Approach, Second Edition. By Andrew Rutherford.

Appendix B

Table B.1 Upper Percentage Points of the *F* Distribution

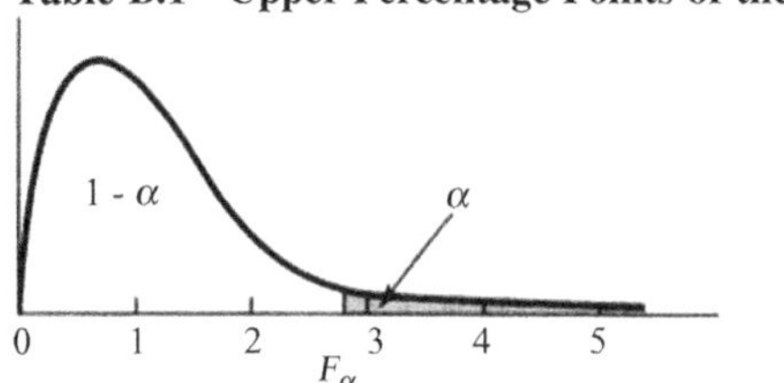

df for Denominator	α	*df* for Numerator 1	2	3	4	5	6	7	8	9	10	11	12
	0.25	5.83	7.50	8.20	8.58	8.82	8.98	9.10	9.19	9.26	9.32	9.36	9.41
1	0.10	39.9	49.5	53.6	55.8	57.2	58.2	58.9	59.4	59.9	60.2	60.5	60.7
	0.05	161	200	216	225	230	234	237	239	241	242	243	244
	0.25	2.57	3.00	3.15	3.23	3.28	3.31	3.34	3.35	3.37	3.38	3.39	3.39
2	0.10	8.53	9.00	9.16	9.24	9.29	9.33	9.35	9.37	9.38	9.39	9.40	9.41
	0.05	18.5	19.0	19.2	19.2	19.3	19.3	19.4	19.4	19.4	19.4	19.4	19.4
	0.01	98.5	99.0	99.2	99.2	99.3	99.3	99.4	99.4	99.4	99.4	99.4	99.4
	0.25	2.02	2.28	2.36	2.39	2.41	2.42	2.43	2.44	2.44	2.44	2.45	2.45
3	0.10	5.54	5.46	5.39	5.34	5.31	5.28	5.27	5.25	5.24	5.23	5.22	5.22
	0.05	10.1	9.55	9.28	9.12	9.01	8.94	8.89	8.85	8.81	8.79	8.76	8.74
	0.01	34.1	30.8	29.5	28.7	28.2	27.9	27.7	27.5	27.3	27.2	27.1	27.1
	0.25	1.81	2.00	2.05	2.06	2.07	2.08	2.08	2.08	2.08	2.08	2.08	2.08
4	0.10	4.54	4.32	4.19	4.11	4.05	4.01	3.98	3.95	3.94	3.92	3.91	3.90
	0.05	7.71	6.94	6.59	6.39	6.26	6.16	6.09	6.04	6.00	5.96	5.94	5.91
	0.01	21.2	18.0	16.7	16.0	15.5	15.2	15.0	14.8	14.7	14.5	14.4	14.4
	0.25	1.69	1.85	1.88	1.89	1.89	1.89	1.89	1.89	1.89	1.89	1.89	1.89
5	0.10	4.06	3.78	3.62	3.52	3.45	3.40	3.37	3.34	3.32	3.30	3.28	3.27
	0.05	6.61	5.79	5.41	5.19	5.05	4.95	4.88	4.82	4.77	4.74	4.71	4.68
	0.01	16.3	13.3	12.1	11.4	11.0	10.7	10.5	10.3	10.2	10.1	9.96	9.89
	0.25	1.62	1.76	1.78	1.79	1.79	1.78	1.78	1.78	1.77	1.77	1.77	1.77
6	0.10	3.78	3.46	3.29	3.18	3.11	3.05	3.01	2.98	2.96	2.94	2.92	2.90
	0.05	5.99	5.14	4.76	4.53	4.39	4.28	4.21	4.15	4.10	4.06	4.03	4.00
	0.01	13.7	10.9	9.78	9.15	8.75	8.47	8.26	8.10	7.98	7.87	7.79	7.72
	0.25	1.57	1.70	1.72	1.72	1.71	1.71	1.70	1.70	1.69	1.69	1.69	1.68
7	0.10	3.59	3.26	3.07	2.96	2.88	2.83	2.78	2.75	2.72	2.70	2.68	2.67
	0.05	5.59	4.74	4.35	4.12	3.97	3.87	3.79	3.73	3.68	3.64	1.60	3.57
	0.01	12.2	9.55	8.45	7.85	7.46	7.19	6.99	6.84	6.72	6.62	6.54	6.47
	0.25	1.54	1.66	1.67	1.66	1.66	1.65	1.64	1.64	1.63	1.63	1.63	1.62
8	0.10	3.46	3.11	2.92	2.81	2.73	2.67	2.62	2.59	2.56	2.54	2.52	2.50
	0.05	5.32	4.46	4.07	3.84	3.69	3.58	3.50	3.44	3.39	3.35	3.31	3.28
	0.01	11.3	8.65	7.59	7.01	6.63	6.37	6.18	6.03	5.91	5.81	5.73	5.67
	0.25	1.51	1.62	1.63	1.63	1.62	1.61	1.60	1.60	1.59	1.59	1.58	1.58
9	0.10	3.36	3.01	2.81	2.69	2.61	2.55	2.51	2.47	2.44	2.42	2.40	2.38
	0.05	5.12	4.26	3.86	3.63	3.48	3.37	3.29	3.23	3.18	3.14	3.10	3.07
	0.01	10.6	8.02	6.99	6.42	6.06	5.80	5.61	5.47	5.35	5.26	5.18	5.11

df for Numerator													
15	20	24	30	40	50	60	100	120	200	500	∞	α	*df* for Denominator
9.49	9.58	9.63	9.67	9.71	9.74	9.76	9.78	9.80	9.82	9.84	9.85	0.25	
61.2	61.7	62.0	62.3	62.5	62.7	62.8	63.0	63.1	63.2	63.3	63.3	0.10	1
246	248	249	250	251	252	252	253	253	254	254	254	0.05	
3.41	3.43	3.43	3.44	3.45	3.45	3.46	3.47	3.47	3.48	3.48	3.48	0.25	
9.42	9.44	9.45	9.46	9.47	9.47	9.47	9.48	9.48	9.49	9.49	9.49	0.10	2
19.4	19.4	19.5	19.5	19.5	19.5	19.5	19.5	19.5	19.5	19.5	19.5	0.05	
99.4	99.4	99.5	99.5	99.5	99.5	99.5	99.5	99.5	99.5	99.5	99.5	0.01	
2.46	2.46	2.46	2.47	2.47	2.47	2.47	2.47	2.47	2.47	2.47	2.47	0.25	
5.20	5.18	5.18	5.17	5.16	5.15	5.15	5.14	5.14	5.14	5.14	5.13	0.10	3
8.70	8.66	8.64	8.62	8.59	8.58	8.57	8.55	8.55	8.54	8.53	8.53	0.05	
26.9	26.7	26.6	26.5	26.4	26.4	26.3	26.2	26.2	26.2	26.1	26.1	0.01	
2.08	2.08	2.08	2.08	2.08	2.08	2.08	2.08	2.08	2.08	2.08	2.08	0.25	
3.87	3.84	3.83	3.82	3.80	3.80	3.79	3.78	3.78	3.77	3.76	3.76	0.10	4
5.86	5.80	5.77	5.75	5.72	5.70	5.69	5.66	5.66	5.65	5.64	5.63	0.05	
14.2	14.0	13.9	13.8	13.7	13.7	13.7	13.6	13.6	13.5	13.5	13.5	0.01	
1.89	1.88	1.88	1.88	1.88	1.83	1.87	1.87	1.87	1.87	1.87	1.87	0.25	
3.24	1.21	3.19	3.17	3.16	3.15	3.14	3.13	3.12	3.12	3.11	3.10	0.10	5
4.62	4.56	4.53	4.50	4.46	4.44	4.43	4.41	4.40	4.39	4.37	4.36	0.05	
9.72	9.55	9.47	9.38	9.29	9.24	9.20	9.13	9.11	9.08	9.04	9.02	0.01	
1.76	1.76	1.75	1.75	1.75	1.75	1.74	1.74	1.74	1.74	1.74	1.74	0.25	
2.87	2.84	2.82	2.80	2.78	2.77	2.76	2.75	2.74	2.73	2.73	2.72	0.10	6
3.94	3.87	3.84	3.81	3.77	3.75	3.74	3.71	3.70	3.69	3.68	3.67	0.05	
7.56	7.40	7.31	7.23	7.14	7.09	7.06	6.99	6.97	6.93	6.90	6.88	0.01	
1.68	1.67	1.67	1.66	1.66	1.66	1.65	1.65	1.65	1.65	1.65	1.65	0.25	
2.63	2.59	2.58	2.56	2.54	2.52	2.51	2.50	2.49	2.48	2.48	2.47	0.10	7
3.51	3.44	3.41	3.38	3.34	3.32	3.30	3.27	3.27	3.25	3.24	3.23	0.05	
6.31	6.16	6.07	5.99	5.91	5.86	5.82	5.75	5.74	5.70	5.67	5.65	0.01	
1.62	1.61	1.60	1.60	1.59	1.59	1.59	1.58	1.58	1.58	1.58	1.58	0.25	
2.46	2.42	2.40	2.38	2.36	2.35	2.34	2.32	2.32	2.31	2.30	2.29	0.10	8
3.22	3.15	3.12	3.08	3.04	3.02	3.01	2.97	2.97	2.95	2.94	2.93	0.05	
5.52	5.36	5.28	5.20	5.12	5.07	5.03	4.96	4.95	4.91	4.88	4.86	0.01	
1.57	1.56	1.56	1.55	1.55	1.54	1.54	1.53	1.53	1.53	1.53	1.53	0.25	
2.34	2.30	2.28	2.25	2.23	2.22	2.21	2.19	2.18	2.17	2.17	2.16	0.10	9
3.01	2.94	2.90	2.86	2.83	2.80	2.79	2.76	2.75	2.73	2.72	2.71	0.05	
4.96	4.81	4.73	4.65	4.57	4.52	4.48	4.42	4.40	4.36	4.33	4.31	0.01	

Table B.1 (*Continued*)

df for Denominator	α	*df* for Numerator											
		1	2	3	4	5	6	7	8	9	10	11	12
10	0.25	1.49	1.60	1.60	1.59	1.59	1.58	1.57	1.56	1.56	1.55	1.55	1.54
	0.10	3.29	2.92	2.73	2.61	2.52	2.46	2.41	2.38	2.35	2.32	2.30	2.28
	0.05	4.96	4.10	3.71	3.48	3.33	3.22	3.14	3.07	3.02	2.98	2.94	2.91
	0.01	10.0	7.56	6.55	5.99	5.64	5.39	5.20	5.06	4.94	4.85	4.77	4.71
11	0.25	1.47	1.58	1.58	1.57	1.56	1.55	1.54	1.53	1.53	1.52	1.52	1.51
	0.10	3.23	2.86	2.66	2.54	2.45	2.39	2.34	2.30	2.27	2.25	2.23	2.21
	0.05	4.84	3.98	3.59	3.36	3.20	3.09	3.01	2.95	2.90	2.85	2.82	2.79
	0.01	9.65	7.21	6.22	5.67	5.32	5.07	4.89	4.74	4.63	4.54	4.46	4.40
12	0.25	1.46	1.56	1.56	1.55	1.54	1.53	1.52	1.51	1.51	1.50	1.50	1.49
	0.10	3.18	2.81	2.61	2.48	2.39	2.33	2.28	2.24	2.21	2.19	2.17	2.15
	0.05	4.75	3.89	3.49	3.26	3.11	3.00	2.91	2.85	2.80	2.75	2.72	2.69
	0.01	9.33	6.93	5.95	5.41	5.06	4.82	4.64	4.50	4.39	4.30	4.22	4.16
13	0.25	1.45	1.55	1.55	1.53	1.52	1.51	1.50	1.49	1.49	1.48	1.47	1.47
	0.10	3.14	2.76	2.56	2.43	2.35	2.28	2.23	2.20	2.16	2.14	2.12	2.10
	0.05	4.67	3.81	3.41	3.18	3.03	2.92	2.83	2.77	2.71	2.67	2.63	2.60
	0.01	9.07	6.70	5.74	5.21	4.86	4.62	4.44	4.30	4.19	4.10	4.02	3.96
14	0.25	1.44	1.53	1.53	1.52	1.51	1.50	1.49	1.48	1.47	1.46	1.46	1.45
	0.10	3.10	2.73	2.52	2.39	2.31	2.24	2.19	2.15	2.12	2.10	2.08	2.05
	0.05	4.60	3.74	3.34	3.11	2.96	2.85	2.76	2.70	2.65	2.60	2.57	2.53
	0.01	8.86	6.51	5.56	5.04	4.69	4.46	4.28	4.14	4.03	3.94	3.86	3.80
15	0.25	1.43	1.52	1.52	1.51	1.49	1.48	1.47	1.46	1.46	1.45	1.44	1.44
	0.10	3.07	2.70	2.49	2.36	2.27	2.21	2.16	2.12	2.09	2.06	2.04	2.02
	0.05	4.54	3.68	3.29	3.06	2.90	2.79	2.71	2.64	2.59	2.54	2.51	2.48
	0.01	8.68	6.36	5.42	4.89	4.56	4.32	4.14	4.00	3.89	3.80	3.73	3.67
16	0.25	1.42	1.51	1.51	1.50	1.48	1.47	1.46	1.45	1.44	1.44	1.44	1.43
	0.10	3.05	2.67	2.46	2.33	2.24	2.18	2.13	2.09	2.06	2.03	2.01	1.99
	0.05	4.49	3.63	3.24	3.01	2.85	2.74	2.66	2.59	2.54	2.49	2.46	2.42
	0.01	8.53	6.2.1	5.29	4.77	4.44	4.20	4.03	3.89	3.78	3.69	3.62	3.55
17	0.25	1.42	1.51	1.50	1.49	1.47	1.46	1.45	1.44	1.43	1.43	1.42	1.41
	0.10	3.03	2.64	2.44	2.31	2.22	2.15	2.10	2.06	2.03	2.00	1.98	1.96
	0.05	4.45	3.59	3.20	2.96	2.81	2.70	2.61	2.55	2.49	2.45	2.41	2.38
	0.01	8.40	6.11	5.18	4.67	4.34	4.10	3.93	3.79	3.68	3.59	3.52	3.46
18	0.25	1.41	1.50	1.49	1.48	1.46	1.45	1.44	1.43	1.42	1.42	1.41	1.40
	0.10	3.01	2.62	2.42	2.29	2.20	2.13	2.08	2.04	2.00	1.98	1.96	1.93
	0.05	4.41	3.55	3.16	2.93	2.77	2.66	2.58	2.51	2.46	2.41	2.37	2.34
	0.01	8.29	6.01	5.09	4.58	4.25	4.01	3.84	3.71	3.60	3.51	3.43	3.37
19	0.25	1.41	1.49	1.49	1.47	1.46	1.44	1.43	1.42	1.41	1.41	1.40	1.40
	0.10	2.99	2.61	2.40	2.27	2.18	2.11	2.06	2.02	1.98	1.96	1.94	1.91
	0.05	4.38	3.52	3.13	2.90	2.74	2.63	2.54	2.48	2.42	2.38	2.34	2.31
	0.01	8.18	5.9.1	5.01	4.50	4.17	3.94	3.77	3.63	3.52	3.43	3.36	3.30
20	0.25	1.40	1.49	1.48	1.46	1.45	1.44	1.43	1.42	1.41	1.40	1.39	1.39
	0.10	2.97	2.59	2.3B	2.25	2.16	2.09	2.04	2.00	1.96	1.94	1.92	1.89
	0.05	4.35	3.49	3.10	2.87	2.71	2.60	2.51	2.45	2.39	2.35	2.31	2.28
	0.01	3.10	5.85	4.94	4.43	4.10	3.87	3.70	3.56	3.46	3.37	3.29	3.23

df for Numerator													*df* for Denominator
15	20	24	30	40	50	60	100	120	200	500	∞	α	
1.53	1.52	1.52	1.51	1.51	1.50	1.50	1.49	1.49	1.49	1.48	1.48	0.25	
2.24	2.20	2.18	2.16	2.13	2.12	2.11	2.09	2.08	2.07	2.06	2.06	0.10	10
2.85	2.77	2.74	2.70	2.66	2.64	2.62	2.59	2.58	2.56	2.55	2.54	0.05	
4.56	4.41	4.33	4.25	4.17	4.12	4.08	4.01	4.00	3.96	3.93	3.91	0.01	
1.50	1.49	1.49	1.48	1.47	1.47	1.47	1.46	1.46	1.46	1.45	1.45	0.25	
2.17	2.12	2.10	2.08	2.05	2.04	2.03	2.00	2.00	1.99	1.98	1.97	0.10	11
2.72	2.65	2.61	2.57	2.53	2.51	2.49	2.46	2.45	2.43	2.42	2.40	0.05	
4.25	4.10	4.02	3.94	3.86	3.81	3.78	3.71	3.69	3.66	3.62	3.60	0.01	
1.48	1.47	1.46	1.45	1.45	1.44	1.44	1.43	1.43	1.43	1.42	1.42	0.25	
2.10	2.06	2.04	2.01	1.99	1.97	1.96	1.94	1.93	1.92	1.91	1.90	0.10	12
2.62	2.54	2.51	2.47	2.43	2.40	2.38	2.35	2.34	2.32	2.31	2.30	0.05	
4.01	3.86	3.78	3.70	3.62	3.57	3.54	3.47	3.45	3.41	3.38	3.36	0.01	
1.46	1.45	1.44	1.43	1.42	1.42	1.42	1.41	1.41	1.40	1.40	1.40	0.25	
2.05	2.01	1.98	1.96	1.93	1.92	1.90	1.88	1.88	1.86	1.85	1.85	0.10	13
2.53	2.46	2.42	2.38	2.34	2.31	2.30	2.26	2.25	2.23	2.22	2.21	0.05	
3.82	3.66	3.59	3.51	3.43	3.38	3.34	3.27	3.25	3.22	3.19	3.17	0.01	
1.44	1.43	1.42	1.41	1.41	1.40	1.40	1.39	1.39	1.39	1.38	1.38	0.25	
2.01	1.96	1.94	1.91	1.89	1.87	1.86	1.83	1.83	1.82	1.80	1.80	0.10	14
2.46	2.39	2.35	2.31	2.27	2.24	2.22	2.19	2.18	2.16	2.14	2.13	0.05	
3.66	3.51	3.43	3.35	3.27	3.22	3.18	3.11	3.09	3.06	3.03	3.00	0.01	
1.43	1.41	1.41	1.40	1.39	1.39	1.38	1.38	1.37	1.37	1.36	1.36	0.25	
1.97	1.92	1.90	1.87	1.85	1.83	1.82	1.79	1.79	1.77	1.76	1.76	0.10	15
2.40	2.33	2.29	2.25	2.20	2.18	2.16	2.12	2.11	2.10	2.08	2.07	0.05	
3.52	3.37	3.29	3.21	3.13	3.08	3.05	2.98	2.96	2.92	2.89	2.87	0.01	
1.41	1.40	1.39	1.38	1.37	1.37	1.36	1.36	1.35	1.35	1.34	1.34	0.25	
1.94	1.89	1.87	1.84	1.81	1.79	1.78	1.76	1.75	1.74	1.73	1.72	0.10	16
2.35	2.28	2.24	2.19	2.15	2.12	2.11	2.07	2.06	2.04	2.02	2.01	0.05	
3.41	3.26	3.18	3.10	3.02	2.97	2.93	2.86	2.84	2.81	2.78	2.75	0.01	
1.40	1.39	1.38	1.37	1.36	1.35	1.35	1.34	1.34	1.34	1.33	1.33	0.25	
1.91	1.86	1.84	1.81	1.78	1.76	1.75	1.73	1.72	1.71	1.69	1.69	0.10	17
2.31	2.23	2.19	2.15	2.10	2.08	2.06	2.02	2.01	1.99	1.97	1.96	0.05	
3.31	3.16	3.08	3.00	2.92	2.87	2.83	2.76	2.75	2.71	2.68	2.65	0.01	
1.39	1.38	1.37	1.36	1.35	1.34	1.34	1.33	1.33	1.32	1.32	1.32	0.25	
1.89	1.84	1.81	1.78	1.75	1.74	1.72	1.70	1.69	1.68	1.67	1.66	0.10	18
2.27	2.19	2.15	2.11	2.06	2.04	2.02	1.98	1.97	1.95	1.93	1.92	0.05	
3.23	3.08	3.00	2.92	2.84	2.78	2.75	2.68	2.66	2.62	2.59	2.57	0.01	
1.38	1.37	1.36	1.35	1.34	1.33	1.33	1.32	1.32	1.31	1.31	1.30	0.25	
1.86	1.81	1.79	1.76	1.73	1.71	1.70	1.67	1.67	1.65	1.64	1.63	0.10	19
2.23	2.16	2.11	2.07	2.03	2.00	1.98	1.94	1.93	1.91	1.89	1.88	0.05	
3.15	3.00	2.92	2.84	2.76	2.71	2.67	2.60	2.58	2.55	2.51	2.49	0.01	
1.37	1.36	1.35	1.34	1.33	1.33	1.32	1.31	1.31	1.30	1.30	1.29	0.25	
1.84	1.79	1.77	1.74	1.71	1.69	1.68	1.65	1.64	1.63	1.62	1.61	0.10	20
2.20	2.12	2.08	2.04	1.99	1.97	1.95	1.41	1.90	1.88	1.86	1.84	0.05	
3.09	2.94	2.86	2.78	2.69	2.64	2.61	2.54	2.52	2.48	2.44	2.42	0.01	

Table B.1 (*Continued*)

df for Denominator	α	*df* for Numerator 1	2	3	4	5	6	7	8	9	10	11	12
	0.25	1.40	1.48	1.47	1.45	1.44	1.42	1.41	1.40	1.39	1.39	1.38	1.37
22	0.10	2.95	2.56	2.35	2.22	2.13	2.06	2.01	1.97	1.93	1.90	1.88	1.86
	0.05	4.30	3.44	3.05	2.82	2.66	2.55	2.46	2.40	2.34	2.30	2.26	2.23
	0.01	7.95	5.72	4.82	4.31	3.99	3.76	3.59	3.45	3.35	3.26	3.18	3.12
	0.25	1.39	1.47	1.46	1.44	1.43	1.41	1.40	1.39	1.38	1.38	1.37	1.36
24	0.10	2.93	2.54	2.33	2.19	2.10	2.04	1.98	1.94	1.91	1.88	1.85	1.83
	0.05	4.26	3.40	3.01	2.78	2.62	2.51	2.42	2.36	2.30	2.25	2.21	2.18
	0.01	7.82	5.61	4.72	4.22	3.90	3.67	3.50	3.36	3.26	3.17	3.09	3.03
	0.25	1.38	1.46	1.45	1.44	1.42	1.41	1.39	1.38	1.37	1.37	1.36	1.35
26	0.10	2.91	2.52	2.31	2.17	2.08	2.01	1.96	1.92	1.88	1.86	1.84	1.81
	0.05	4.23	3.37	2.98	2.74	2.59	2.47	2.39	2.32	2.27	2.22	2.18	2.15
	0.01	7.72	5.53	4.64	4.14	3.82	3.59	3.42	3.29	3.18	3.09	3.02	2.96
	0.25	1.38	1.46	1.45	1.43	1.41	1.40	1.39	1.38	1.37	1.36	1.35	1.34
28	0.10	2.89	2.50	2.29	2.16	2.06	2.00	1.94	1.90	1.87	1.84	1.81	1.79
	0.05	4.20	3.34	2.95	2.71	2.56	2.45	2.36	2.29	2.24	2.19	2.15	2.12
	0.01	7.64	5.45	4.57	4.07	3.75	3.53	3.36	3.23	3.12	3.03	2.96	2.90
	0.25	1.38	1.45	1.44	1.42	1.41	1.39	1.38	1.37	1.36	1.35	1.35	1.34
30	0.10	2.88	2.49	2.28	2.14	2.05	1.98	1.93	1.88	1.85	1.82	1.79	1.77
	0.05	4.17	3.32	2.92	2.69	2.53	2.42	2.33	2.27	2.21	2.16	2.13	2.09
	0.01	7.56	5.39	4.51	4.02	3.70	3.47	3.30	3.17	3.07	2.98	2.91	2.84
	0.25	1.36	1.44	1.42	1.40	1.39	1.37	1.36	1.35	1.34	1.33	1.32	1.31
40	0.10	2.84	2.44	2.23	2.09	2.00	1.93	1.87	1.83	1.79	1.76	1.73	1.71
	0.05	4.08	3.23	2.84	2.61	2.45	2.34	2.25	2.18	2.12	2.08	2.04	2.00
	0.01	7.31	5.18	4.31	3.83	3.51	3.29	3.12	2.99	2.89	2.80	2.73	2.66
	0.25	1.35	1.42	1.41	1.38	1.37	1.35	1.33	1.32	1.31	1.30	1.29	1.29
60	0.10	2.79	2.39	2.18	2.04	1.95	1.87	1.82	1.77	1.74	1.71	1.68	1.66
	0.05	4.00	3.15	2.76	2.53	2.37	2.25	2.17	2.10	2.04	1.99	1.95	1.92
	0.01	7.08	4.98	4.13	3.65	3.34	3.12	2.95	2.82	2.72	2.63	2.56	2.50
	0.25	1.34	1.40	1.39	1.37	1.35	1.33	1.31	1.30	1.29	1.28	1.27	1.26
120	0.10	2.75	2.35	2.13	1.99	1.90	1.82	1.77	1.72	1.68	1.65	1.62	1.60
	0.05	3.92	3.07	2.68	2.45	2.29	2.17	2.09	2.02	1.96	1.91	1.87	1.83
	0.01	6.85	4.79	3.95	3.48	3.17	2.96	2.79	2.66	2.56	2.47	2.40	2.34
	0.25	1.33	1.39	1.38	1.36	1.34	1.32	1.31	1.29	1.28	1.27	1.26	1.25
200	0.10	2.73	2.33	2.11	1.97	1.88	1.80	1.75	1.70	1.66	1.63	1.60	1.57
	0.05	3.89	3.04	2.65	2.42	2.26	2.14	2.06	1.98	1.93	1.88	1.84	1.80
	0.01	6.76	4.71	3.88	3.41	3.11	2.89	2.73	2.60	2.50	2.41	2.34	2.27
	0.25	1.32	1.39	1.37	1.35	1.33	1.31	1.29	1.28	1.27	1.25	1.24	1.24
∞	0.10	2.71	2.30	2.08	1.94	1.85	1.77	1.72	1.67	1.63	1.60	1.57	1.55
	0.05	3.84	3.00	2.60	2.37	2.21	2.10	2.01	1.94	1.88	1.83	1.79	1.75
	0.01	6.63	4.61	3.78	3.32	3.02	2.80	2.64	2.51	2.41	2.32	2.25	2.18

Table B.1 is abridged from Table 18 in Pearson E.S. & Hartley H.O. (eds) (1958). *Biometrika Tables for Statisticians*. Vol. 1, 2nd ed. New York: Wiley. Reproduced with permission of the editors and the trustees of *Biometrika*.

df for Numerator													
15	20	24	30	40	50	60	100	120	200	500	∞	α	*df* for Denominator
1.36	1.34	1.33	1.32	1.31	1.31	1.30	1.30	1.30	1.29	1.29	1.28	0.25	
1.81	1.76	1.73	1.70	1.67	1.65	1.64	1.61	1.60	1.59	1.58	1.57	0.10	22
2.15	2.07	2.03	1.98	1.94	1.91	1.89	1.85	1.84	1.82	1.80	1.78	0.05	
2.98	2.83	2.75	2.67	2.58	2.53	2.50	2.42	2.40	2.36	2.33	2.31	0.01	
1.35	1.33	1.32	1.31	1.30	1.29	1.29	1.28	1.28	1.27	1.27	1.26	0.25	
1.78	1.73	1.70	1.67	1.64	1.62	1.61	1.58	1.57	1.56	1.54	1.53	0.10	24
2.11	2.03	1.98	1.94	1.89	1.86	1.84	1.80	1.79	1.77	1.75	1.73	0.05	
2.89	2.74	2.66	2.58	2.49	2.44	2.40	2.33	2.31	2.27	2.24	2.21	0.01	
1.34	1.32	1.31	1.30	1.29	1.28	1.28	1.26	1.26	1.26	1.25	1.25	0.25	
1.76	1.71	1.68	1.65	1.61	1.59	1.58	1.55	1.54	1.53	1.51	1.50	0.10	26
2.07	1.99	1.95	1.90	1.85	1.82	1.80	1.76	1.75	1.73	1.71	1.69	0.05	
2.81	2.66	2.58	2.50	2.42	2.36	2.33	2.25	2.23	2.19	2.16	2.13	0.01	
1.33	1.31	1.30	1.29	1.28	1.27	1.27	1.26	1.25	1.25	1.24	1.24	0.25	
1.74	1.69	1.66	1.63	1.59	1.57	1.56	1.53	1.52	1.50	1.49	1.48	0.10	28
2.04	1.96	1.91	1.87	1.82	1.79	1.77	1.73	1.71	1.69	1.67	1.65	0.05	
2.75	2.60	2.52	2.44	2.35	2.30	2.26	2.19	2.17	2.13	2.09	2.06	0.01	
1.32	1.30	1.29	1.28	1.27	1.26	1.26	1.25	1.24	1.24	1.23	1.23	0.25	
1.72	1.67	1.64	1.61	1.57	1.55	1.54	1.51	1.50	1.48	1.47	1.46	0.10	30
2.01	1.93	1.89	1.84	1.79	1.76	1.74	1.70	1.68	1.66	1.64	1.62	0.05	
2.70	2.55	2.47	2.39	2.30	2.25	2.21	2.13	2.11	2.07	2.03	2.01	0.01	
1.30	1.28	1.26	1.25	1.24	1.23	1.22	1.21	1.21	1.20	1.19	1.19	0.25	
1.66	1.61	1.57	1.54	1.51	1.48	1.47	1.43	1.42	1.41	1.39	1.38	0.10	40
1.92	1.84	1.79	1.74	1.69	1.66	1.64	1.59	1.58	1.55	1.53	1.51	0.05	
2.52	2.37	2.29	2.20	2.11	2.06	2.02	1.94	1.92	1.87	1.83	1.80	0.01	
1.27	1.25	1.24	1.22	1.21	1.20	1.19	1.17	1.17	1.16	1.15	1.15	0.25	
1.60	1.54	1.51	1.48	1.44	1.41	1.40	1.36	1.35	1.33	1.31	1.29	0.10	60
1.84	1.75	1.70	1.65	1.59	1.56	1.53	1.48	1.47	1.44	1.41	1.39	0.05	
2.35	2.20	1.12	2.03	1.94	1.88	1.84	1.75	1.73	1.68	1.63	1.60	0.01	
1.24	1.22	1.21	1.19	1.18	1.17	1.16	1.14	1.13	1.12	1.11	1.10	0.25	
1.55	1.48	1.45	1.41	1.37	1.34	1.32	1.27	1.26	1.24	1.21	1.19	0.10	120
1.75	1.66	1.61	1.55	1.50	1.46	1.43	1.37	1.35	1.32	1.28	1.25	0.05	
2.19	2.03	1.95	1.86	1.76	1.70	1.66	1.56	1.53	1.48	1.42	1.38	0.01	
1.23	1.21	1.20	1.18	1.16	1.14	1.12	1.11	1.10	1.09	1.08	1.06	0.25	
1.52	1.46	1.42	1.38	1.34	1.31	1.28	1.24	1.22	1.20	1.17	1.14	0.10	200
1.72	1.62	1.57	1.52	1.46	1.41	1.39	1.32	1.29	1.26	1.22	1.19	0.05	
2.13	1.97	1.89	1.79	1.69	1.63	1.58	1.48	1.44	1.39	1.33	1.28	0.01	
1.22	1.19	1.18	1.16	1.14	1.13	1.12	1.09	1.08	1.07	1.04	1.00	0.25	
1.49	1.42	1.38	1.34	1.30	1.26	1.24	1.18	1.17	1.13	1.08	1.00	0.10	∞
1.67	1.57	1.52	1.46	1.39	1.35	1.32	1.24	1.22	1.17	1.11	1.00	0.05	
2.04	1.88	1.79	1.70	1.59	1.52	1.47	1.36	1.32	1.25	1.15	1.00	0.01	

Appendix C

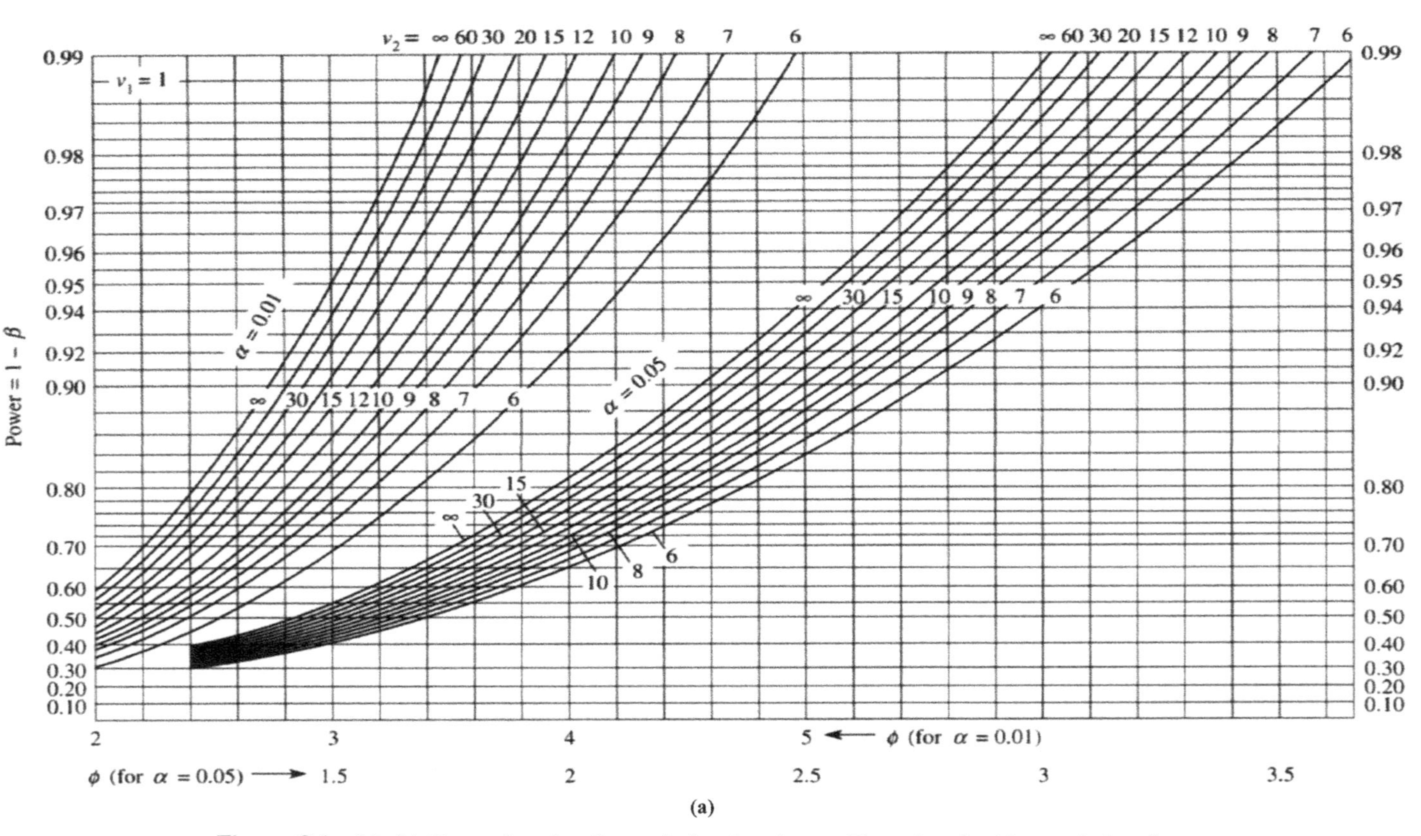

(a)

Figure C.1 (a)–(h) Power function for analysis of variance. (Reproduced with permission from Pearson E.S. and Hartley H.O. (1951). Charts of the power function for analysis of variance tests, derived from the non-central *F*-distribution. *Biometrika*, *38*, 112–130.)

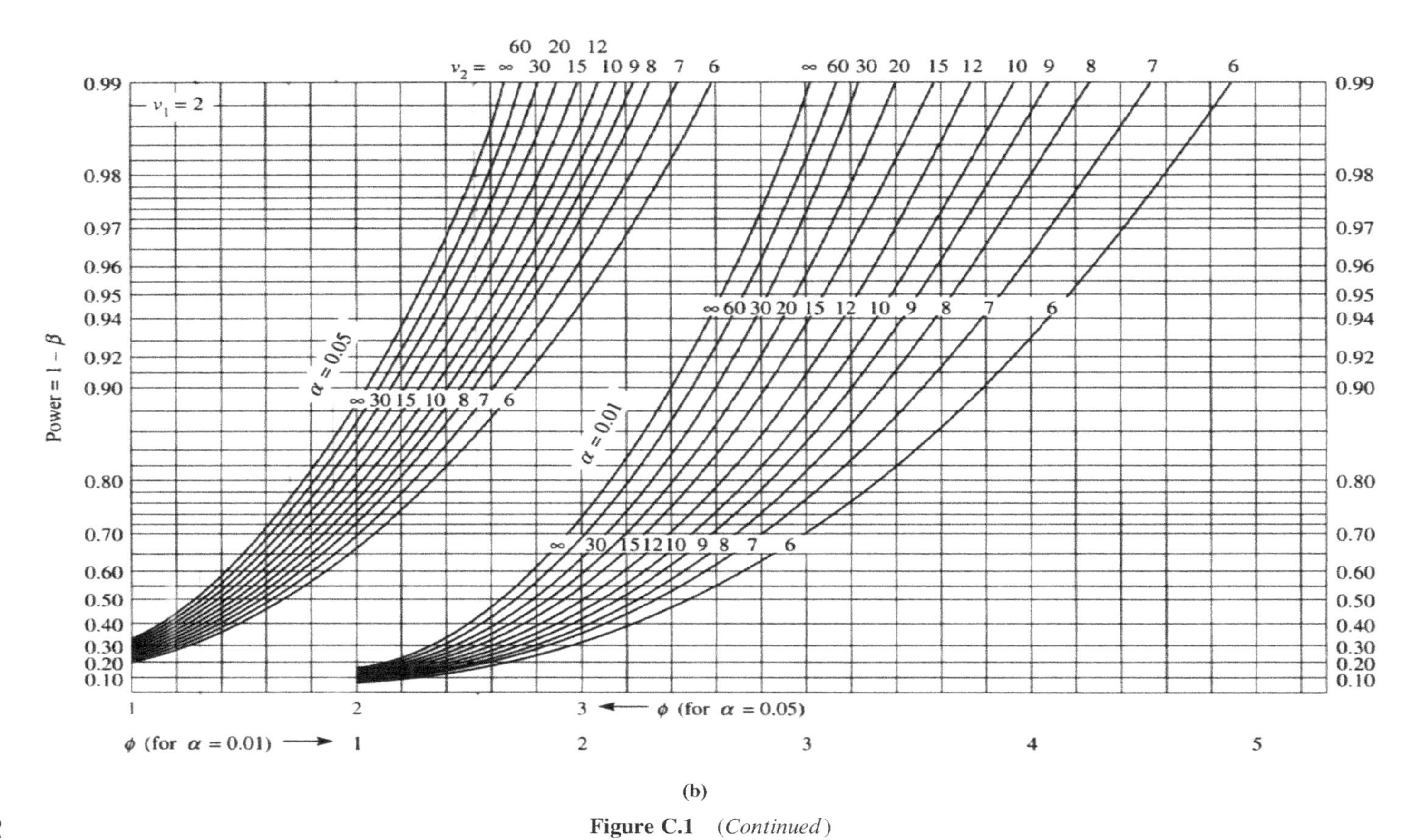

(b)

Figure C.1 *(Continued)*

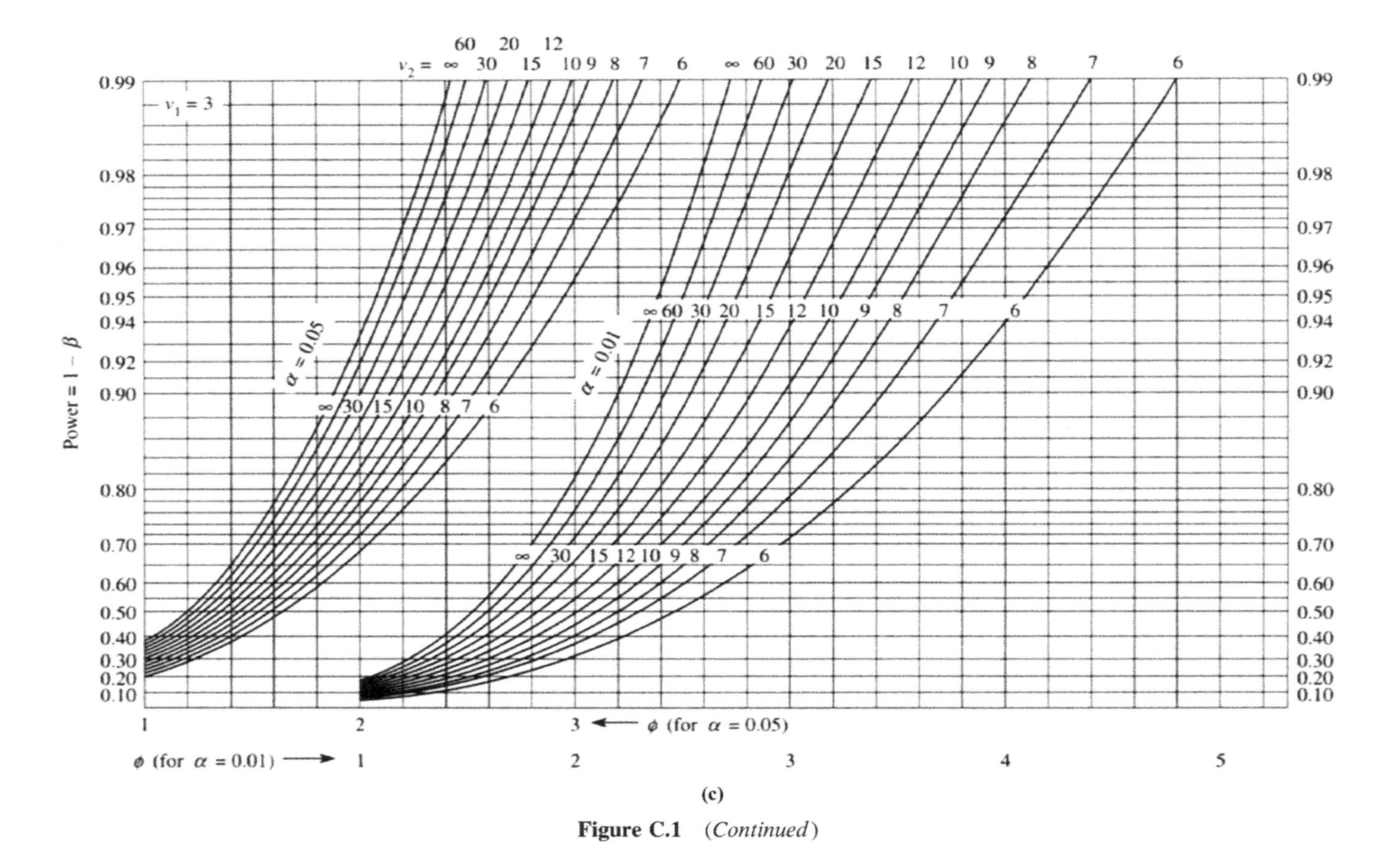

(c)

Figure C.1 *(Continued)*

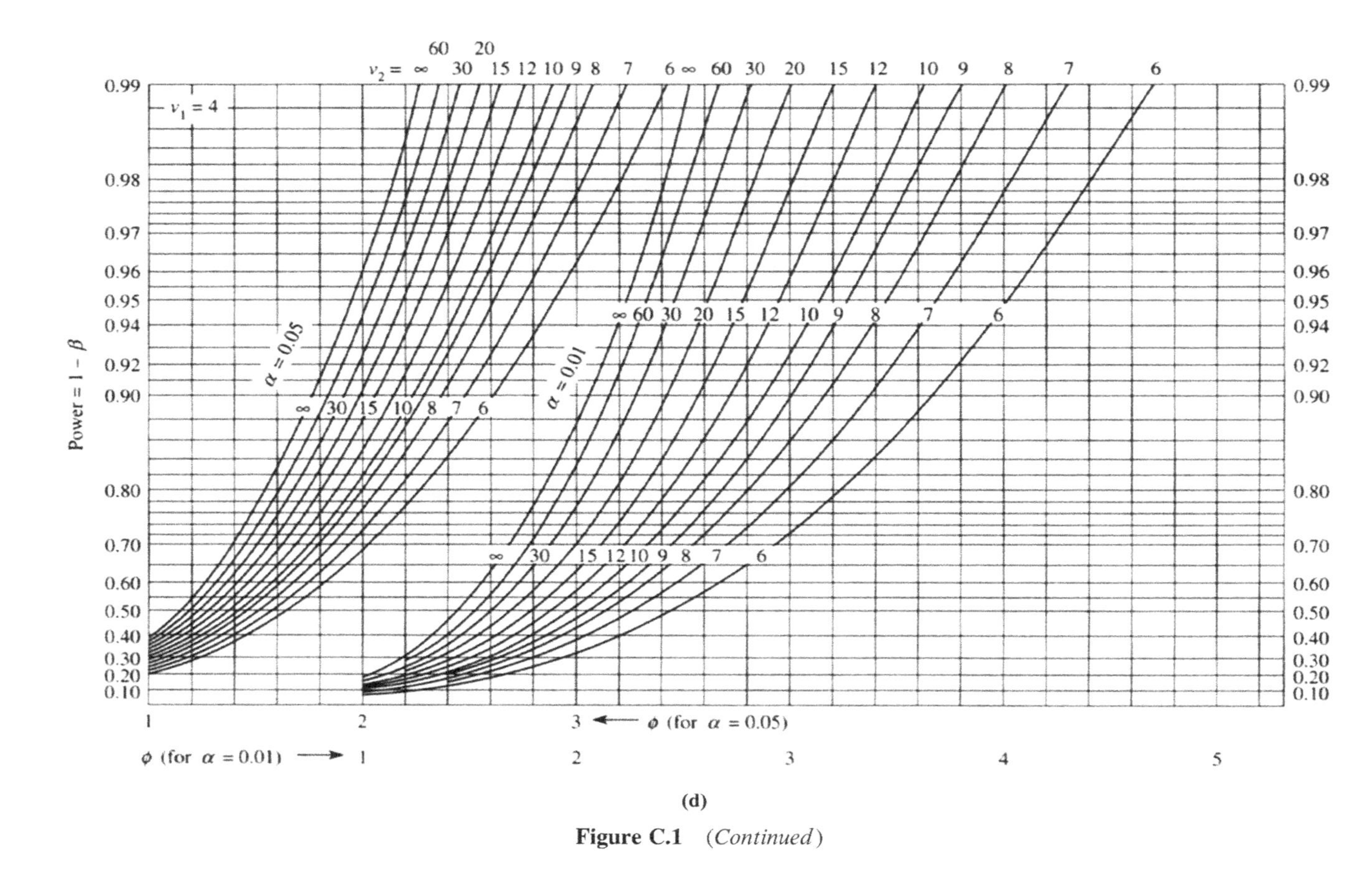

(d)

Figure C.1 *(Continued)*

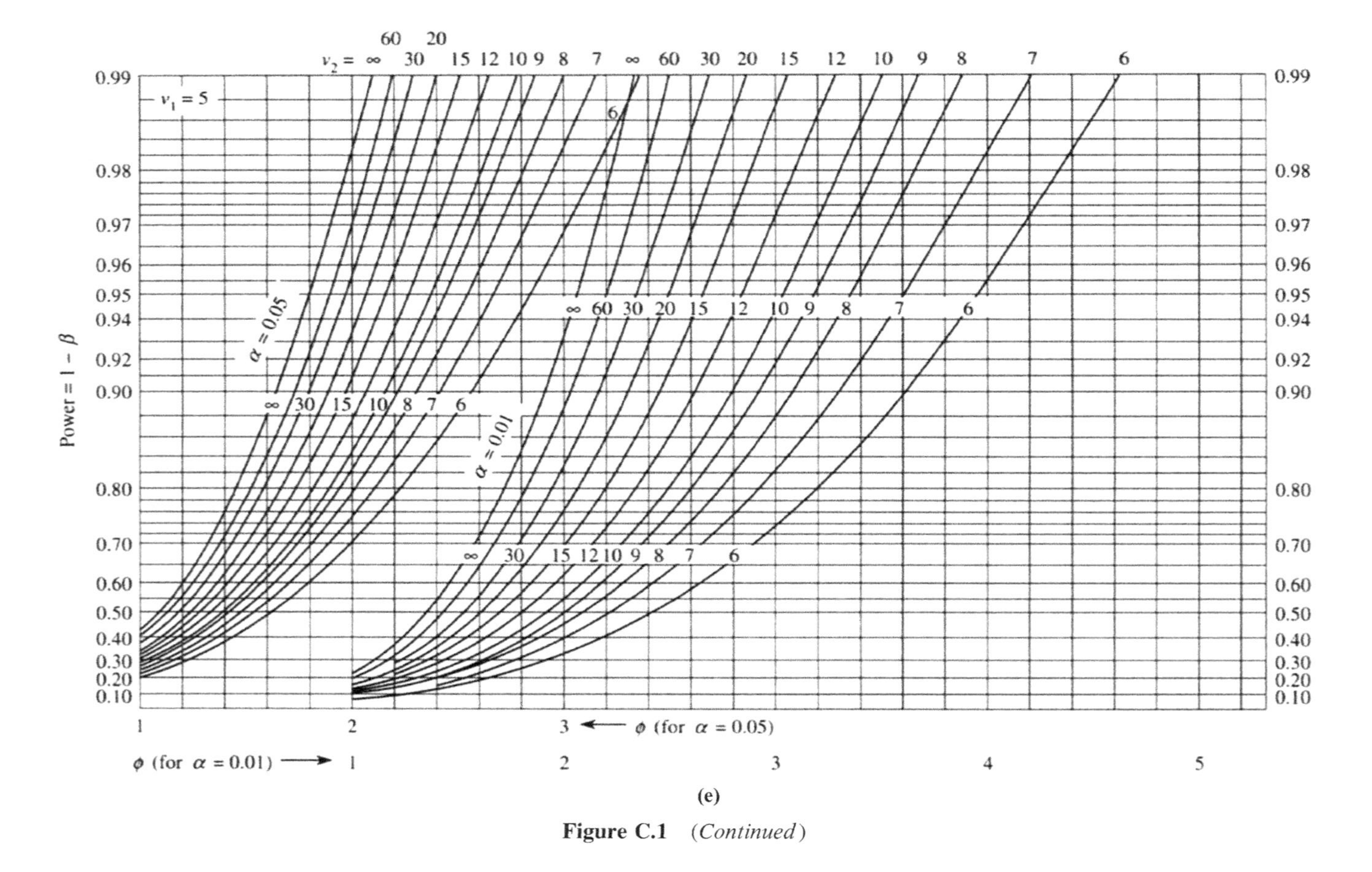

(e)

Figure C.1 *(Continued)*

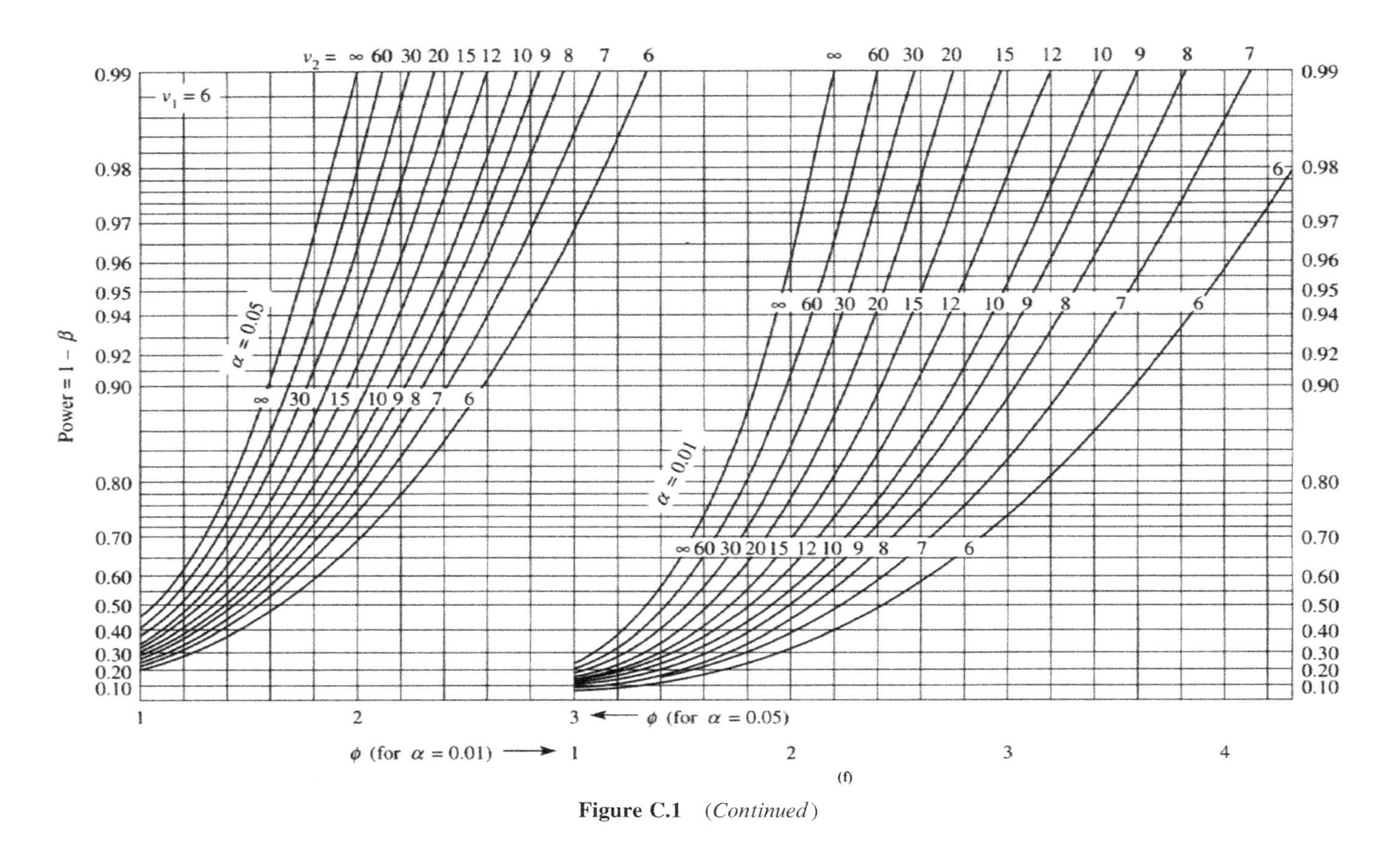

Figure C.1 *(Continued)*

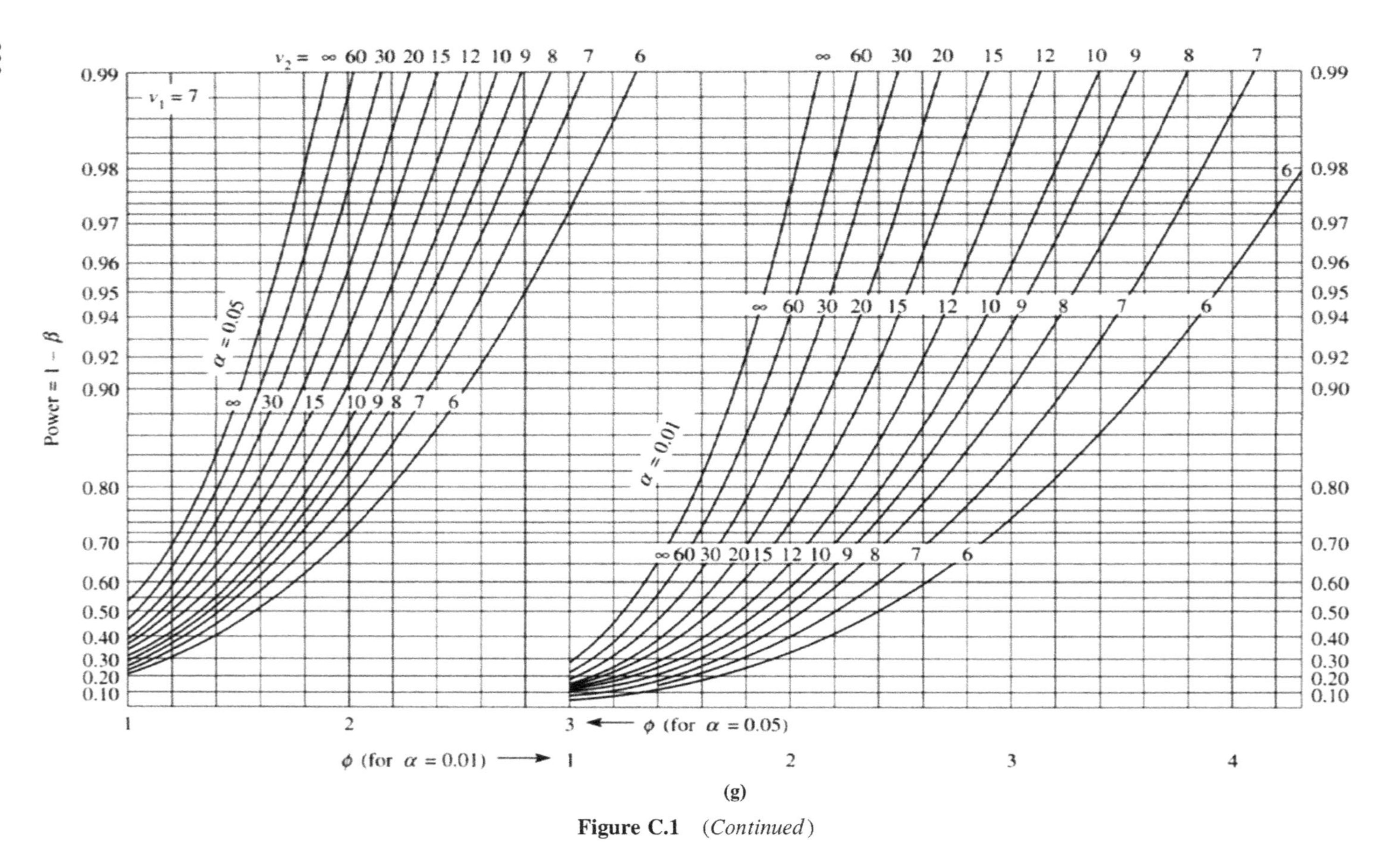

(g)

Figure C.1 (*Continued*)

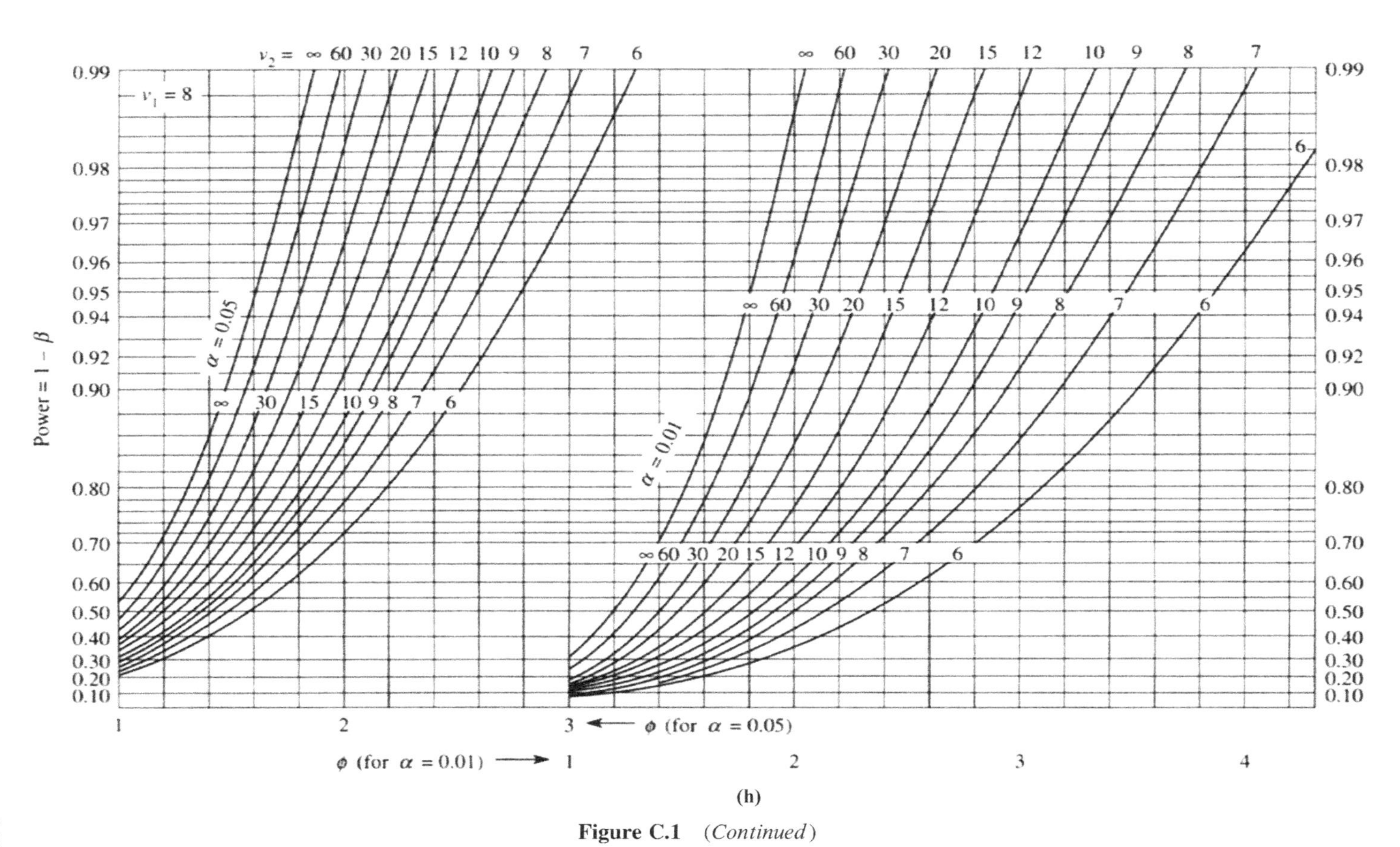

(h)

Figure C.1 *(Continued)*

References

Aickin, M. (1999). Other method for adjustment of multiple testing exists. *British Medical Journal, 318*, 127.

Arnold, B.C. (1970). Hypothesis testing incorporating a preliminary test of significance. *Journal of the American Statistical Association, 65*, 1590–1596.

Atiqullah, M. (1964). The robustness of the covariance analysis of a one-way classification. *Biometrika, 49*, 83–92.

Bakan, D. (1966). The test of significance in psychological research. *Psychological Bulletin, 66*, 423–437.

Bancroft, T.A. (1964). Analysis and inference for incompletely specified models involving the use of preliminary test(s) of significance. *Biometrics, 20*, 427–442.

Bender, R. & Lange, S. (1999). Multiple test procedures other than Bonferroni's deserve wider use. *British Medical Journal, 318*, 600–601.

Bender, R. & Lange, S. (2001). Adjusting for multiple testing: when and how? *Journal of Clinical Epidemiology, 54*, 343–349.

Benjamani, Y. & Hochberg, Y. (1995). Controlling the false discovery rate: a practical and powerful approach to multiple testing. *Journal of the Royal Statistical Society, Series B, 57*, 289–300.

Benjamani, Y. & Hochberg, Y. (2000). On the adaptive control of the false discovery rate in multiple testing with independent statistics. *Journal of Educational and Behavioral Statistics, 25*, 60–83.

Benjamini, Y. & Yekutieli, D. (2001). The control of the false discovery rate in multiple testing under dependency. *Annals of Statistics, 29*, 1165–1188.

Bentler, P.M. (1980). Multivariate analysis with latent variables: causal modelling. *Annual Review of Psychology, 31*, 419–456.

Berkson, J. (1938). Some difficulties of interpretation encountered in the application of the chi-square test. *Journal of the American Statistical Association, 33*, 526–536.

ANOVA and ANCOVA: A GLM Approach, Second Edition. By Andrew Rutherford.

Bernhardson, C.S. (1975). Type 1 error rates when multiple comparison procedures follow a significant *F* test of ANOVA. *Biometrics*, *31*, 229–232.

Bingham, C. & Fienberg, S.E. (1982). Textbook analysis of covariance: is it correct? *Biometrics*, *38*, 747–753.

Block, H.W., Savits, T.H., & Wang, J. (2008). Negative dependence and the Simes inequality. *Journal of Statistical Planning and Inference*, *138*, 4107–4110.

Boik, R.J. (1981). A priori tests in repeated measures designs: effects of nonsphericity. *Psychometrika*, *46*, 241–255.

Bollen, K.A. (1989). *Structural Equations with Latent Variables*. New York: Wiley.

Bolker, B.M., Brooks, M.E., Clark, C.J., Geange, S.W., Poulsen, J.R., Stevens, M.H.H., & White, J-S.S. (2009). Generalized linear mixed models: a practical guide for ecology and evolution. *Trends in Ecology and Evolution*, *24*, 127–135.

Bonferroni, C.E. (1936). Teoria statistica delle classi e calcolo delle probabilità. *Pubblicazioni del R Istituto Superiore di Scienze Economiche e Commerciali di Firenze*, *8*, 3–62.

Boole, G. (1854). *An Investigation of the Laws of Thought on Which are Founded the Mathematical Theories of Logic and Probabilities*. London: Macmillan (Reprinted: New York: Dover Publications).

Box, G.E.P. (1954). Some theorems on quadratic forms applied in the study of analysis of variance problems. II. Effects of inequality of variance and of correlation between errors in the two-way classification. *Annals of Mathematical Statistics*, *25*, 484–498.

Box, G.E.P. & Jenkins, G.M. (1976). *Time Series Analysis: Forecasting and Control*. San Francisco, CA: Holden-Day.

Bradley, J.V. (1978). Robustness? *British Journal of Mathematical and Statistical Psychology*, *31*, 144–152.

Braun, H.I. (1994). *The Collected works of John Tukey*. Vol. VIII. *Multiple Comparisons: 1948–1983*. New York: Chapman & Hall.

Brewer, M. (2000). Research design and issues of validity. In H. Reis & C. Judd (eds), *Handbook of Research Methods in Social and Personality Psychology*. Cambridge: Cambridge University Press.

Carroll, R.M. & Nordholm, L.A. (1975). Sampling characteristics of Kelley's ε and Hay's ω. *Educational and Psychological Measurement*, *35*, 541–554.

Chambers, J.M., Cleveland, W.S., Kleiner, B., & Tukey, P.A. (1983). *Graphical Methods for Data Analysis*. Belmont, CA: Wadsworth.

Chang, C.K., Rom, D.M., & Sarkar, S.K. (1996). *Modified Bonferroni Procedure for Repeated Significance Testing*. Technical Report 96–01, Temple University.

Clark, H.H. (1976). Reply to Wike and Church. *Journal of Verbal Learning and Verbal Behavior*, *15*, 257–261.

Clinch, J.J. & Keselman, H.J. (1982). Parametric alternatives to the analysis of variance. *Journal of Educational Statistics*, *7*, 207–214.

Cochran, W.G. (1957). Analysis of covariance: its nature and uses. *Biometrics*, *13*, 261–281.

Cohen, J. (1962). The statistical power of abnormal-social psychological research. *Journal of Abnormal and Social Psychology*, *65*, 145–153.

Cohen, J. (1969). *Statistical Power Analysis for the Behavioral Sciences*. 1st edition. Hillsdale, NJ: LEA.

Cohen, J. (1983). The cost of dichotomization. *Applied Psychological Measurement, 7,* 249–253.

Cohen, J. (1988). *Statistical Power Analysis for the Behavioral Sciences.* 2nd edition. Hillsdale, NJ: LEA.

Cohen, J. (1992a). A power primer. *Psychological Bulletin, 112,* 155–159.

Cohen, J. (1992b). Statistical power analysis. *Current Directions in Psychological Science, 1,* 98–101.

Cohen, J. & Cohen, P. (1975). *Applied Multiple Regression/Correlation Analysis for the Behavioral Sciences.* Hillsdale, NJ: LEA.

Cohen, J. & Cohen, P. (1983). *Applied Multiple Regression/Correlation Analysis for the Behavioral Sciences.* 2nd edition. Hillsdale, NJ: LEA.

Cohen, J., Cohen, P., West, S.G., & Aiken, L.S. (2003). *Applied Multiple Regression/ Correlation Analysis for the Behavioral Sciences.* 3rd edition. Hillsdale, NJ: LEA.

Collier, R.O. & Hummel, T.J. (1977). *Experimental Design and Interpretation.* Berkeley, CA: McCutchan.

Cook, R.D. & Weisberg, S. (1983). Diagnostics for heteroscedasticity in regression. *Biometrika, 70,* 1–10.

Cox, D.R. & McCullagh, P. (1982). Some aspects of analysis of covariance. *Biometrics, 38,* 541–561.

Daniel, C. & Wood, F.S. (1980). *Fitting Equations to Data.* 2nd edition. New York: Chapman & Hall.

Darlington, R.B. (1968). Multiple regression in psychological research. *Psychological Bulletin, 69,* 161–182.

Davidson, M.L. (1972). Univariate versus multivariate tests in repeated measures experiments. *Psychological Bulletin, 77,* 446–452.

Dienes, Z. (2008). *Understanding Psychology as a Science: An Introduction to Scientific and Statistical Inference.* Basingstoke: Palgrave/Macmillan.

Dodd, D.H. & Schultz, R.F. (1973). Computational procedures for estimating magnitude of effect for some analysis of variance designs. *Psychological Bulletin, 79,* 391–395.

Donner, A. & Koval, J.J. (1980). The estimation of the intraclass correlation in the analysis of family data. *Biometrics, 36,* 19–25.

Draper, N.R. & Smith, H. (1998). *Applied Regression Analysis.* 3rd edition. New York: Wiley.

Dudoit, S., Shaffer, J.P., & Boldrick, J.C. (2003). Multiple hypothesis testing in microarray experiments. *Statistical Science, 18,* 71–103.

Duncan, D.B. (1955). Multiple range and multiple *F* tests. *Biometrics, 11,* 1–42.

Dunn, O.J. (1961). Multiple comparisons among means. *Journal of the American Statistical Association, 56,* 52–64.

Dunnett, C.W. (1955). A multiple comparison procedure for comparing several treatments with a control. *Journal of the American Statistical Association, 50,* 1096–1121.

Dunnett, C.W. & Tamhane, A.C. (1993). Power comparisons of some step-up multiple test procedures. *Statistics & Probability Letters, 16,* 55–58.

Edgeworth, F.Y. (1886). Progressive means. *Journal of the Royal Statistical Society, 49,* 469–475.

Elashoff, J.D. (1969). Analysis of covariance: a delicate instrument. *American Educational Research Journal, 6,* 383–401.

Emerson, J.D. (1991). In D.C. Hoaglin, F. Mosteller, & J.W. Tukey (eds), *Fundamentals of Exploratory Analysis of Variance*. Hoboken, NJ: Wiley.

Evans, S.H. & Anastasio, E.J. (1968). Misuse of analysis of covariance when treatment effect and covariate are confounded. *Psychological Bulletin*, *69*, 225–234.

Farcomeni, A. (2008). A review of modern multiple hypothesis testing, with particular attention to the false discovery proportion. *Statistical Methods in Medical Research*, *17*, 347–388.

Faul, F., Erdfelder, E., Lang, A.-G., & Buchner, A. (2007). G*Power 3: a flexible statistical power analysis program for the social, behavioural and biomedical sciences. *Behavior Research Methods*, *39*, 175–191.

Feise, R.J. (2002). Do multiple outcome measures require *p*-value adjustment? *BMC Medical Research Methodology*, *2*, 8.

Fisher, R.A. (1924). *International Mathematical Conference,* Toronto.

Fisher, R.A. (1925). Theory of statistical estimation. *Proceedings of the Cambridge Philosophical Society*, *22*, 700–725.

Fisher, R. A. (1926). The arrangement of field experiments. *Journal of the Ministry of Agriculture, Great Britain*, *33*, 503–513.

Fisher, R.A. (1932). *Statistical Methods for Research Workers*. Edinburgh: Oliver & Boyd.

Fisher, R. A. (1925). *Statistical Methods for Research Workers*. Edinburgh: Oliver & Boyd.

Fisher, R.A. (1934). Two new properties of mathematical likelihood. *Proceedings of the Royal Society A*, *144*, 285–307.

Fisher, R.A. (1935a). The logic of inductive inference. *Journal of the Royal Statistical Society*, *98*, 39–54.

Fisher, R.A. (1935b). *The Design of Experiments*. Edinburgh: Oliver & Boyd.

Galton, F. (1886). Regression toward mediocrity in hereditary stature. *Journal of the Anthropological Institute*, *15*, 246–263.

Galton, F. (1888). Co-relations and their measurement, chiefly from anthropometric data. *Proceedings of the Royal Society*, *15*, 135–145.

Games, P.A. (1983). Curvilinear transformation of the dependent variable. *Psychological Bulletin*, *93*, 382–387.

Games, P.A. (1984). Data transformations, power and skew: a rebuttal to Levine and Dunlap. *Psychological Bulletin*, *95*, 345–347.

Gamst, G., Meyers, L.S., & Guarino, A.J. (2008). *Analysis of Variance Designs: A Conceptual and Computational Approach with SPSS and SAS*. New York: Cambridge University Press.

Geisser, S. & Greenhouse, S.W. (1958). An extension of Box's results on the use of the *F* distribution in multivariate analysis. *Annals of Mathematical Statistics*, *29*, 885–891.

Glass, G.V., Peckham, P.D., & Saunders, J.R. (1972). Consequences of failure to meet assumptions underlying the fixed effects analysis of variance and covariance. *Review of Educational Research*, *42*, 237–288.

Gordon, R.A. (1968). Issues in multiple regression. *American Journal of Sociology*, *73*, 592–616.

Gosset, W.S. (1908). The probable error of the mean. *Biometrika*, *6*, 1–25.

Greenwald, A.G. (1976). Within-subject designs: to use or not to use? *Psychological Bulletin*, *83*, 314–320.

Grissom, R.J. (2000). Heterogeneity of variance in clinical data. *Journal of Consulting & Clinical Psychology*, *68*, 155–165.

Grissom, R.J. & Kim, J.J. (2005). *Effect Sizes for Research: A Broad Practical Approach*. New York: LEA.

Hamilton, B.L. (1976). A Monte Carlo test of the robustness of parametric and nonparametric analysis of covariance against unequal regression slopes. *Journal of the American Statistical Association, 71*, 864–869.

Hand, D.H. & Crowder, M. (1996). *Practical Longitudinal Data Analysis*. London: Chapman & Hall.

Hand, D.J. & Taylor, C.C. (1987). *Multivariate Analysis of Variance and Repeated Measures: A Practical Approach for Behavioural Scientists*. London: Chapman & Hall.

Harter, H.L. (1980). History of multiple comparisons. In P.R. Krishnaiah (ed.), *Handbook of Statistics*. Vol. 1. Amsterdam: North Holland. pp. 617–622.

Hays, W. L. (1963). *Statistics*. New York: Holt, Rinehart & Winston.

Hays, W.L. (1994). *Statistics*. 5th edition. Fort Worth, TX: Harcourt Brace.

Hayter, A.J. (1986). The maximum familywise error rate of Fisher's least significant difference test. *Journal of the American Statistical Association, 81*, 1000–1004.

Hinkelman, K. & Kempthorne, O. (2008). *Design and Analysis of Experiments*. Vol. 1. *Introduction to Experimental Design*. Hoboken, NJ: Wiley.

Hoaglin, D.C., Mosteller, F., & Tukey, J.W. (1985). Introduction to more refined estimators. In D.C. Hoaglin, F. Mosteller, & J.W. Tukey (eds), *Understanding Robust and Exploratory Data Analysis*. New York: Wiley. pp. 283–296.

Hochberg, Y. (1988). A sharper Bonferroni procedure for multiple tests of significance. *Biometrika, 75*, 800–802.

Hochberg, Y. & Rom, D. (1995). Extensions of multiple testing procedures based on Simes' test. *Journal of Statistical Planning and Inference, 48*, 141–152.

Hochberg, Y. & Tamhane, A.C. (1987). *Multiple Comparison Procedures*. New York: Wiley.

Hoenig, J.M. & Heisey, D.M. (2001). The abuse of power: the pervasive fallacy of power calculations for data analysis. *The American Statistician, 55*, 19–24.

Hoffman, L., & Rovine, M.J. (2007). Multilevel models for the experimental psychologist: Foundations and illustrative examples. *Behavior Research Methods, 39*, 101–117.

Hollingsworth, H.H. (1976). An analytical investigation of the robustness and power of ANCOVA with the presence of heterogeneous regression slopes. Paper presented at the *Annual Meeting of the American Educational Research Association* Washington, DC.

Holm, S. (1979). A simple sequentially rejective multiple test procedure. *Scandinavian Journal of Statistics, 6*, 65–70.

Hommel, G. (1983). Tests of the overall hypothesis for arbitrary dependence structures. *Biometrical Journal*, 25, 423–430.

Hommel, G. (1988). A stagewise rejective multiple test procedure based on a modified Bonferroni test. *Biometrika, 75*, 383–386.

Horton, R.L. (1978). *The General Linear Model: Data Analysis in the Social and Behavioral Sciences*. New York: McGraw-Hill.

Howell, D.C. (2010). *Statistical Methods for Psychology*. 5th edition. Belmont, CA: Wadsworth, Cengage Learning.

Huitema, B.E. (1980). *The Analysis of Covariance and Alternatives*. New York: Wiley.

Hüttener, H.J.M. & van den Eeden, P. (1995). *The Multilevel Design. A Guide with an Annotated Bibliography, 1980–1993*. Westport, CT: Greenwood Press.

Huynh, H. & Feldt, L.S. (1970). Conditions under which mean square ratios in repeated measurements designs have exact *F*-distributions. *Journal of the American Statistical Association, 65*, 1582–1589.

Huynh, H. & Feldt, L.S. (1976). Estimation of the Box correction for degrees of freedom from sample data in randomized block and split-plot designs. *Journal of Educational Statistics, 1*, 69–82.

Jaeger, R.G. & Halliday, T.R. (1998). On confirmatory versus exploratory research. *Herpetologica, 54*, S64–S66.

Jöreskog, K.G. (1969). A general approach to confirmatory maximum likelihood factor analysis. *Psychometrika, 34*, 183–202.

Jöreskog, K.G. & Sorbom, D. (1993). *LISREL 8: Structural Equation Modelling with the SIMPLIS Command Language*. Hillsdale, NJ: LEA.

Judd, C.M. & McClelland, G.H. (1989). *Data Analysis: A Model Comparison Approach*. San Diego, CA: Harcourt Brace Jovanovich.

Kelley, T.L. (1935). An unbiased correlation ratio measure. *Proceedings of the National Academy of Sciences of the United States of America, 21*, 554–559.

Kempthorne, O. (1980). The term design matrix. *The American Statistician, 34*, 249.

Kendall, M.G. (1948). *The Advanced Theory of Statistics*. Vol. 2. London: Charles Griffin & Company.

Kenny, D.A. & Judd, C.M. (1986). Consequences of violating the independence assumption in analysis of variance. *Psychological Bulletin, 99*, 422–431.

Keppel, G. & Wickens, T.D. (2004). *Design and Analysis: A Researcher's Handbook*. 4th edition. Upper Saddle River, NJ: Pearson.

Keppel, G. & Zedeck, S. (1989). *Data Analysis for Research Designs: Analysis of Variance and Multiple Regression/Correlation Approaches*. New York: Freeman.

Keselman, H.J., Algina J., & Kowalchuk, R.K. (2001). The analysis of repeated measures designs: a review. *British Journal of Mathematical and Statistical Psychology, 54*, 1–20.

Keselman, H.J., Algina, J., Kowalchuk, R.K., & Wolfinger, R.D. (1998). A comparison of two approaches for selecting covariance structures in the analysis of repeated measurements. *Communications in Statistics: Simulation and Computation, 27*, 591–604.

Keselman, H.J., Holland, B., & Cribbie, R.A. (2005). Multiple comparison procedures. In B.S. Everitt & D.C. Howell (eds), *Encyclopedia of Statistics in the Behavioral Sciences*. Chichester: Wiley.

Keselman, J.C., Lix, L.M., & Keselman, H.J. (1996). The analysis of repeated measurements: a quantitative research synthesis. *British Journal of Mathematical and Statistical Psychology, 49*, 275–298.

Keuls, M. (1952). The use of the "Studentized range" in connection with an analysis of variance. *Euphytica, 1*, 112–122.

Kirby, K.N. (1993). *Advanced Data Analysis with SYSTAT*. New York: Van Nostrand Reinhold.

Kirk, R.E. (1968). *Experimental Design: Procedures for the Behavioral Sciences*. Monterey, CA: Brookes/Cole.

Kirk, R.E. (1982). *Experimental Design: Procedures for the Behavioral Sciences*. 2nd edition. Monterey, CA: Brookes/Cole.

Kirk, R.E. (1994). Choosing a multiple comparison procedure. In B. Thompson (ed.), *Advances in Social Science Methodology*. Vol. 3. Greenwich, CT: JAI Press. pp. 77–121.

Kirk, R.E. (1995). *Experimental Design: Procedures for the Behavioral Sciences*. 3rd edition. Pacific Grove, CA: Brookes/Cole.

Kmenta, J. (1971). *Elements of Econometrics*. New York: Macmillan.

Kramer, C.Y. (1956). Extension of multiple range test to group means with unequal number of replications. *Biometrics*, *12*, 307–310.

Kutner, M.H., Nachtsheim, C.J., Neter, J., & Li, W. (2005). *Applied Linear Statistical Models*. 5th edition. New York: McGraw Hill.

Laird, N.M. & Ware, J.H. (1982). Random-effects models for longitudinal data. *Biometrics*, *38*, 963–974.

Lane, P.W. & Nelder, J.A. (1982). Analysis of covariance and standardization as instances of prediction. *Biometrics*, *38*, 613–621.

Lehman, E.L. (1993). The Fisher, Neyman-Pearson theories of testing hypotheses: one theory or two? *Journal of the American Statistical Association*, *88*, 1242–1249.

Lehmann, E.L., Romano, J.P., & Shaffer, J.P. (2005). On optimality of stepdown and stepup multiple test procedures. *The Annals of Statistics*, *33*, 1084–1108.

Levene, H. (1960). Robust tests for equality of variances. In I. Olkin, S.G. Ghurye, W. Hoeffding, W.G. Madow, & H.B. Mann (eds), *Contributions to Probability and Statistics: Essays in Honor of Harold Hotelling*. London: London University Press. pp. 278–292.

Levine, D.W. & Dunlap, W.P. (1982). Power of the *F* test with skewed data: should one transform or not? *Psychological Bulletin*, *92*, 272–280.

Levine, D.W. & Dunlap, W.P. (1983). Data transformation, power and skew: a rejoinder to games. *Psychological Bulletin*, *93*, 596–599.

Lilliefors, H.W. (1967). On the Kolmogorov–Smirnov test for normality with mean and variance unknown. *Journal of the American Statistical Association*, *62*, 399–402.

Lix, L.M. & Keselman, H.J. (1998). To trim or not to trim: tests of location equality under heteroscedasticity and nonnormality. *Educational and Psychological Measurement*, *58*, 409–429.

Longford, N.T. (1993). *Random Coefficient Models*. New York: Oxford University Press.

Lovie, A.D. (1991a). A short history of statistics in twentieth century psychology. In P. Lovie & A.D. Lovie (eds), *New Developments in Statistics for Psychology and the Social Sciences*. London: Routledge/British Psychological Society.

Lovie, P. (1991b). Regression diagnostics: a rough guide to safer regression. In P. Lovie & A.D. Lovie (eds), *New Developments in Statistics for Psychology and the Social Sciences*. London: Routledge/British Psychological Society.

Ludbrook, J. (1998). Multiple comparison procedures updated. *Clinical and Experimental Pharmacology and Physiology*, *25*, 1032–1037.

Macmillan, N.A. & Creelman, C.D. (2005). *Detection Theory: A User's Guide*. 2nd edition. Mahwah, NJ: LEA.

MacRae, A.W. (1995). Descriptive and inferential statistics. In A.M. Coleman (ed.), *Psychological Research Methods and Statistics*. London: Longman.

Marcus, R., Peritz, E., & Gabriel, K.R. (1976). On closed testing procedures with special reference to ordered analysis of variance. *Biometrika*, *63*, 655–660.

Maxwell, S.E. (2004). The persistence of underpowered studies in psychological research: causes, consequences and remedies. *Psychological Methods, 9*, 147–163.

Maxwell, S.E., Camp, C.J., & Arvey, R.D. (1981). Measures of strength of association: a comparative examination. *Journal of Applied Psychology, 66*, 525–534.

Maxwell, S.E. & Delaney, H.D. (1993). Bivariate median splits and spurious statistical significance. *Psychological Bulletin, 113*, 181–190.

Maxwell, S.E. & Delaney, H.D. (2004). *Designing Experiments and Analysing Data: A Model Comparison Perspective*. 2nd edition. Mahwah, NJ: LEA.

Maxwell, S.E., Delaney, H.D., & Manheimer, J.M. (1985). ANOVA of residuals and ANCOVA: correcting an illusion by using model comparisons and graphs. *Journal of Educational Statistics, 10*, 197–209.

Maxwell, S.E., Kelly, K., & Rausch, J.R. (2008). Sample size planning for statistical power and accuracy in parameter estimation. *Annual Review of Psychology, 59*, 537–563.

McClelland, G.H. (1997). Optimal design in psychological research. *Psychological Methods, 2*, 3–19.

McCullagh, P. & Nelder, J.A. (1989). *Generalised Linear Models*. 2nd edition. London: Chapman & Hall.

Micceri, T. (1989). The unicorn, the normal curve and other improbable creatures. *Psychological Bulletin, 105*, 156–166.

Miles, J. & Shevlin, M. (2001). *Applying Regression and Correlation*. London: Sage.

Miller, R.G. (1966). *Simultaneous Statistical Inference*. New York: Wiley.

Miller, R.G. (1981). *Simultaneous Statistical Inference*. 2nd edition. New York: Springer.

Mitchell, J. (1986). Measurement scales and statistics: a clash of paradigms. *Psychological Bulletin, 100*, 398–407.

Montgomery, D.C. & Peck, E.A. (1982). *Introduction to Linear Regression Analysis*. New York: Wiley.

Moser, B.K. & Stevens, G.R. (1992). Homogeneity of variance in the two-sample means test. *The American Statistician, 46*, 19–21.

Moser, B.K., Stevens, G.R., & Watts, C.L. (1989). The two-sample *t* test versus Satterthwaite's approximate *F*-test. *Communication in Statistics: Theory and Methods, 18*, 3963–3975.

Mosteller, F. & Tukey, J.W. (1977). *Data Analysis and Regression*. New York: Addison-Wesley.

Murphy, K.R. & Myors, B. (2004). *Statistical Power Analysis: A Simple and General Model for Traditional and Modern Hypothesis Testing*. 2nd edition. Mahwah, NJ: LEA.

Myers, J.L., Well, A.D., & Lorch, R.F. (2010). *Research Design and Statistical Analysis*. 3rd edition. Routledge: Hove.

Nelder, J.A. (1977). A reformulation of linear models. *Journal of the Royal Statistical Society, Series A, 140*, 48–77.

Newman, D. (1939). The distribution of the range in samples from a normal population, expressed in terms of an independent estimate of the standard deviation. *Biometrika, 31*, 20–30.

Neyman, J. & Pearson, E. S. (1928). On the use and interpretation of certain test criteria for purposes of statistical inference. *Biometrika, 20A*, 175–240, 263–294.

Neyman, J. & Pearson, E.S. (1933). On the problem of the most efficient tests of statistical hypotheses. *Philosophical Transactions of the Royal Society of London, Series A, 231*, 289–337.

Norusis, M.J. (1990). *SPSS/PC + Statistics™ 4.0*. Chicago, IL: SPSS Inc.

O'Brien, P.C. (1983). The appropriateness of analysis of variance and multiple comparison procedures. *Biometrika, 39*, 787–788.

O'Brien, R.G. & Kaiser, M.K. (1985). MANOVA method for analysing repeated measures designs: an extensive primer. *Psychological Bulletin, 97*, 316–333.

Olejnik, S., Li, J., Supattathum, S., & Huberty, C.J. (1997). Multiple testing and statistical power with modified Bonferroni procedures. *Journal of Educational and Behavioural Statistics, 22*, 389–406.

O'Neil, R. & Wetherill, G.B. (1971). The present state of multiple comparison methods. *Journal of the Royal Statistical Society, Series B, 33*, 218–241.

Overall, J.E. & Woodward, J.A. (1977). Nonrandom assignment and the analysis of covariance. *Psychological Bulletin, 84*, 588–594.

Pearson, K. (1896). Regression, heredity and panmixia. *Philosophical Transactions of the Royal Society of London, Series A, 187*, 253–267.

Pearson, K. (1905). *Mathematical Contributions to the Theory of Evolution. XIV. On the General Theory of Skew Correlation and Non-linear Regression.* Drapers' Company Research Memoirs, Biometric Series II. London: Dulau.

Pedhazur, E.J. (1982). *Multiple Regression in Behavioral Research.* 2nd edition. New York: Holt, Rinehart & Winston Inc.

Pedhazur, E.J. (1997). *Multiple Regression in Behavioral Research.* 3rd edition. Fort Worth, TX: Harcourt Brace.

Permutt, T. (1990). Testing for imbalance of covariates in controlled experiments. *Statistics in Medicine, 9*, 1455–1462.

Perneger, T.V. (1998). What's wrong with Bonferroni adjustments. *British Medical Journal, 316*, 1236–1238.

Plackett, R.L. (1972). Studies in the history of probability and statistics. XXIX. The discovery of the method of least squares. *Biometrika, 59*, 239–251.

Preece, D.A. (1982). The design and analysis of experiments: what has gone wrong? *Utilitas Mathematica, Series A, 21*, 201–244.

Ramsey, P.H. (1993). Multiple comparisons of independent means. In L.K. Edwards (ed.), *Applied Analysis of Variance in Behavioral Science.* New York: Marcel Dekker.

Ramsey, P.H. (2002). Comparison of closed testing procedures for pairwise testing of means. *Psychological Methods, 7*, 504–523.

Rao, C.R. (1965). *Linear Statistical Inference and Its Applications.* New York: Wiley.

Rao, C.V. & Saxena, K.P. (1981). On approximation of power of a test procedure based on preliminary tests of significance. *Communication in Statistics: Theory and Methods, 10*, 1305–1321.

Richardson, J.T. (1996). Measures of effect size. *Behavior Research Methods, Instruments & Computers, 28*, 12–22.

Rodgers, J.L., Nicewander, W.A., & Toothaker, L. (1984). Linearly independent, orthogonal and uncorrelated variables. *The American Statistician, 38*, 133–134.

Rodland, E.A. (2006). Simes' procedure is 'valid on average'. *Biometrika, 93*, 742–746.

Rogosa, D.R. (1980). Comparing non-parallel regression lines. *Psychological Bulletin, 88*, 307–321.

Rom, D.M. (1990). A sequentially rejective test procedure based on a modified Bonferroni inequality. *Biometrika, 77*, 663–665.

Rosenthal, R. (1987). *Judgement Studies: Design, Analysis and Meta Analysis*. Cambridge, UK: Cambridge University Press.

Rosnow, R.L. & Rosenthal, R. (1989). Statistical procedures and the justification of knowledge in psychological science. *American Psychologist*, *44*, 1276–1284.

Rothman, K.J. (1990). No adjustments are needed for multiple comparisons. *Epidemiology*, *1*, 43–46.

Rouanet, H. & Lépine, D. (1970). Comparison between treatments in a repeated measures design: ANOVA and multivariate methods. *British Journal of Mathematical and Statistical Psychology*, *23*, 147–163.

Rubin, D.B. (1977). Assignment to treatment group on the basis of a covariate. *Journal of Educational Statistics*, *2*, 1–26.

Rutherford, A. (1992). Alternatives to traditional analysis of covariance. *British Journal of Mathematical and Statistical Psychology*, *45*, 197–223.

Ryan, T.A. (1959). Multiple comparisons in psychological research. *Psychological Bulletin*, *56*, 26–47.

Ryan, T.A. (1960). Significance tests for multiple comparison of proportion, variance, and other statistics. *Psychological Bulletin*, *57*, 318–328.

Saleh, A.K. & Sen, P.K. (1983). Asymptotic properties of tests of hypotheses following a preliminary test. *Statistical Decisions*, *1*, 455–477.

Samuel-Cahn, E. (1996). Is the Simes improved Bonferroni procedure conservative? *Biometrika*, *83*, 928–933.

Sarkar, S.K. (1998). Some probability inequalities for ordered MTP_2 random variables: a proof of the Simes conjecture. *The Annals of Statistics*, *26*, 494–504.

Sarkar, S.K. (2002). Some results on false discovery rate in stepwise multiple testing procedures. *Annals of Statistics*, *30*, 239–257.

Sarker, S.K. & Chang, C.K. (1997). The Simes method for multiple hypothesis testing with positively dependent test statistics. *Journal of the American Statistical Association*, *92*, 1601–1608.

Satterthwaite, F.E. (1946). An approximate distribution of estimates of variance components. *Biometrics Bulletin*, *2*, 110–114.

Saville, D.J. (1990). Multiple comparison procedures: the practical solution. *American Statistician*, *44*, 174–180.

Savitz, D.A. & Olshan, A.F. (1995). Multiple comparisons and related issues in the interpretation of epidemiologic data. *American Journal of Epidemiology*, *142*, 904–908.

Savitz, D.A. & Olshan, A.F. (1998). Describing data requires no adjustment for multiple comparisons: a reply from Savitz and Olshan. *American Journal of Epidemiology*, *147*, 813–814.

Scheffe, H. (1953). A method for judging all contrasts in the analysis of variance. *Biometrika*, *40*, 87–104.

Scheffe, H. (1959). *The Analysis of Variance*. New York: Wiley.

Seaman, M.A., Levin, J.R., & Serlin, R.C. (1991). New developments in pairwise multiple comparisons: some powerful and practicable procedures. *Psychological Bulletin*, *110*, 577–586.

Searle, S.R. (1979). Alternative covariance models for the 2-way crossed classification. *Communications in Statistics: Theory and Methods*, *8*, 799–818.

Searle, S.R. (1987). *Linear Models for Unbalanced Data*. New York: Wiley.

Searle, S.R. (1997). *Linear Models*. New York: Wiley.

Searle, S.R. Casella, G., & McCulloch, C.E. (1992). *Variance Components*. New York: Wiley.

Seber, G.A.F. (1977). *Linear Regression Analysis*. New York: Wiley.

Seneta, E. (1993). Probability inequalities and Dunnett's test. In F.M. Hoppe (ed.), *Multiple Comparisons, Selections and Applications in Biometry*. New York: Marcel Dekker. Chapter 3, pp. 29–45.

Senn, S.J. (1989). Covariate imbalance and random allocation in clinical trials. *Statistics in Medicine*, *8*, 467–475.

Shadish, W., Cook, T., & Campbell, D. (2002). *Experimental and Quasi-Experimental Designs for Generalized Causal Inference*. 2nd edition. Boston: Houghton Mifflin.

Shaffer, J.P. (1979). Comparison of means: an F-test followed by a modified multiple range procedure. *Journal of Educational Statistics*, *4*, 14–23.

Shaffer, J.P. (1986). Modified sequentially rejective multiple test procedures. *Journal of the American Statistical Association*, *81*, 826–831.

Shaffer, J.P. (1995). Multiple hypothesis testing. *Annual Review of Psychology*, *46*, 561–584.

Shapiro, S.S. & Wilk, M.B. (1965). An analysis of variance test for normality (complete samples). *Biometrika*, *52*, 591–611.

Shirley, E.A.C. & Newnham, P. (1984). The choice between analysis of variance and analysis of covariance with special reference to the analysis of organ weights in toxicology studies. *Statistics in Medicine*, *3*, 85–91.

Sidak, Z. (1967). Rectangular confidence regions for the means of multivariate normal distributions. *Journal of the American Statistical Association*, *62*, 626–633.

Siegel, S. & Castellan, N.J. (1988). *Nonparametric Statistics for the Behavioral Sciences*. New York: McGraw-Hill.

Simes, R.J. (1986). An improved Bonferroni procedure for multiple tests of significance. *Biometrika*, *73*, 751–754.

Smith, H.F. (1957). Interpretation of adjusted treatment means and regressions in analysis of covariance. *Biometrics*, *13*, 282–308.

Snedecor, G.W. (1934). *Analysis of Variance and Covariance*. Ames, IO: Iowa State University Press.

Snedecor, G.W. & Cochran, W.G. (1980). *Statistical Methods*. 7th edition. Ames, IA: Iowa State University.

Snijders, T. & Bosker, R. (1999). *Multilevel Analysis: An Introduction to Basic and Advanced Multilevel Modeling*. London: Sage.

Stevens, S.S. (1951). Mathematics, measurement and psychophysics. In S.S. Stevens (ed.), *Handbook of Experimental Psychology*. New York: Wiley.

Stigler, S.M. (1986). *The History of Statistics: The Measurement of Uncertainty Before 1900*. Cambridge, MA: Belknap Press.

Stigler, S.M. (2002). *Statistics on the Table: The History of Statistical Concepts and Methods*. Cambridge, MA: Harvard University Press.

Suppes, P. & Zinnes, J.L. (1963). Basic measurement theory. In R.D. Luce, R.R. Bush, & E. Galanter (eds), *Handbook of Mathematical Psychology*. Vol. 1. New York: Wiley.

Tabachnick, B.G. & Fidell, L.S. (2007). *Experimental Design Using ANOVA*. Belmont, CA: Duxbury, Thompson Brookes/Cole.

Tan, W.Y. (1982). Sampling distributions and robustness of *t, F* and variance ratio in two samples and ANOVA models with respect to departure from normality. *Communications in Statistics: Theory and Methods*, *11*, 486–511.

Taylor, M.J. & Innocenti, M.S. (1993). Why covariance? A rationale for using analysis of covariance procedures in randomized studies. *Journal of Early Intervention*, *17*, 455–466.

Tomarken, A.J. & Serlin, R.C. (1986). Comparison of ANOVA alternatives under variance heterogeneity and specific noncentrality structures. *Psychological Bulletin*, *99*, 90–99.

Toothaker, L.E. (1991). *Multiple Comparisons for Researchers*. Thousand Oaks, CA: Sage.

Townsend, J.T. & Ashby, F.G. (1984). Measurement scales and statistics: the misconception misconceived. *Psychological Bulletin*, *96*, 394–401.

Tukey, J.W. (1949). One degree of freedom for non-additivity. *Biometrics*, *5*, 232–249.

Tukey, J.W. (1953). The problem of multiple comparisons. Unpublished manuscript. See Braun (1994), pp. 1–300.

Tukey, J.W. (1955). Queries. *Biometrics*, *11*, 111–113.

Tukey, J.W. (1977). *Exploratory Data Analysis*. Reading, MA: Addison-Wesley.

Turk, D.C., Dworkin, R.H., McDermott, M.P., Bellamy, N., Burke, L.B., et al. (2008). Analyzing multiple endpoints in clinical trials of pain treatments: IMMPACT recommendations. *Pain*, *139*, 485–493.

Urquhart, N.S. (1982). Adjustment in covariance when one factor affects the covariate. *Biometrics*, *38*, 651–660.

Vargha, A., Rudas, T., Delaney, H.D., & Maxwell, S.E. (1996). Dichotomization, partial correlation and conditional independence. *Journal of Educational and Behavioural Statistics*, *21*, 264–282.

Weisberg, S. (1985). *Applied Linear Regression*. 2nd edition. New York: Wiley.

Wells, C.S. & Hintze, J.M. (2007). Dealing with assumptions underlying statistical tests. *Psychology in the Schools*, *44*, 495–502.

Westfall, P.H., Tobias, R.D., Rom, D., Wolfinger, R.D., & Hochberg, Y. (1999). *Multiple Comparisons and Multiple Tests Using SAS*. Cary, NC: SAS Institute Inc.

Westfall, P.H. & Young, S.S. (1993). *Resampling Based Multiple Testing*. New York: Wiley.

Wherry, R.J. (1931). A new formula for predicting the shrinkage of the coefficient of multiple correlation. *Annals of Mathematical Statistics*, *2*, 240–457.

Wilcox, R.R. (1987). New designs in analysis of variance. *Annual Review of Psychology*, *38*, 29–60.

Wilcox, R.R. (1998). How many discoveries have been lost by ignoring modern statistical methods? *American Psychologist*, *53*, 300–314.

Wilcox, R.R. (2003). *Applying Contemporary Statistical Techniques*. San Diego, CA: Academic Press.

Wilkinson, G.N. & Rogers, C.E. (1973). Symbolic description of factorial models for analysis of variance. *Applied Statistics*, *22*, 392–399.

Wilkinson, L. & Task Force on Statistical Inference (1999). Statistical methods in psychology journals: guidelines and explanations. *American Psychologist*, *54*, 594–604.

Winer, B.J. (1962). *Statistical Principles in Experimental Design*. New York: McGraw-Hill.

Winer, B.J. (1971). *Statistical Principles in Experimental Design.* 2nd edition. New York: McGraw-Hill.

Winer, B.J., Brown, D.R., & Michels, K.M. (1991). *Statistical Principles in Experimental Design.* 3rd edition. New York: McGraw-Hill.

Wonnacott, R.J. & Wonnacott, T.H. (1970). *Econometrics.* New York: Wiley.

Wright, D.B. (1998). People, materials and situations. In J. Nunn (ed.), *Laboratory Psychology.* Hove: Psychology Press.

Wright, D.B. & London, K. (2009). Multilevel modelling: Beyond the basic applications. *British Journal of Mathematical and Statistical Psychology, 62*, 439–456.

Yule, G.U. (1907). On the theory of correlation for any number of variables treated by a new system of notation. *Proceedings of the Royal Society, Series A, 79*, 182–193.

Zimmerman, D.W. (1996). Some properties of preliminary tests of equality of variances in the two-sample location problem. *The Journal of General Psychology, 123*, 217–231.

Zimmerman, D.W. (2004). A note on preliminary tests of equality of variances. *British Journal of Mathematical and Statistical Psychology, 57*, 173–181.

Index

ANOVA and ANCOVA: A GLM Approach, Second Edition. By Andrew Rutherford.

Printed and bound by CPI Group (UK) Ltd, Croydon, CR0 4YY

16/04/2025

14658367-0002